AF842972

HANDBOOK OF HAZARDOUS WASTE INCINERATION

To Claire, my wife, for her love and her patience.

HANDBOOK OF HAZARDOUS WASTE INCINERATION

CALVIN R. BRUNNER, P.E.

TPR
TAB Professional and Reference Books
Division of TAB BOOKS Inc.
Blue Ridge Summit, PA
1989

FIRST EDITION
FIRST PRINTING

Copyright © 1989 by TAB BOOKS Inc.
Printed in the United States of America

Reproduction or publication of the content in any manner, without express permission of the publisher, is prohibited. No liability is assumed with respect to the use of the information herein.

Library of Congress Cataloging in Publication Data

Brunner, Calvin R.
Handbook of hazardous waste incineration.

Includes index.
1. Hazardous wastes—Incineration. I. Title.
TD811.5.B78 1988 628.4′4 88-2249
ISBN 0-8306-9318-1

TAB BOOKS Inc. offers software for sale. For information and a catalog, please contact TAB Software Department, Blue Ridge Summit, PA 17294-0850.

Questions regarding the content of this book should be addressed to:

Reader Inquiry Branch
TAB BOOKS Inc.
Blue Ridge Summit, PA 17294-0214

Edited by Stephen Kamm

Contents

Preface

Hazardous waste has been dumped indiscriminately, it has been burned uncontrollably, and it has been buried irresponsibly. This book describes methods of disposal of hazardous waste through incineration in a responsible, environmentally secure manner.

The purpose of this book is to present the entire field of hazardous-waste incineration to the reader. Information on regulations, site clean-up, air emissions control, and current and emerging thermal treatment technologies are of relevance to the lay reader as well as the engineering professional. Non-technical people with an interest in this subject matter include supervisory personnel in government and industry, members of public interest groups, members of the news media, and students of environmental technology. The technical material in the text, including calculations of system and equipment parameters and air emissions discharges, should be of interest to the engineer and designer.

The high profile of the problems of hazardous-waste accumulations and disposal has resulted in the mobilization of regulatory disciplines, environmental engineers, environmental scientists, industrial generators, remediation contractors, equipment manufacturers, environmental organizations, and concerned citizens across the country. They are mobilized to reduce the amount of hazardous waste generated and to help ensure that the disposal of these wastes is made in an appropriate manner. This book directly addresses these concerns.

The disposal of hazardous wastes is controlled by regulatory procedures that are described at the beginning of this text. The types of systems available for the incineration of hazardous wastes are detailed in three chapters, followed by specific applications. Three additional chapters are included for these applications: site clean-up, incineration in high-temperature industrial processes, and ocean incineration technology. A chapter is devoted to European systems, which have a longer history than American systems.

Three chapters are devoted to emissions forecasting and control, and the last four chapters cover incinerator calculations, from heat balance techniques to energy recovery. The text is supplemented with appendices containing the basic information necessary to perform incineration calculations. A comprehensive glossary is also included.

1

Introduction

The demands of present society create wastes that appear to increase in quantity as the standard of living increases. This is true not only in the United States but in the rest of the world as well. The ability of the land, water, and air to absorb these wastes is limited; reminders of this are all too apparent. Poisoned rivers, darkened skies, and wastelands are endemic to the industrialized world. Resistance to this trend, coupled with the start of vigorous attempts to reverse inadequate waste disposal practices has been only a recent occurrence.

Over the past fifteen years the annual expenditure on pollution control in the United States has risen from less than one percent to almost six percent of industrial capital equipment investment. By the end of the century this figure, which does not include pollution control for automobiles or the cost of disposal of municipal solid waste or sewage, is expected to more than double.

The most effective means of dealing with this problem, dramatically illustrated in Fig. 1-1, is to reduce the amount of wastes generated. In industry, this means process modification or the establishment of new markets for generated streams. In the public sector, waste reduction translates to source separation, recycling, and resource recovery.

Waste reduction is a goal that will be driven forward by economic conditions. As the cost of waste disposal increases, new methods of waste

Fig. 1-1. Hazardous-waste dump. Source: USEPA.

reduction will become economically attractive, and will be implemented. It is difficult to imagine; however, a time when there will be no waste generation. For the foreseeable future, despite effective reduction practices, there will still be hundreds of millions of tons of waste that must be destroyed or controlled to protect the public health and the sanctity of the environment.

HAZARDOUS WASTE

The first step in the control of waste disposal is the establishment of waste definitions. There are many different types of wastes from sources throughout the spectrum of industrial, agricultural, mining, and domestic activity, as listed in Table 1-1. The majority of these wastes, those from agriculture and livestock, generally do not pose a severe threat to life. Many other wastes are not life threatening either, but they all must be disposed of in an environmentally sound manner.

A top priority in the restoration and maintenance of a safe and tenable environment is the disposition of a particular classification of waste materials: hazardous wastes. The federal government has established a set of definitions which identify wastes that are of a particular concern. Chapter 2 discusses these wastes in detail.

The term *hazardous waste* has a specific statutory definition. However, in common usage, this term tends to include other wastes. The Toxic Substances Control Act (TSCA) regulates wastes containing PCBs; the Nuclear Regulatory Act deals with wastes having a radioactive content.

Table 1-1. Solid Waste Generation in the United States, 1978.

SOURCE		Million Tons/Year	Percent
Municipal		230	5.2
Industrial		140	3.1
Mineral		1,700	38.2
Animal Wastes		1,740	39.1
Crop Wastes		640	14.4
	TOTAL	4,450	100.0

Source: Reference 1.

The wastes discussed in this book will include wastes designated by the Resource Conservation and Recovery Act (RCRA) and those wastes included under TSCA. When the term "hazardous waste" is used, PCBs and related wastes under TSCA will also be included in this designation.

If wastes have been defined as hazardous, the regulations described in Chapter 2 address their storage, transportation, treatment and disposal.

INCINERATION

While industry is struggling with the mechanisms of waste reduction, the avenues previously open for waste disposal are closing, particularly in the area of hazardous wastes. Incineration is becoming more attractive as an alternative disposal method. Features of the incineration option include the following:

- The hazardous component of the waste is destroyed.
- The volume and weight of the waste is reduced to a fraction of its original size.
- Waste reduction is immediate; it does not require long-term residence in a biological treatment pond or other land disposal system.
- Waste can be incinerated on-site, without being transported to a distant area.
- Air discharges can be effectively controlled for minimal impact on the atmospheric environment.
- The ash residue may be subject to de-classification as a hazardous waste. In this case, incineration becomes, essentially, a final disposal method as well as a treatment method for hazardous waste.
- Incineration requires a relatively small disposal area, not the acres and acres needed for lagoons or other land disposal methods.
- Incineration is easily terminated. The cessation of incineration activity will remove any liability for the generator or operator. With land burial, the liabilities are indefinite and uncertain.

- Through heat recovery techniques, the cost of operation can be reduced or offset by the use or sale of energy.

Although incineration is becoming increasingly attractive as a waste disposal option, it is not universally applicable to waste disposal. Note the following considerations:

- Some materials, such as highly aqueous wastes or non-combustible soils, are not incinerable.
- Certain organic materials, particularly those with a chlorine component, tend to generate products of incomplete combustion that are often toxic and present their own disposal problems.
- Incineration represents a high capital cost.
- Skilled operators are required.
- Supplemental fuel is required to bring up an incinerator to operating temperature and, with some materials, to maintain combustion temperatures.

As noted previously, the incineration option is, on balance, an attractive option for a growing number of waste streams. The trend today is an increase in the use of incineration for a wider profile of wastes.

TRANSPORTATION

An issue directly related to the disposal of hazardous waste is the transportation of hazardous waste. Although the federal government has normally taken the lead role in the regulation of transportation activities, the fear of hazardous product and waste incidents is bringing state and local governments into the regulation of these activities. An indication of the seriousness of this issue is presented in Table 1-2, which is a list of the

Table 1-2. Reported Incidents Involving Transportation of Hazardous Materials, 1973-1983.

Transport Mode	Incidents	Deaths	Injuries	Damages[1] ($ Million)
Highway	10,289	19	419	8.15
Rail	975	4	222	4.67
Water	26	0	3	0.07
Air	150	< 1	9	0.43
Freight forwarder	2	0	2	< 0.01
Other	20	0	8	0.01
Total	11,462	24	663	13.33

[1]Property damage reported within 15 days after an accident.

Source: Reference 3.

annual average number of incidents of transport anomalies over a ten year period.

It is interesting to note from this table that the injuries from rail transport are higher per incident than any other named occurrence. While only 8½ percent of the total reported incidents of transportation accidents involving hazardous wastes is associated with rail transport, it results in over one-third of the reported injuries. When a rail car has an accident, the population at large surrounding the track is vulnerable, more so than in any other transport mode. In most instances of highway accidents, the truck driver is normally the only one at risk; the surrounding community is not threatened to the extent typical of rail accidents. These statistics explain the growing concern of state legislatures regarding the proper control of hazardous materials passing through their state.

States that have passed or are considering legislation requiring the registration of hazardous-wastes transport are indicated in Table 1-3 and Fig. 1-2. Twenty-six of these states require hazardous-waste transporters to register with the state and impose fees ranging from $25 to $500. These fees may cover only a single trip, or extend over a whole year. In some states additional registration requirements are mandated, such as inspections in California, Connecticut, and Rhode Island.

One result of the attention that these states are giving to the transportation of hazardous wastes is that the licensing, registration, and permit requirements vary widely, particularly at the state and local level.

The evaluation of incineration compared to other methods of disposal should include considerations of transportation from the generator to the treatment facility.

HAZARDOUS-WASTE INCINERATION

The Federal Government maintains a log of hazardous-waste treatment facilities in the *EPA/DOE Hazardous Waste Control Technology Data Base*. This includes all treatment methods, including incineration. The Department of Energy at the Oak Ridge National Laboratory in Tennessee is the agency responsible for maintenance of this data base.

As of March, 1987, 37 industrial hazardous-waste incinerators and 8 PCB incinerators have been permitted out of a total of 199 such incinerators extant in the United States. The majority of the non-permitted incinerators are either in the process of being permitted, are R&D (used for research and development, generally laboratory installations) incinerators requiring special consideration, or have been registered with EPA but are exempt from hazardous-waste regulations.

The data base contains not only the name and use of the incinerator facilities that have applied for permits, but also technical information on

Table 1-3. States with Proposed or Existing Hazardous Waste Transportation Requirements.

State	Company Registration	Company Fee	Vehicle Registration	Vehicle Fee	Driver Training	Spill Provisions
Alabama	Yes	—	—	—	—	—
Arkansas	Yes	Yes	—	—	—	—
California	Yes	Yes	—	—	Yes	—
Connecticut	Yes	Yes	Yes	—	Yes	Yes
Delaware	Yes	—	Yes	—	—	—
Georgia	Yes	Yes	—	—	—	—
Idaho	Yes	Yes	—	—	—	—
Illinois	Yes	—	Yes	—	—	—
Indiana	Yes	Yes	Yes	—	—	—
Kansas	Yes	Yes	—	—	—	—
Kentucky	Yes	Yes	—	—	—	—
Louisiana	Yes	No	—	—	—	—
Maine	Yes	Yes	Yes	Yes	Yes	—
Maryland	Yes	—	Yes	Yes	Yes	Yes
Massachusetts	Yes	Yes	Yes	Yes	—	Yes
Michigan	Yes	Yes	Yes	Yes	—	—
Minnesota	Yes	—	—	—	—	—
Missouri	Yes	—	—	Yes	—	—
New Hampshire	—	—	Yes	Yes	—	—
New Jersey	—	—	Yes	Yes	—	—
New York	—	Yes	—	—	—	—
North Dakota	Yes	—	—	—	—	—
Ohio	Yes	Yes	—	Yes	—	—
Oklahoma	Yes	—	—	—	—	—
Oregon	Yes	—	—	—	—	—
Pennsylvania	Yes	Yes	—	—	—	Yes
Rhode Island	Yes	—	Yes	Yes	—	—
South Carolina	Yes	—	—	—	—	—
Tennessee	Yes	Yes	—	—	—	—
Texas	Yes	Yes	—	—	—	—
Vermont	Yes	—	Yes	Yes	—	—
Virginia	Yes	—	—	—	—	—
Wisconsin	Yes	—	—	—	—	—

Source: Reference 2.

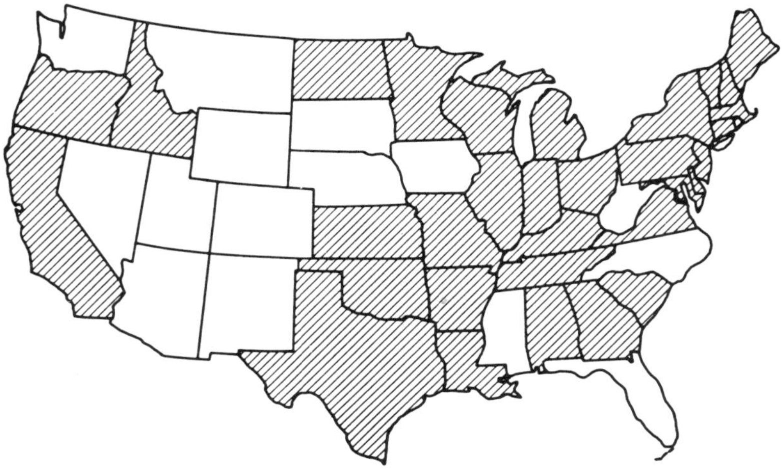

Fig. 1-2. States with proposed or existing hazardous-waste transportation regulations. Source: Reference 2.

each of them. Such information includes waste quantity, hazardous constituents in the waste, heat value of the waste stream, and results of the test burn performed for permit compliance.

REFERENCES

1. *Report to the U.S. Congress—Resource Recovery and Source Reduction,* United States Environmental Protection Agency, Report SW-118, February 1978.

2. *Hazardous Materials: State and Local Activities,* U.S. Office of Technology Assessment, OTA-SET-301, March 1986.

3. *Annual Report on Hazardous Materials Transportation, Calendar Year 1983,* U.S. Department of Transportation.

2

Regulatory Requirements

The federal government has taken the lead in developing a strategy to prevent the discharge or accumulation of hazardous wastes in the environment. The two main regulatory mechanisms for the identification and control of hazardous wastes, RCRA and TSCA, will be discussed in this chapter.

RCRA AND TSCA

The Toxic Substances Control Act (TSCA) addresses the control of *polychlorinated biphenyls* (PCBs) in the environment. It was the first comprehensive piece of legislation designed in response to growing public awareness of and concern about the discharge of industrial pollutants into the air, the water, and the earth. The Resource Conservation and Recovery Act (RCRA) represents the next generation of legislative concern about the discharge of dangerous materials into the environment.

RCRA defines hazardous wastes and describes the methods required for the control of these wastes. It includes wastes from industry, institutions, the public at large, and all other segments of the economy where wastes with a potential negative impact on the quality of life may be generated.

Both RCRA and TSCA regulations require that hazardous-waste generators, transporters, depositories, treatment facilities, and ultimate repositories obtain permits. The RCRA permit is more universal than that required by TSCA. RCRA permits can be issued by the state or the federal government, as described later in this chapter. Authority for the issuance of TSCA permits resides in the federal government, namely the EPA. Eventually, the TSCA requirements and permitting authority, will be included within the RCRA statutes, which will simplify permitting and reporting requirements. At this time, however, TSCA must be complied with when a waste contains a significant PCB component (50 ppm by weight or greater) and RCRA must be complied with for other hazardous waste.

The RCRA permitting process is described in this chapter. Incinerator regulations for both RCRA and TSCA are also discussed.

RCRA

The Resource Conservation and Recovery Act (RCRA) is the basic mechanism utilized by the federal government for protection of the public welfare with regard to the disposition of hazardous wastes. RCRA first became law in 1976 and had a number of major revisions, most recently in the fall of 1984.

RCRA consists of eight parts (subtitles), identified as A through H. Subtitle C, Hazardous Waste Management, incorporates 13 sections, as listed in Table 2-1. This subtitle describes what the EPA must do to control hazardous waste handling and disposal and provides the EPA with the authority to carry out the provisions of this act.

Table 2-1. RCRA Subtitle C Hazardous Waste Management.

Section 3001	Identification and listing of hazardous wastes.
Section 3002	Standards applicable to generators of hazardous waste.
Section 3003	Standards applicable to transporters of hazardous waste.
Section 3004	Standards applicable to owners and operators of hazardous waste treatment, storage and disposal facilities.
Section 3005	Permits for treatment, storage, or disposal of hazardous waste.
Section 3006	Authorized state hazardous waste programs.
Section 3007	Inspections.
Section 3008	Federal enforcement.
Section 3009	Retention of State authority.
Section 3010	Effective date.
Section 3011	Authorization of assistance to states.
Section 3012	Restrictions on recycled oil.
Section 3013	Monitoring, analysis and testing.

Table 2-2. Parts of Title 40 of the CFR.

PART	TITLE
40 CFR 124	Procedures for Decision making.
40 CFR 260	Hazardous Waste Management System: General.
40 CFR 261	Identification and Listing of Hazardous Waste.
40 CFR 262	Standards Applicable to Generation of Hazardous Waste.
40 CFR 263	Standards Applicable to Transporters of Hazardous Waste.
40 CFR 264	Standards for Owners and Operators of Hazardous Waste Treatment, Storage and Disposal Facilities.
40 CFR 265	Interim Status Standards of Owners and Operators of Hazardous Waste Treatment, Storage and Disposal Facilities.
40 CFR 266	Reserved.
40 CFR 267	Interim Standards for Owners and Operators of New Hazardous Waste Land Disposal Facilities.
40 CFR 270	EPA Administered Permit Programs: The Hazardous Waste Permit Program.
40 CFR 271	Requirements for Authorization of State Hazardous Waste Programs.

Hazardous waste regulations are contained in Title 40 of the Code of Federal Regulations (CFR). Within Title 40, the hazardous waste permitting, handling and disposal regulations are contained in Part 124 and Parts 260 through 271. Table 2-2 identifies these sections of the regulations. Of these sections, the regulations of particular concern to a hazardous waste incineration facility are described in Table 2-3. Subpart O of Part 264 of these regulations detail incinerator permitting, operating and reporting requirements. Table 2-4 lists the sections of this standard.

STATE AUTHORITY

Initially RCRA placed authority over the entire permit process with the EPA regional office. Each regional office retains this authority until or unless a state program is approved to replace it. This approval process is phased as follows:

Phase I	Waste identification, standards for generators and transporters, and interim status requirements for *TSD* (transport, storage, disposal) facilities.
Phase II	
Component A	Permits for storage and treatment in tanks, surface impoundments and waste piles and the use of containers.

Table 2-3. RCRA Incinerator Regulations Summary.

Part 124	Specifies the administrative aspects of the permit procedures such as when public hearings are required and the steps included in the general decision making process.
Part 260	Provides an overview of the regulations and defines terms such as "INCINERATOR - an enclosed device using controlled flame combustion, the primary purpose of which is to thermally break down hazardous waste".
Part 261	Characterizes hazardous wastes by ignitability, corrosivity, reactivity and EP toxicity, and lists hazardous-waste constituents.
Part 262	Applies to generators and may apply to a hazardous-waste incinerator if the ash, residue and solids and sludges that remain after incineration cannot be delisted as a hazardous waste.
Part 264	Applies to permitted hazardous-waste facilities and presents general facility requirements as well as the specific performance standards and operating requirements that apply to incinerators.
Part 265	Applies to interim status facilities, general facility and unit operating standards similar, though less specific than those standards within Part 264.
Part 270	Defines the EPA permit process for hazardous-waste facilities. It lists the general informational requirements of the permit and cites specific data that must be supplied for an incinerator. It also cites permit conditions that are applicable to all permitted facilities and provides information on items such as changes to and termination of the final RCRA permit.
Part 271	Specifies the procedures that EPA will follow in approving state programs.

Table 2-4. 40 CFR 264 Subpart O - Incinerators.

264.340	Applicability
264.341	Waste analysis
264.342	Principal organic hazardous constituents (POHCs)
264.344	Hazardous-waste permits
264.345	Operating requirements
264.346	Reserved for future use
264.347	Monitoring and inspections
264.348–.350	Reserved for future use
264.351	Closure
264.352–.339	Reserved for future use

Component B	Permits for incinerators.
Component C	Permits for land disposal facilities.

This phased process is an interim process. Eventually each state will be given authority to approve and issue all hazardous waste permits. As of March 15, 1987, half the states have received either final authorization or Phase II, Component B (incinerator) interim permitting authorization.

RCRA PERMIT

The RCRA permit itself consists of two parts, Part A and Part B. Figure 2-1, Incinerator Part B Permit Process, illustrates this process which is relatively complex, costly, and time-consuming.

For a new facility, i.e., one that began operation or was constructed after November 19, 1980, both parts are submitted together. If the facility was in operation before this date, it is classified as an existing facility and each part is handled separately. Existing facilities were required to submit a Part A application to the regional EPA office prior to November 19, 1980.

The Part A permit consists of two forms. Form One requests general information, such as the name and address of the facility; the name, address, and telephone number of both the owner and the operator; the permits or construction approvals received or applied for under various regulatory programs, and a brief description of the nature of the business requesting the permit. Form Two requires specific information of the hazardous waste being processed, including the process design capacity of each storage, treatment, or disposal operation; the EPA waste hazard code (identified later in this chapter) and annual quantity of each waste generated or accepted; and the process used for each waste, as well as drawings and photographs of the facility.

Once a facility had filed the two forms that comprise Part A of the RCRA application, it was granted interim status. Under interim status, the facility is treated as having been issued a RCRA permit until the EPA or authorized state agency makes a final determination on the complete permit application. This would occur after the Part B application has been called, submitted by the facility, and processed by the cognizant RCRA permitting authority.

Owners and operators of new facilities cannot, by definition, qualify for interim status. They must submit both the Part A and Part B applications together and are granted only a final permit.

The second and more extensive portion of the RCRA permit application is Part B. Unlike Part A, there is no standard form for Part B. The permit applicant must describe in narrative form the activities and procedures that must be undertaken at the facility to demonstrate proper protection of human health and the environment. For existing facilities operating under interim status, the Part B application must be submitted within 6 months after it is requested (or "called in") by the permitting agency. For new facilities, it must be submitted 180 days before physical construction is scheduled to begin.

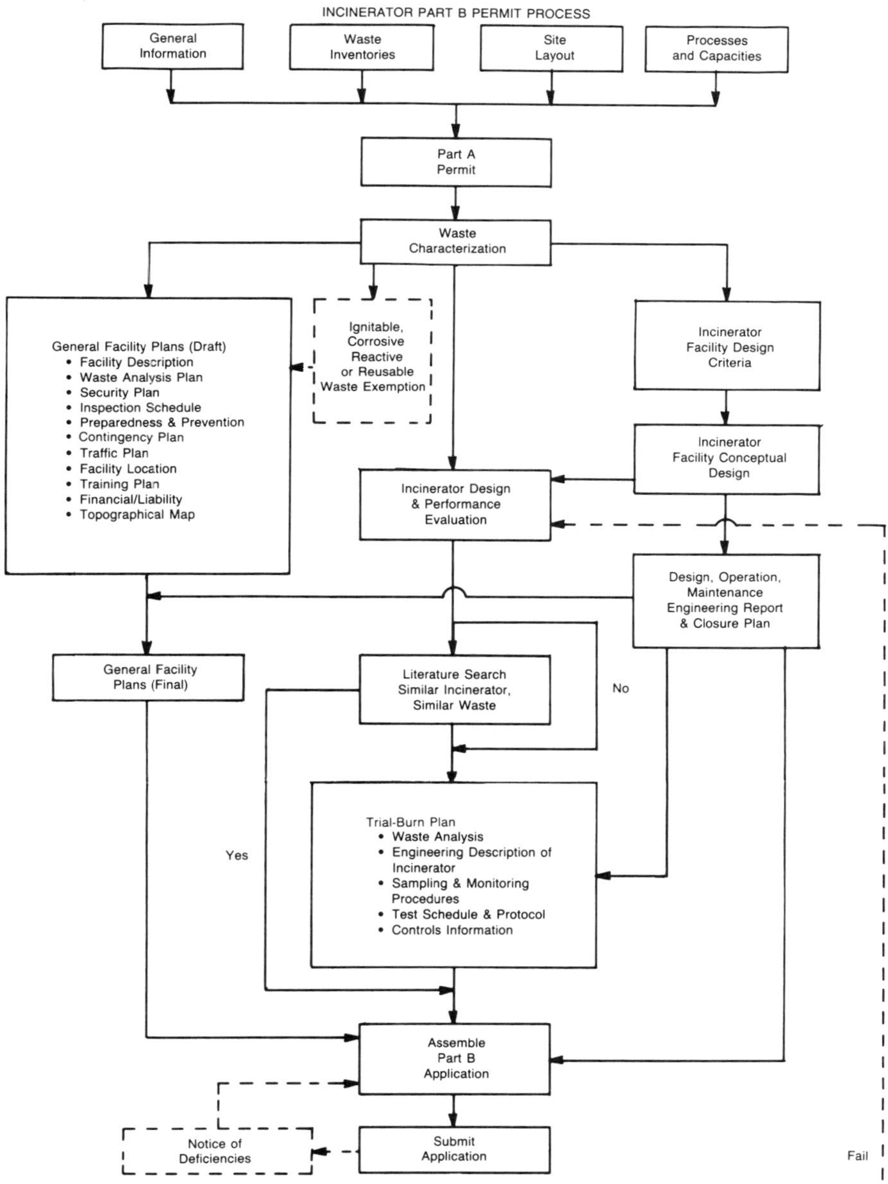

Fig. 2-1. Incinerator Part B permit process.

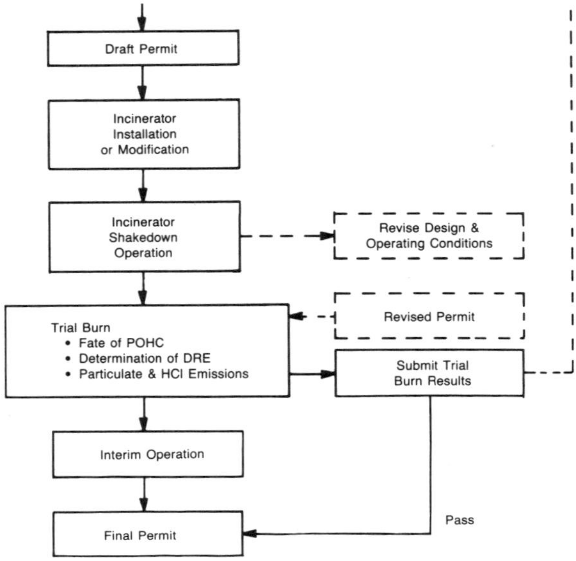

Fig. 2-1. Cont.

Two different types of information must be supplied in the Part B application:

- general information that is required of all permit applicants, and
- facility-specific information that is applicable only to incinerators.

These requirements are detailed in Subparts 264 and 270 of the regulations. Key submittals under these requirements include the following:

- General facility description. This must list the type, size, and location of the facility as well as the types of activities conducted and the waste types stored, treated, or disposed of.
- Waste analysis plan. This must contain sufficient information to characterize the waste and specify parameters to be monitored, as well as sampling methods, analysis methods and sampling frequency.
- Security procedures. This section must state how unauthorized entry will be prevented or minimized through the use of such measures as 24-hour surveillance or physical barriers with controlled access.
- Inspection schedule. This schedule must identify each item to be inspected with corresponding inspection frequencies.
- Preparedness and prevention. This section must address how the facility is designed and operated to minimize hazards. It must

describe the equipment and procedures to be used in case of emergencies.

- Contingency plan. This section must also address procedures for dealing with emergencies. While the preparedness and prevention section describes emergencies dealing specifically with the incineration process, this section should include external considerations such as fires, explosions, and unplanned releases of hazardous wastes. It should also include evacuation plans and incident reporting procedures.
- Traffic plans. This must describe general traffic patterns and the pattern of traffic moving hazardous wastes within the facility.
- Facility location. This section must describe the location of the facility with respect to seismic faults, the 100-year flood plain, etc.
- Training plan. This must describe the nature and frequency of on-site training. It must be shown that facility personnel are trained to adequately respond to emergencies. This plan must also include procedures for repairing and replacing emergency and monitoring equipment.
- Closure plan. This must describe how and when the facility will be closed and the cost of closure.
- Cost estimate. A cost estimate of facility closure must be included. A financial assurance plan must also be included to demonstrate that there are financial resources dedicated to closing the facility.
- Topographic map. A topographic map must be included which describes the land for a distance of 1,000 feet outside the perimeter of the facility.

HAZARDOUS-WASTE CLASSIFICATION

A set of definitions of hazardous wastes have been established in 40 CFR 261 as follows:

1. Ignitable waste (ignitability), hazard code "I". This waste will have at least one of the following properties:

    - A liquid having a flash point less that 140° F. An aqueous solution containing less than 20 percent alcohol by volume is excluded from this definition.
    - A substance, other than a liquid, which can cause fire through friction or through absorption of moisture, or can ignite into fire under spontaneous chemical change under standard temperature and pressure. When this substance burns, it does so vigorously and persistently.
    - An ignitable compressed gas (see 49 CFR 173.300 for further definition).

- An oxidizer (see 40 CFR 173.151 for further definition).

An ignitable waste which is not listed elsewhere as a hazardous waste is given an EPA hazardous waste number of D001.

2. Corrosive waste (corrosivity), hazard code ''C''. This waste will have either or both of the following properties:

- An aqueous waste with a pH equal to or less than 2.0 (acidic) or a pH equal to or greater than 12.5 (basic).
- A liquid that corrodes carbon steel (grade SAE 1030) at a rate greater than 0.250 inches per year.

A corrosive waste which is not listed elsewhere as a hazardous waste is given an EPA hazardous-waste number of D002.

3. Reactive waste (reactivity), hazard code ''R''. This waste will have at least one of the following characteristics.

- A substance which is normally unstable and undergoes violent physical or chemical change without detonating.
- A substance that reacts violently with water.
- A waste that forms a potentially explosive mixture when wetted with water.
- A substance that can generate harmful gases, vapors, or fumes when mixed with water.
- A cyanide or sulfide bearing waste which can generate harmful gases, vapors, or fumes when exposed to pH conditions between 2.0 and 12.5.
- A waste which, when subjected to a strong initiating source or when heated in confinement, will detonate or generate an explosive reaction.
- A substance which is readily capable of detonation at standard temperature and pressure.
- An explosive listed as Class A, Class B, or ''forbidden'' in accordance with 49 CFR 173.

A reactive waste which is not listed elsewhere as a hazardous waste is given an EPA hazardous-waste number of D003.

4. EP toxic waste (EP toxicity), hazard code ''E''. If the extract from a representative sample of this waste (EP, extract procedure) contains contamination in excess of that allowed in Table 2-5, it is classified as a hazardous waste.

Table 2-5. Allowable Contaminants.

EPA hazardous waste number	Contaminant	Maximum concentration (milligrams per liter)
D004	Arsenic	5.0
D005	Barium	100.0
D006	Cadmium	1.0
D007	Chromium	5.0
D008	Lead	5.0
D009	Mercury	0.2
D010	Selenium	1.0
D011	Silver	5.0
D012	Endrin (1,2,3,4,10,10-hexachloro-1,7-epoxy-1,4,4a,5,6,7,8,8a-octahydro-1,4-endo, endo-5,8-dimethano-naphthalene.	0.02
D013	Lindane (1,2,3,4,5,6-hexa- chlorocyclohexane, gamma isomer.	0.4
D014	Methoxychlor (1,1,1-Trichloro-2,2-bis [p-methoxyphenyl]ethane).	10.0
D015	Toxaphene ($C_{10}H_{10}Cl_8$, Technical chlorinated camphene, 67–69 percent chlorine).	0.5
D016	2,4-D, (2,4-Dichlorophenoxyacetic acid).	10.0
D017	2,4,5-TP Silvex (2,4,5-Trichlorophenoxypropionic acid).	1.0

An EP toxic waste which is not listed elsewhere as a hazardous waste is given an EPA hazardous-waste number corresponding to the number of the contaminant listed in Table 2-5 causing it to be hazardous.

5. Acute hazardous waste, hazard code "H". A substance that has been found to be fatal to humans in low doses, or in the absence of data on human toxicity, has been found to be fatal in corresponding human concentrations in small animals.
6. Toxic waste, hazard code "T". Wastes that have been found, through laboratory studies, to have a *carcinogenic, mutagenic,* or *teratogenic* effect on human or other life forms. Definitions of these terms are as follows:

- Carcinogenic: Producing or tending to produce cancer.
- Mutagenic: Capable of inducing mutations in future offspring.
- Teratogenic: Producing abnormal growth in fetuses.

HAZARDOUS-WASTE LISTINGS

A set of lists based on the above hazardous-waste classifications have been included in the RCRA regulations identifying hazardous wastes:

1. Nonspecific sources. The list in Table 2-6 identifies *nonspecific sources*, the wastes of which are hazardous. The hazard code at the right of each item refers to the quality of the waste which is responsible for its hazardous classification.

2. Specific sources. Table 2-7 lists specific processes which generate wastes classified as hazardous. The waste's hazard code identifies the reason(s) for its hazardous classification.

3. Acute hazardous wastes. The substances identified in Table 2-8, when discarded, are classified as hazardous wastes.

4. Toxic wastes. The substances listed in Table 2-9, when discarded, are classified as toxic hazardous wastes.

For a more comprehensive and up-to-date discussion and listing of hazardous wastes, see 40 CFR 261. These lists change from time to time, with the addition of new wastes and, occasionally, the deletion of wastes previously listed.

ANCILLARY MATERIALS

The container or container liner in contact with a hazardous waste is also hazardous. In addition, clothing, debris, soil, etc. that has become contaminated with a hazardous waste is considered hazardous. These hazardous-waste definitions are subject to the small quantity exclusion noted subsequently in this chapter.

NON-HAZARDOUS WASTES

A number of wastes are specifically excluded from classification as hazardous under RCRA, including the following:

1. domestic sewage;
2. irrigation return flows;
3. nuclear waste;
4. household waste;
5. wastes generated from the growing of crops and the raising of animals (manure) and which are returned to the soil as fertilizers;
6. mining overburden returned to the mine site;
7. fly ash, bottom ash, slag waste, and waste from flue gas emissions control systems when generated from the burning of coal or other fossil fuels; and
8. wastes associated with the exploration, development or production of crude oil, natural gas, or geothermal energy such as drilling fluids and oil-laden wastes.

Table 2-6. Hazardous Waste from Non-Specific Sources.

Industry and EPA hazardous waste No.	Hazardous waste	Hazard code
Generic:		
F001	The following spent halogenated solvents used in degreasing: tetrachloroethylene, trichloroethylene, methylene chloride, 1,1,1-trichloroethane, carbon tetrachloride, and chlorinated fluorocarbons; and sludges from the recovery of these solvents in degreasing operations.	(T)
F002	The following spent halogenated solvents: tetrachloroethylene, methylene chloride, trichloroethylene, 1,1,1-trichloroethane, chlorobenzene, 1,1,2-trichloro-1,2,2-trifluoroethane, ortho-dichlorobenzene, and trichlorofluoromethane; and the still bottoms from the recovery of these solvents.	(T)
F003	The following spent non-halogenated solvents: xylene, acetone, ethyl acetate, ethyl benzene, ethyl ether, methyl isobutyl ketone, n-butyl alcohol, cyclohexanone, and methanol; and the still bottoms from the recovery of these solvents.	(I)
F004	The following spent non-halogenated solvents: cresols and cresylic acid, and nitrobenzene; and the still bottoms from the recovery of these solvents.	(T)
F005	The following spent non-halogenated solvents: toluene, methyl ethyl ketone, carbon disulfide, isobutanol, and pyridine; and the still bottoms from the recovery of these solvents.	(I, T)
F006	Wastewater treatment sludges from electroplating operations except from the following processes: (1) sulfuric acid anodizing of aluminum; (2) tin plating on carbon steel; (3) zinc plating (segregated basis) on carbon steel; (4) aluminum or zinc-aluminum plating on carbon steel; (5) cleaning/stripping associated with tin, zinc and aluminum plating on carbon steel; and (6) chemical etching and milling of aluminum.	(T)
F019	Wastewater treatment sludges from the chemical conversion coating of aluminum	(T)
F007	Spent cyanide plating bath solutions from electroplating operations	(R, T)
F008	Plating bath residues from the bottom of plating baths from electroplating operations where cyanides are used in the process.	(R, T)
F009	Spent stripping and cleaning bath solutions from electroplating operations where cyanides are used in the process.	(R, T)
F010	Quenching bath residues from oil baths from metal heat treating operations where cyanides are used in the process.	(R, T)
F011	Spent cyanide solutions from salt bath pot cleaning from metal heat treating operations.	(R, T)
F012	Quenching waste water treatment sludges from metal heat treating operations where cyanides are used in the process.	(T)
F024	Wastes, including but not limited to, distillation residues, heavy ends, tars, and reactor clean-out wastes from the production of chlorinated aliphatic hydrocarbons, having carbon content from one to five, utilizing free radical catalyzed processes. [This listing does not include light ends, spent filters and filter aids, spent dessicants, wastewater, wastewater treatment sludges, spent catalysts, and wastes listed in § 261.32.].	(T)
FO20	Wastes (except wastewater and spent carbon from hydrogen chloride purification) from the production or manufacturing use (as a reactant, chemical intermediate, or component in a formulating process) of tri- or tetrachlorophenol, or of intermediates used to produce their pesticide derivatives. (This listing does not include wastes from the production of Hexachlorophene from highly purified 2,4,5-trichlorophenol.).	(H)
FO21	Wastes (except wastewater and spent carbon from hydrogen chloride purification) from the production or manufacturing use (as a reactant, chemical intermediate, or component in a formulating process) of pentachlorophenol, or of intermediates used to produce its derivatives.	(H)
FO22	Wastes (except wastewater and spent carbon from hydrogen chloride purification) from the manufacturing use (as a reactant, chemical intermediate, or component in a formulating process) of tetra-, penta-, or hexachlorobenzenes under alkaline conditions.	(H)
FO23	Wastes (except wastewater and spent carbon from hydrogen chloride purification) from the production of materials on equipment previously used for the production or manufacturing use (as a reactant, chemical intermediate, or component in a formulating process) of tri- and tetrachlorophenols. (This listing does not include wastes from equipment used only for the production or use of Hexachlorophene from highly purified 2,4,5-trichlorophenol.).	(H)
FO26	Wastes (except wastewater and spent carbon from hydrogen chloride purification) from the production of materials on equipment previously used for the manufacturing use (as a reactant, chemical intermediate, or component in a formulating process) of tetra-, penta-, or hexachlorobenzene under alkaline conditions.	(H)
FO27	Discarded unused formulations containing tri-, tetra-, or pentachlorophenol or discarded unused formulations containing compounds derived from these chlorophenols. (This listing does not include formulations containing Hexachlorophene sythesized from prepurified 2,4,5-trichlorophenol as the sole component.).	(H)
FO28	Residues resulting from the incineration or thermal treatment of soil contaminated with EPA Hazardous Waste Nos. FO20, FO21, FO22, FO23, FO26, and FO27.	(T)

Table 2-7. Hazardous Waste from Specific Sources.

Industry and EPA hazardous waste No.	Hazardous waste	Hazard code
Wood preservation: K001	Bottom sediment sludge from the treatment of wastewaters from wood preserving processes that use creosote and/or pentachlorophenol.	(T)
Inorganic pigments:		
K002	Wastewater treatment sludge from the production of chrome yellow and orange pigments.	(T)
K003	Wastewater treatment sludge from the production of molybdate orange pigments	(T)
K004	Wastewater treatment sludge from the production of zinc yellow pigments	(T)
K005	Wastewater treatment sludge from the production of chrome green pigments	(T)
K006	Wastewater treatment sludge from the production of chrome oxide green pigments (anhydrous and hydrated).	(T)
K007	Wastewater treatment sludge from the production of iron blue pigments	(T)
K008	Oven residue from the production of chrome oxide green pigments	(T)
Organic chemicals:		
K009	Distillation bottoms from the production of acetaldehyde from ethylene	(T)
K010	Distillation side cuts from the production of acetaldehyde from ethylene	(T)
K011	Bottom stream from the wastewater stripper in the production of acrylonitrile	(R, T)
K013	Bottom stream from the acetonitrile column in the production of acrylonitrile	(R, T)
K014	Bottoms from the acetonitrile purification column in the production of acrylonitrile	(T)
K015	Still bottoms from the distillation of benzyl chloride	(T)
K016	Heavy ends or distillation residues from the production of carbon tetrachloride	(T)
K017	Heavy ends (still bottoms) from the purification column in the production of epichlorohydrin.	(T)
K018	Heavy ends from the fractionation column in ethyl chloride production	(T)
K019	Heavy ends from the distillation of ethylene dichloride in ethylene dichloride production.	(T)
K020	Heavy ends from the distillation of vinyl chloride in vinyl chloride monomer production.	(T)
K021	Aqueous spent antimony catalyst waste from fluoromethanes production	(T)
K022	Distillation bottom tars from the production of phenol/acetone from cumene	(T)
K023	Distillation light ends from the production of phthalic anhydride from naphthalene	(T)
K024	Distillation bottoms from the production of phthalic anhydride from naphthalene	(T)
K093	Distillation light ends from the production of phthalic anhydride from ortho-xylene	(T)
K094	Distillation bottoms from the production of phthalic anhydride from ortho-xylene	(T)
K025	Distillation bottoms from the production of nitrobenzene by the nitration of benzene	(T)
K026	Stripping still tails from the production of methy ethyl pyridines	(T)
K027	Centrifuge and distillation residues from toluene diisocyanate production	(R, T)
K028	Spent catalyst from the hydrochlorinator reactor in the production of 1,1,1-trichloroethane.	(T)
K029	Waste from the product steam stripper in the production of 1,1,1-trichloroethane	(T)
K095	Distillation bottoms from the production of 1,1,1-trichloroethane	(T)
K096	Heavy ends from the heavy ends column from the production of 1,1,1-trichloroethane.	(T)
K030	Column bottoms or heavy ends from the combined production of trichloroethylene and perchloroethylene.	(T)
K083	Distillation bottoms from aniline production	(T)
K103	Process residues from aniline extraction from the production of aniline	(T)
K104	Combined wastewater streams generated from nitrobenzene/aniline production	(T)
K085	Distillation or fractionation column bottoms from the production of chlorobenzenes	(T)
K105	Separated aqueous stream from the reactor product washing step in the production of chlorobenzenes.	(T)
Inorganic chemicals:		
K071	Brine purification muds from the mercury cell process in chlorine production, where separately prepurified brine is not used.	(T)
K073	Chlorinated hydrocarbon waste from the purification step of the diaphragm cell process using graphite anodes in chlorine production.	(T)
K106	Wastewater treatment sludge from the mercury cell process in chlorine production	(T)
Pesticides:		
K031	By-product salts generated in the production of MSMA and cacodylic acid	(T)
K032	Wastewater treatment sludge from the production of chlordane	(T)
K033	Wastewater and scrub water from the chlorination of cyclopentadiene in the production of chlordane.	(T)
K034	Filter solids from the filtration of hexachlorocyclopentadiene in the production of chlordane.	(T)
K097	Vacuum stripper discharge from the chlordane chlorinator in the production of chlordane.	(T)
K035	Wastewater treatment sludges generated in the production of creosote	(T)
K036	Still bottoms from toluene reclamation distillation in the production of disulfoton	(T)
K037	Wastewater treatment sludges from the production of disulfoton	(T)
K038	Wastewater from the washing and stripping of phorate production	(T)

Table 2-7. Continued.

K039	Filter cake from the filtration of diethylphosphorodithioic acid in the production of phorate.	(T)
K040	Wastewater treatment sludge from the production of phorate	(T)
K041	Wastewater treatment sludge from the production of toxaphene	(T)
K098	Untreated process wastewater from the production of toxaphene	(T)
K042	Heavy ends or distillation residues from the distillation of tetrachlorobenzene in the production of 2,4,5-T.	(T)
K043	2,6-Dichlorophenol waste from the production of 2,4-D	(T)
K099	Untreated wastewater from the production of 2,4-D	(T)
Explosives:		
K044	Wastewater treatment sludges from the manufacturing and processing of explosives	(R)
K045	Spent carbon from the treatment of wastewater containing explosives	(R)
K046	Wastewater treatment sludges from the manufacturing, formulation and loading of lead-based initiating compounds.	(T)
K047	Pink/red water from TNT operations	(R)
Petroleum refining:		
K048	Dissolved air flotation (DAF) float from the petroleum refining industry	(T)
K049	Slop oil emulsion solids from the petroleum refining industry	(T)
K050	Heat exchanger bundle cleaning sludge from the petroleum refining industry	(T)
K051	API separator sludge from the petroleum refining industry	(T)
K052	Tank bottoms (leaded) from the petroleum refining industry	(T)
Iron and steel:		
K061	Emission control dust/sludge from the primary production of steel in electric furnaces.	(T)
K062	Spent pickle liquor from steel finishing operations	(C, T)
Secondary lead:		
K069	Emission control dust/sludge from secondary lead smelting	(T)
K100	Waste leaching solution from acid leaching of emission control dust/sludge from secondary lead smelting.	(T)
Veterinary pharmaceuticals:		
K084	Wastewater treatment sludges generated during the production of veterinary pharmaceuticals from arsenic or organo-arsenic compounds.	(T)
K101	Distillation tar residues from the distillation of aniline-based compounds in the production of veterinary pharmaceuticals from arsenic or organo-arsenic compounds.	(T)
K102	Residue from the use of activated carbon for decolorization in the production of veterinary pharmaceuticals from arsenic or organo-arsenic compounds.	(T)
Ink formulation: K086	Solvent washes and sludges, caustic washes and sludges, or water washes and sludges from cleaning tubs and equipment used in the formulation of ink from pigments, driers, soaps, and stabilizers containing chromium and lead.	(T)
Coking:		
K060	Ammonia still lime sludge from coking operations	(T)
K087	Decanter tank tar sludge from coking operations	(T)

Table 2-8. Acute Hazardous Wastes.

Hazardous waste No.	Substance	Hazardous waste No.	Substance
P023	Acetaldehyde, chloro-	P038	Arsine, diethyl-
P002	Acetamide, N-(aminothioxomethyl)-	P054	Aziridine
P057	Acetamide, 2-fluoro-	P013	Barium cyanide
P058	Acetic acid, fluoro-, sodium salt	P024	Benzenamine, 4-chloro-
P066	Acetimidic acid, N-[(methylcarbamoyl)oxy]thio-, methyl ester	P077	Benzenamine, 4-nitro-
		P028	Benzene, (chloromethyl)-
P001	3-(alpha-Acetonylbenzyl)-4-hydroxycoumarin and salts, when present at concentrations greater than 0.3%	P042	1,2-Benzenediol, 4-[1-hydroxy-2-(methylamino)ethyl]-
		P014	Benzenethiol
P002	1-Acetyl-2-thiourea	P028	Benzyl chloride
P003	Acrolein	P015	Beryllium dust
P070	Aldicarb	P016	Bis(chloromethyl) ether
P004	Aldrin	P017	Bromoacetone
P005	Allyl alcohol	P018	Brucine
P006	Aluminum phosphide	P021	Calcium cyanide
P007	5-(Aminomethyl)-3-isoxazolol	P123	Camphene, octachloro-
P008	4-aAminopyridine	P103	Carbamimidoselenoic acid
P009	Ammonium picrate (R)	P022	Carbon bisulfide
P119	Ammonium vanadate	P022	Carbon disulfide
P010	Arsenic acid	P095	Carbonyl chloride
P012	Arsenic (III) oxide	P033	Chlorine cyanide
P011	Arsenic (V) oxide	P023	Chloroacetaldehyde
P011	Arsenic pentoxide	P024	p-Chloroaniline
P012	Arsenic trioxide	P026	1-(o-Chlorophenyl)thiourea

Hazardous waste No.	Substance
P027	3-Chloropropionitrile
P029	Copper cyanides
P030	Cyanides (soluble cyanide salts), not elsewhere specified
P031	Cyanogen
P033	Cyanogen chloride
P036	Dichlorophenylarsine
P037	Dieldrin
P038	Diethylarsine
P039	O,O-Diethyl S-[2-(ethylthio)ethyl] phosphorodithioate
P041	Diethyl-p-nitrophenyl phosphate
P040	O,O-Diethyl O-pyrazinyl phosphorothioate
P043	Diisopropyl fluorophosphate
P044	Dimethoate
P045	3,3-Dimethyl-1-(methylthio)-2-butanone, O-[(methylamino)carbonyl] oxime
P071	O,O-Dimethyl O-p-nitrophenyl phosphorothioate
P082	Dimethylnitrosamine
P046	alpha, alpha-Dimethylphenethylamine
P047	4,6-Dinitro-o-cresol and salts
P034	4,6-Dinitro-o-cyclohexylphenol
P048	2,4-Dinitrophenol
P020	Dinoseb
P085	Diphosphoramide, octamethyl-
P039	Disulfoton
P049	2,4-Dithiobiuret
P109	Dithiopyrophosphoric acid, tetraethyl ester
P050	Endosulfan
P088	Endothall
P051	Endrin
P042	Epinephrine
P046	Ethanamine, 1,1-dimethyl-2-phenyl-
P084	Ethenamine, N-methyl-N-nitroso-
P101	Ethyl cyanide
P054	Ethylenimine
P097	Famphur
P056	Fluorine
P057	Fluoroacetamide
P058	Fluoroacetic acid, sodium salt
P065	Fulminic acid, mercury(II) salt (R,T)
P059	Heptachlor
P051	1,2,3,4,10,10-Hexachloro-6,7-epoxy-1,4,4a,5,6,7,8,8a-octahydro-endo,endo-1,4:5,8-dimethanonaphthalene
P037	1,2,3,4,10,10-Hexachloro-6,7-epoxy-1,4,4a,5,6,7,8,8a-octahydro-endo,exo-1,4:5,8-demethanonaphthalene
P060	1,2,3,4,10,10-Hexachloro-1,4,4a,5,8,8a-hexahydro-1,4:5,8-endo, endo-dimeth- anonaphthalene
P004	1,2,3,4,10,10-Hexachloro-1,4,4a,5,8,8a-hexahydro-1,4:5,8-endo,exo-dimethanonaphthalene
P060	Hexachlorohexahydro-exo,exo-dimethanonaphthalene
P062	Hexaethyl tetraphosphate
P116	Hydrazinecarbothioamide
P068	Hydrazine, methyl-
P063	Hydrocyanic acid
P063	Hydrogen cyanide
P096	Hydrogen phosphide
P064	Isocyanic acid, methyl ester
P007	3(2H)-Isoxazolone, 5-(aminomethyl)-
P092	Mercury, (acetato-O)phenyl-
P065	Mercury fulminate (R,T)
P016	Methane, oxybis(chloro-
P112	Methane, tetranitro- (R)
P118	Methanethiol, trichloro-
P059	4,7-Methano-1H-indene, 1,4,5,6,7,8,8-heptachloro-3a,4,7,7a-tetrahydro-
P066	Methomyl
P067	2-Methylaziridine
P068	Methyl hydrazine
P064	Methyl isocyanate
P069	2-Methyllactonitrile
P071	Methyl parathion
P072	alpha-Naphthylthiourea
P073	Nickel carbonyl
P074	Nickel cyanide
P074	Nickel(II) cyanide
P073	Nickel tetracarbonyl
P075	Nicotine and salts
P076	Nitric oxide
P077	p-Nitroaniline
P078	Nitrogen dioxide
P076	Nitrogen(II) oxide
P078	Nitrogen(IV) oxide
P081	Nitroglycerine (R)
P082	N-Nitrosodimethylamine
P084	N-Nitrosomethylvinylamine
P050	5-Norbornene-2,3-dimethanol, 1,4,5,6,7,7-hexachloro, cyclic sulfite
P085	Octamethylpyrophosphoramide
P087	Osmium oxide
P087	Osmium tetroxide
P088	7-Oxabicyclo[2.2.1]heptane-2,3-dicarboxylic acid
P089	Parathion
P034	Phenol, 2-cyclohexyl-4,6-dinitro-
P048	Phenol, 2,4-dinitro-
P047	Phenol, 2,4-dinitro-6-methyl-
P020	Phenol, 2,4-dinitro-6-(1-methylpropyl)-
P009	Phenol, 2,4,6-trinitro-, ammonium salt (R)
P036	Phenyl dichloroarsine
P092	Phenylmercuric acetate
P093	N-Phenylthiourea
P094	Phorate
P095	Phosgene
P096	Phosphine
P041	Phosphoric acid, diethyl p-nitrophenyl ester
P044	Phosphorodithioic acid, O,O-dimethyl S-[2-(methylamino)-2-oxoethyl]ester
P043	Phosphorofluoric acid, bis(1-methylethyl)-ester
P094	Phosphorothioic acid, O,O-diethyl S-(ethylthio)methyl ester
P089	Phosphorothioci acid, O,O-diethyl O-(p-nitrophenyl) ester
P040	Phosphorothioic acid, O,O-diethyl O- pyrazinyl ester
P097	Phosphorothioic acid, O,O-dimethyl O-[p-((dimethylamino)-sulfonyl)phenyl]ester
P110	Plumbane, tetraethyl-
P098	Potassium cyanide
P099	Potassium silver cyanide
P070	Propanal, 2-methyl-2-(methylthio)-, O-[(methylamino)carbonyl]oxime
P101	Propanenitrile
P027	Propanenitrile, 3-chloro-
P069	Propanenitrile, 2-hydroxy-2-methyl-
P081	1,2,3-Propanetriol, trinitrate- (R)
P017	2-Propanone, 1-bromo-
P102	Propargyl alcohol
P003	2-Propenal
P005	2-Propen-1-ol
P067	1,2-Propylenimine
P102	2-Propyn-1-ol
P008	4-Pyridinamine
P075	Pyridine, (S)-3-(1-methyl-2-pyrrolidinyl)-, and salts
P111	Pyrophosphoric acid, tetraethyl ester
P103	Selenourea

Hazardous waste No.	Substance	Hazardous waste No.	Substance
P104	Silver cyanide	P049	Thioimidodicarbonic diamide
P105	Sodium azide	P014	Thiophenol
P106	Sodium cyanide	P116	Thiosemicarbazide
P107	Strontium sulfide	P026	Thiourea, (2-chlorophenyl)-
P108	Strychnidin-10-one, and salts	P072	Thiourea, 1-naphthalenyl-
P018	Strychnidin-10-one, 2,3-dimethoxy-	P093	Thiourea, phenyl-
P108	Strychnine and salts	P123	Toxaphene
P115	Sulfuric acid, thallium(I) salt	P118	Trichloromethanethiol
P109	Tetraethyldithiopyrophosphate	P119	Vanadic acid, ammonium salt
P110	Tetraethyl lead	P120	Vanadium pentoxide
P111	Tetraethylpyrophosphate	P120	Vanadium(V) oxide
P112	Tetranitromethane (R)	P001	Warfarin, when present at concentrations greater than 0.3%
P062	Tetraphosphoric acid, hexaethyl ester		
P113	Thallic oxide	P121	Zinc cyanide
P113	Thallium(III) oxide	P122	Zinc phosphide (R,T)
P114	Thallium(I) selenite	P122	Zinc phosphide, when present at concentrations greater than 10%
P115	Thallium(I) sulfate		
P045	Thiofanox		

USABLE HAZARDOUS WASTE

Regulations applicable to the disposal (incineration) of hazardous wastes are not currently applicable to a hazardous waste that meets either of the following criteria:

- The waste is being recycled or reclaimed.
- The waste is in storage or is being treated, prior to its reclamation.

Reclamation includes the use of a waste for the generation of heat energy. Note, however, that the waste must have significant heating value, that is, a minimum of 6,000 to 8,000 Btu/lb as fired. The regulations state specifically that a waste cannot be fired to avoid regulation unless it can legitimately be used for heat generation and recovery.

BOILER EXCLUSION

Section 40 CFR 260.32 allows certain controlled-flame combustion chambers that might not appear to be conventional boilers to be classified as boilers. A qualifying combustion device is not classified as a hazardous waste incinerator. Test burns and other requirements of a hazardous waste incinerator are not required for combustion chambers classified as boilers. In particular, a boiler requires no RCRA permit.

To qualify as a boiler, a combustion chamber must meet all of the following requirements:

- The unit must have physical provisions for recovering and exporting thermal energy in the form of steam, hot water (or other liquid), or heated gases.
- The unit's combustion chamber and primary energy recovery section(s) must be of integral design. This is defined as the combustion chamber and the primary energy recovery section (such as a

Table 2-9. Toxic Wastes.

Hazardous Waste No.	Substance
U001	Acetaldehyde (I)
U034	Acetaldehyde, trichloro-
U187	Acetamide, N-(4-ethoxyphenyl)-
U005	Acetamide, N-9H-fluoren-2-yl
U112	Acetic acid, ethyl ester (I)
U144	Acetic acid, lead salt
U214	Acetic acid, thallium(I) salt
U002	Acetone (I)
U003	Acetonitrile (I,T)
U248	3-(alpha-Acetonylbenzyl)-4-hydroxycoumarin and salts, when present at concentrations of 0.3% or less
U004	Acetophenone
U005	2-Acetylaminofluorene
U006	Acetyl chloride (C,R,T)
U007	Acrylamide
U008	Acrylic acid (I)
U009	Acrylonitrile
U150	Alanine, 3-[p-bis(2-chloroethyl)amino] phenyl-, L-
U011	Amitrole
U012	Aniline (I,T)
U014	Auramine
U015	Azaserine
U010	Azirino(2′,3′:3,4)pyrrolo(1,2-a)indole-4,7-dione, 6-amino-8-[((aminocarbonyl) oxy)methyl]-1,1a,2,8,8a,8b-hexahydro-8a-methoxy-5-methyl-,
U157	Benz[j]aceanthrylene, 1,2-dihydro-3-methyl-
U016	Benz[c]acridine
U016	3,4-Benzacridine
U017	Benzal chloride
U018	Benz[a]anthracene
U018	1,2-Benzanthracene
U094	1,2-Benzanthracene, 7,12-dimethyl-
U012	Benzenamine (I,T)
U014	Benzenamine, 4,4′-carbonimidoylbis(N,N-dimethyl-
U049	Benzenamine, 4-chloro-2-methyl-
U093	Benzenamine, N,N′-dimethyl-4-phenylazo-
U158	Benzenamine, 4,4′-methylenebis(2-chloro-
U222	Benzenamine, 2-methyl-, hydrochloride
U181	Benzenamine, 2-methyl-5-nitro
U019	Benzene (I,T)
U038	Benzeneacetic acid, 4-chloro-alpha-(4-chlorophenyl)-alpha-hydroxy, ethyl ester
U030	Benzene, 1-bromo-4-phenoxy-
U037	Benzene, chloro-
U190	1,2-Benzenedicarboxylic acid anhydride
U028	1,2-Benzenedicarboxylic acid, [bis(2-ethylhexyl)] ester
U069	1,2-Benzenedicarboxylic acid, dibutyl ester
U088	1,2-Benzenedicarboxylic acid, diethyl ester
U102	1,2-Benzenedicarboxylic acid, dimethyl ester
U107	1,2-Benzenedicarboxylic acid, di-n-octyl ester
U070	Benzene, 1,2-dichloro-
U071	Benzene, 1,3-dichloro-
U072	Benzene, 1,4-dichloro-
U017	Benzene, (dichloromethyl)-
U223	Benzene, 1,3-diisocyanatomethyl- (R,T)
U239	Benzene, dimethyl-(I,T)
U201	1,3-Benzenediol
U127	Benzene, hexachloro-
U056	Benzene, hexahydro- (I)
U188	Benzene, hydroxy-
U220	Benzene, methyl-
U105	Benzene, 1-methyl-1-2,4-dinitro-
U106	Benzene, 1-methyl-2,6-dinitro-
U203	Benzene, 1,2-methylenedioxy-4-allyl-
U141	Benzene, 1,2-methylenedioxy-4-propenyl-
U090	Benzene, 1,2-methylenedioxy-4-propyl-
U055	Benzene, (1-methylethyl)- (I)
U169	Benzene, nitro- (I,T)
U183	Benzene, pentachloro-
U185	Benzene, pentachloro-nitro-
U020	Benzenesulfonic acid chloride (C,R)
U020	Benzenesulfonyl chloride (C,R)
U207	Benzene, 1,2,4,5-tetrachloro-
U023	Benzene, (trichloromethyl)-(C,R,T)
0234	Benzene, 1,3,5-trinitro- (R,T)
U021	Benzidine
U202	1,2-Benzisothiazolin-3-one, 1,1-dioxide
U120	Benzo[j,k]fluorene
U022	Benzo[a]pyrene
U022	3,4-Benzopyrene
U197	p-Benzoquinone
U023	Benzotrichloride (C,R,T)
U050	1,2-Benzphenanthrene
U085	2,2′-Bioxirane (I,T)
U021	(1,1′-Biphenyl)-4,4′-diamine
U073	(1,1′-Biphenyl)-4,4′-diamine, 3,3′-dichloro-
U091	(1,1′-Biphenyl)-4,4′-diamine, 3,3′-dimethoxy-
U095	(1,1′-Biphenyl)-4,4′-diamine, 3,3′-dimethyl-
U024	Bis(2-chloroethoxy) methane
U027	Bis(2-chloroisopropyl) ether
U244	Bis(dimethylthiocarbamoyl) disulfide
U028	Bis(2-ethylhexyl) phthalate
U246	Bromine cyanide
U225	Bromoform
U030	4-Bromophenyl phenyl ether
U128	1,3-Butadiene, 1,1,2,3,4,4-hexachloro-
U172	1-Butanamine, N-butyl-N-nitroso-
U035	Butanoic acid, 4-[Bis(2-chloroethyl)amino] benzene-
U031	1-Butanol (I)
U159	2-Butanone (I,T)
U160	2-Butanone peroxide (R,T)
U053	2-Butenal
U074	2-Butene, 1,4-dichloro- (I,T)
U031	n-Butyl alchohol (I)
U136	Cacodylic acid
U032	Calcium chromate
U238	Carbamic acid, ethyl ester
U178	Carbamic acid, methylnitroso-, ethyl ester
U176	Carbamide, N-ethyl-N-nitroso-
U177	Carbamide, N-methyl-N-nitroso-
U219	Carbamide, thio-
U097	Carbamoyl chloride, dimethyl-
U215	Carbonic acid, dithallium(I) salt
U156	Carbonochloridic acid, methyl ester (I,T)
U033	Carbon oxyfluoride (R,T)
U211	Carbon tetrachloride
U033	Carbonyl fluoride (R,T)
U034	Chloral
U035	Chlorambucil
U036	Chlordane, technical
U026	Chlornaphazine
U037	Chlorobenzene
U039	4-Chloro-m-cresol
U041	1-Chloro-2,3-epoxypropane
U042	2-Chloroethyl vinyl ether
U044	Chloroform
U046	Chloromethyl methyl ether
U047	beta-Chloronaphthalene
U048	o-Chlorophenol
U049	4-Chloro-o-toluidine, hydrochloride
U032	Chromic acid, calcium salt
U050	Chrysene
U051	Creosote
U052	Cresols

Hazardous Waste No.	Substance
U052	Cresylic acid
U053	Crotonaldehyde
U055	Cumene (I)
U246	Cyanogen bromide
U197	1,4-Cyclohexadienedione
U056	Cyclohexane (I)
U057	Cyclohexanone (I)
U130	1,3-Cyclopentadiene, 1,2,3,4,5,5-hexa- chloro-
U058	Cyclophosphamide
U240	2,44-D, salts and esters
U059	Daunomycin
U060	DDD
U061	DDT
U142	Decachlorooctahydro-1,3,4-metheno-2H-cyclobuta[c,d]-pentalen-2-one
U062	Diallate
U133	Diamine (R,T)
U221	Diaminotoluene
U063	Dibenz[a,h]anthracene
U063	1,2:5,6-Dibenzanthracene
U064	1,2:7,8-Dibenzopyrene
U064	Dibenz[a,i]pyrene
U066	1,2-Dibromo-3-chloropropane
U069	Dibutyl phthalate
U062	S-(2,3-Dichloroallyl) diisopropylthiocarbamate
U070	o-Dichlorobenzene
U071	m-Dichlorobenzene
U072	p-Dichlorobenzene
U073	3,3′-Dichlorobenzidine
U074	1,4-Dichloro-2-butene (I,T)
U075	Dichlorodifluoromethane
U192	3,5-Dichloro-N-(1,1-dimethyl-2-propynyl) benzamide
U060	Dichloro diphenyl dichloroethane
U061	Dichloro diphenyl trichloroethane
U078	1,1-Dichloroethylene
U079	1,2-Dichloroethylene
U025	Dichloroethyl ether
U081	2,4-Dichlorophenol
U082	2,6-Dichlorophenol
U240	2,4-Dichlorophenoxyacetic acid, salts and esters
U083	1,2-Dichloropropane
U084	1,3-Dichloropropene
U085	1,2:3,4-Diepoxybutane (I,T)
U108	1,4-Diethylene dioxide
U086	N,N-Diethylhydrazine
U087	O,O-Diethyl-S-methyl-dithiophosphate
U088	Diethyl phthalate
U089	Diethylstilbestrol
U148	1,2-Dihydro-3,6-pyradizinedione
U090	Dihydrosafrole
U091	3,3′-Dimethoxybenzidine
U092	Dimethylamine (I)
U093	Dimethylaminoazobenzene
U094	7,12-Dimethylbenz[a]anthracene
U095	3,3′-Dimethylbenzidine
U096	alpha,alpha-Dimethylbenzylhydroperoxide (R)
U097	Dimethylcarbamoyl chloride
U098	1,1-Dimethylhydrazine
U099	1,2-Dimethylhydrazine
U101	2,4-Dimethylphenol
U102	Dimethyl phthalate
U103	Dimethyl sulfate
U105	2,4-Dinitrotoluene
U106	2,6-Dinitrotoluene
U107	Di-n-octyl phthalate
U108	1,4-Dioxane
U109	1,2- Diphenylhydrazine
U110	Dipropylamine (I)
U111	Di-N-propylnitrosamine
U001	Ethanal (I)
U174	Ethanamine, N-ethyl-N-nitroso-
U067	Ethane, 1,2-dibromo-
U076	Ethane, 1,1-dichloro-
U077	Ethane, 1,2-dichloro-
U114	1,2-Ethanediylbiscarbamodithioic acid
U131	Ethane, 1,1,1,2,2,2-hexachloro-
U024	Ethane, 1,1′-[methylenebis(oxy)]bis[2-chloro-
U003	Ethanenitrile (I, T)
U117	Ethane,1,1′-oxybis- (I)
U025	Ethane, 1,1′-oxybis[2-chloro-
U184	Ethane, pentachloro-
U208	Ethane, 1,1,1,2-tetrachloro-
U209	Ethane, 1,1,2,2-tetrachloro-
U218	Ethanethioamide
U247	Ethane, 1,1,1,-trichloro-2,2-bis(p-methoxy-phenyl).
U227	Ethane, 1,1,2-trichloro-
U043	Ethene, chloro-
U042	Ethene, 2-chloroethoxy-
U078	Ethene, 1,1-dichloro-
U079	Ethene, trans-1,2-dichloro-
U210	Ethene, 1,1,2,2-tetrachloro-
U173	Ethanol, 2,2′-(nitrosoimino)bis-
U004	Ethanone, 1-phenyl-
U006	Ethanoyl chloride (C,R,T)
U112	Ethyl acetate (I)
U113	Ethyl acrylate (I)
U238	Ethyl carbamate (urethan)
U038	Ethyl 4,4′-dichlorobenzilate
U114	Ethylenebis(dithiocarbamic acid)
U067	Etylene dibromide
U077	Ethylene dichloride
U115	Ethlene oxide (I,T)
U116	Ethylene thiourea
U117	Ethyl ether (I)
U076	Ethylidene dichloride
U118	Ethylmethacrylate
U119	Ethyl methanesulfonate
U139	Ferric dextran
U120	Fluoranthene
U122	Formaldehyde
U123	Formic acid (C,T)
U124	Furan (I)
U125	2-Furancarboxaldehyde (I)
U147	2,5-Furandione
U213	Furan, tetrahydro- (I)
U125	Furfural (I)
U124	Furfuran (I)
U206	D-Glucopyranose, 2-deoxy-2(3-methyl-3-nitrosoureido)-
U126	Glycidylaldehyde
U163	Guanidine, N-nitroso-N-methyl-N′nitro-
U127	Hexachlorobenzene
U128	Hexachlorobutadiene
U129	Hexachlorocyclohexane (gamma isomer)
U130	Hexachlorocyclopentadiene
U131	Hexachloroethane
U132	Hexachlorophene
U243	Hexachloropropene
U133	Hydrazine (R,T)
U086	Hydrazine, 1,2-diethyl-
U098	Hydrazine, 1,1-dimethyl-
U099	Hydrazine, 1,2-dimethyl-
U109	Hydrazine, 1,2-diphenyl-
U134	Hydrofluoric acid (C,T)
U134	Hydrogen fluoride (C,T)
U135	Hydrogen sulfide
U096	Hydroperoxide, 1-methyl-1-phenylethyl- (R)
U136	Hydroxydimethylarsine oxide
U116	2-Imidazolidinethione
U137	Indeno[1,2,3-cd]pyrene
U139	Iron dextran
U140	Isobutyl alcohol (I,T)
U141	Isosafrole
U142	Kepone
U143	Lasiocarpine

Hazardous Waste No.	Substance
U144	Lead acetate
U145	Lead phosphate
U146	Lead subacetate
U129	Lindane
U147	Maleic anhydride
U148	Maleic hydrazide
U149	Malononitrile
U150	Melphalan
U151	Mercury
U152	Methacrylonitrile (I,T)
U092	Methanamine, N-methyl- (I)
U029	Methane, bromo-
U045	Methane, chloro- (I,T)
U046	Methane, chloromethoxy-
U068	Methane, dibromo-
U080	Methane, dichloro-
U075	Methane, dichlorodifluoro-
U138	Methane, iodo-
U119	Methanesulfonic acid, ethyl ester
U211	Methane, tetrachloro-
U121	Methane, trichlorofluoro-
U153	Methanethiol (I,T)
U225	Methane, tribromo-
U044	Methane, trichloro-
U121	Methane, trichlorofluoro-
U123	Methanoic acid (C,T)
U036	4,7-Methanoindan, 1,2,4,5,6,7,8,8-octachloro-3a,4,7,7a-tetrahydro-
U154	Methanol (I)
U155	Methapyrilene
U247	Methoxychlor.
U154	Methyl alcohol (I)
U029	Methyl bromide
U186	1-Methylbutadiene (I)
U045	Methyl chloride (I,T)
U156	Methyl chlorocarbonate (I,T)
U226	Methylchloroform
U157	3-Methylcholanthrene
U158	4,4′-Methylenebis(2-chloroaniline)
U132	2,2′-Methylenebis(3,4,6-trichlorophenol)
U068	Methylene bromide
U080	Methylene chloride
U122	Methylene oxide
U159	Methyl ethyl ketone (I,T)
U160	Methyl ethyl ketone peroxide (R,T)
U138	Methyl iodide
U161	Methyl isobutyl ketone (I)
U162	Methyl methacrylate (I,T)
U163	N-Methyl-N′-nitro-N-nitrosoguanidine
U161	4-Methyl-2-pentanone (I)
U164	Methylthiouracil
U010	Mitomycin C
U059	5,12-Naphthacenedione, (8S-cis)-8-acetyl-10-[(3-amino-2,3,6-trideoxy-alpha-L-lyxo-hexopyranosyl)oxyl]-7,8,9,10-tetrahydro-6,8,11-trihydroxy-1-methoxy-
U165	Naphthalene
U047	Naphthalene, 2-chloro-
U166	1,4-Naphthalenedione
U236	2,7-Naphthalenedisulfonic acid, 3,3′-[(3,3′-dimethyl-(1,1′-biphenyl)-4,4′diyl)]-bis(azo)bis(5-amino-4-hydroxy)-,tetrasodium salt
U166	1,4,Naphthaquinone
U167	1-Naphthylamine
U168	2-Naphthylamine
U167	alpha-Naphthylamine
U168	beta-Naphthylamine
U026	2-Naphthylamine, N,N′-bis(2-chloromethyl)-
U169	Nitrobenzene (I,T)
U170	p-Nitrophenol
U171	2-Nitropropane (I)
U172	N-Nitrosodi-n-butylamine
U173	N-Nitrosodiethanolamine
U174	N-Nitrosodiethylamine
U111	N-Nitroso-N-propylamine
U176	N-Nitroso-N-ethylurea
U177	N-Nitroso-N-methylurea
U178	N-Nitroso-N-methylurethane
U179	N-Nitrosopiperidine
U180	N-Nitrosopyrrolidine
U181	5-Nitro-o-toluidine
U193	1,2-Oxathiolane, 2,2-dioxide
U058	2H-1,3,2-Oxazaphosphorine, 2-[bis(2-chloroethyl)amino]tetrahydro-, oxide 2-
U115	Oxirane (I,T)
U041	Oxirane, 2-(chloromethyl)-
U182	Paraldehyde
U183	Pentachlorobenzene
U184	Pentachloroethane
U185	Pentachloronitrobenzene
See F027	Pentachlorophenol
U186	1,3-Pentadiene (I)
U187	Phenacetin
U188	Phenol
U048	Phenol, 2-chloro-
U039	Phenol, 4-chloro-3-methyl-
U081	Phenol, 2,4-dichloro-
U082	Phenol, 2,6-dichloro-
U101	Phenol, 2,4-dimethyl-
U170	Phenol, 4-nitro-
See F027	Phenol, pentachloro-
Do	Phenol, 2,3,4,6-tetrachloro-
Do	Phenol, 2,4,5-trichloro-
Do	Phenol, 2,4,6-trichloro-
U137	1,10-(1,2-phenylene)pyrene
U145	Phosphoric acid, Lead salt
U087	Phosphorodithioic acid, 0,0-diethyl-, S-methylester
U189	Phosphorous sulfide (R)
U190	Phthalic anhydride
U191	2-Picoline
U192	Pronamide
U194	1-Propanamine (I,T)
U110	1-Propanamine, N-propyl- (I)
U066	Propane, 1,2-dibromo-3-chloro-
U149	Propanedinitrile
U171	Propane, 2-nitro- (I)
U027	Propane, 2,2′oxybis[2-chloro-
U193	1,3-Propane sultone
U235	1-Propanol, 2,3-dibromo-, phosphate (3:1)
U126	1-Propanol, 2,3-epoxy-
U140	1-Propanol, 2-methyl- (I,T)
U002	2-Propanone (I)
U007	2-Propenamide
U084	Propene, 1,3-dichloro-
U243	1-Propene, 1,1,2,3,3,3-hexachloro-
U009	2-Propenenitrile
U152	2-Propenenitrile, 2-methyl- (I,T)
U008	2-Propenoic acid (I)
U113	2-Propenoic acid, ethyl ester (I)
U118	2-Propenoic acid, 2-methyl-, ethyl ester
U162	2-Propenoic acid, 2-methyl-, methyl ester (I,T)
See F027	Propionic acid, 2-(2,4,5-trichlorophenoxy)-
U194	n-Propylamine (I,T)
U083	Propylene dichloride
U196	Pyridine
U155	Pyridine, 2-[(2-(dimethylamino)-2-thenylamino]-
U179	Pyridine, hexahydro-N-nitroso-
U191	Pyridine, 2-methyl-
U164	4(1H)-Pyrimidinone, 2,3-dihydro-6-methyl-2-thioxo-
U180	Pyrrole, tetrahydro-N-nitroso-
U200	Reserpine
U201	Resorcinol
U202	Saccharin and salts
U203	Safrole
U204	Selenious acid
U204	Selenium dioxide

Hazardous Waste No.	Substance	Hazardous waste No.	Substance
U205	Selenium disulfide (R,T)	U223	Toluene diisocyanate (R,T)
U015	L-Serine, diazoacetate (ester)	U222	O-Toluidine hydrochloride
See F027	Silvex	U011	1H-1,2,4-Triazol-3-amine
U089	4,4'-Stilbenediol, alpha,alpha'-diethyl-	U226	1,1,1-Trichloroethane
U206	Streptozotocin	U227	1,1,2-Trichloroethane
U135	Sulfur hydride	U228	Trichloroethene
U103	Sulfuric acid, dimethyl ester	U228	Trichloroethylene
U189	Sulfur phosphide (R)	U121	Trichloromonofluoromethane
U205	Sulfur selenide (R,T)	See F027	2,4,5-Trichlorophenol
See F027	2,4,5-T	Do	2,4,6-Trichlorophenol
U207	1,2,4,5-Tetrachlorobenzene	Do	2,4,5-Trichlorophenoxyacetic acid
U208	1,1,1,2-Tetrachloroethane	U234	sym-Trinitrobenzene (R,T)
U209	1,1,2,2-Tetrachloroethane	U182	1,3,5-Trioxane, 2,4,5-trimethyl-
U210	Tetrachloroethylene	U235	Tris(2,3-dibromopropyl) phosphate
See F027	2,3,4,6-Tetrachlorophenol	U236	Trypan blue
U213	Tetrahydrofuran (I)	U237	Uracil, 5[bis(2-chloromethyl)amino]-
U214	Thallium(I) acetate	U237	Uracil mustard
U215	Thallium(I) carbonate	U043	Vinyl chloride
U216	Thallium(I) chloride	U248	Warfarin, when present at concentrations of 0.3% or less
U217	Thallium(I) nitrate		
U218	Thioacetamide	U239	Xylene (I)
U153	Thiomethanol (I,T)	U200	Yohimban-16-carboxylic acid, 11,17-dimethoxy-18-[(3,4,5-trimethoxy-benzoyl)oxy]-, methyl ester
U219	Thiourea		
U244	Thiram		
U220	Toluene	U249	Zinc phosphide, when present at concentrations of 10% or less.
U221	Toluenediamine		

waterwall) being physically formed into one manufactured or assembled unit. A unit in which the combustion chamber and the primary energy recovery section(s) are joined by flues or breechings is not integrally designed. Process heaters (units that transfer energy directly to a process stream) and fluid bed combustion units are not precluded from being of integral design.

- While in operation, the unit must maintain a thermal energy recovery efficiency of at least 60 percent calculated in terms of the recovered energy compared with the thermal value of the fuel.
- The unit must export at least 75 percent of the recovered energy, calculated on an annual basis. In this calculation no credit shall be taken for recovered heat used internally such as combustion air preheat or turbine-driving of fans or pumps by generated steam.

QUANTITY EXCLUSION

Provisions are included in the RCRA regulations to exempt small quantity hazardous-waste generators from the rigorous procedures necessary for compliance with the disposal statutes. The small generator exclusion includes the following provisions:

1. Generation of less than 220 pounds (100 kilograms) per month of hazardous wastes, unless otherwise specified
2. Accumulation of less that 2200 pounds (1000 kilograms) of hazardous waste in any one calendar month, unless otherwise specified

3. A waste containing no more than 2.2 pounds (one kilogram) of any product or chemical listed in Table 2-8, Acute Hazardous Waste
4. Containers not greater than 5.3 gallons (20 liters) or liners no greater than 22 pounds (10 kilograms) in weight, holding materials listed in Table 2-8, Acute Hazardous Waste
5. No greater than 220 pounds (100 kilograms) of residue, soil contaminated from a spill, etc., from a material listed in Table 2-8, Acute Hazardous Waste

The regulations allow that a hazardous waste subject to the small quantity exclusion may be mixed with non-hazardous waste and remain subject to this exclusion even though the resultant mixture exceeds the small quantity limitation.

GENERAL CONSIDERATIONS

The waste generator is responsible to determine if his waste is hazardous, and to classify it with an appropriate hazardous-waste code. The generator is then obligated to report the existence of this waste, its transport, and method of disposal, using a manifest which is described in 40 CFR 260.

These regulations will undoubtedly be subject to intense review as they are applied. Litigation and subsequent revisions are to be expected; these regulations should be regularly monitored because of these expected changes.

HAZARDOUS-WASTE INCINERATION REGULATIONS

If a hazardous waste is hazardous only because it has characteristics of ignitability or corrosivity, or if the waste will be re-used, reclaimed, or used for heat recovery, its incineration need not comply with hazardous-waste incineration criteria. In addition, certain wastes which are hazardous solely because of reactivity may be incinerated without compliance with these criteria.

These criteria require that a trial burn be run on the *principal organic hazardous constituent* (POHC). The POHC is any compound listed in Table 2-10 that is present in the waste. If more than one compound from Table 2-10 can be identified as being present in the waste stream, the selection of the POHC will be a matter of judgment based on the difficulty of incineration and the quantity of that compound present. More than one of those compounds present may be designated as the POHC(s). The final determination of the compound(s) designated as the POHC(s) is reserved for the permit writer, which is either the EPA or the designated state agency. A recommended method for determination of the POHC is described in the following section.

Table 2-10. Hazardous Constituents (Appendix VIII).

Acetaldehyde
(Acetato)phenylmercury
Acetonitrile
3-(alpha-Acetonylbenzyl)-4-hydroxycoumarin and salts
2-Acetylaminofluorene
Acetyl chloride
1-Acetyl-2-thiourea
Acrolein
Acrylamide
Acrylonitrile
Aflatoxins
Aldrin
Allyl alcohol
Aluminum phosphide
4-Aminobiphenyl
6-Amino-1,1a,2,8,8a,8b-hexahydro-8-(hydroxymethyl)-8a-methoxy-5-methylcarbamate azirino(2',3':3,4) pyrrolo(1,2-a)indole-4,7-dione (ester) (Mitomycin C)
5-(Aminomethyl)-3-isoxazolol
4-Aminopyridine
Amitrole
Antimony and compounds, N.O.S.[1]
Aramite
Arsenic and compounds, N.O.S.
Arsenic acid
Arsenic pentoxide
Arsenic trioxide
Auramine
Azaserine
Barium and compounds, N.O.S.
Barium cyanide
Benz[c]acridine
Benz[a]anthracene
Benzene
Benzenearsonic acid
Benzenethiol
Benzidine
Benzo[a]anthracene
Benzo[b]fluoranthene
Benzo[j]fluoranthene
Benzo[a]pyrene
Benzotrichloride
Benzyl chloride
Beryllium and compounds, N.O.S.
Bis(2-chloroethoxy)methane
Bis(2-chloroethyl) ether
N,N-Bis(2-chloroethyl)-2-naphthylamine
Bis(2-chloroisopropyl) ether
Bis(chloromethyl) ether
Bis(2-ethylhexyl) phthalate
Bromoacetone
Bromomethane
4-Bromophenyl phenyl ether
Brucine
2-Butanone peroxide
Butyl benzyl phthalate
2-sec-Butyl-4,6-dinitrophenol (DNBP)
Cadmium and compounds, N.O.S.
Calcium chromate
Calcium cyanide
Carbon disulfide
Chlorambucil
Chlordane (alpha and gamma isomers)
Chlorinated benzenes, N.O.S.
Chlorinated ethane, N.O.S.
Chlorinated naphthalene, N.O.S.
Chlorinated phenol, N.O.S.
Chloroacetaldehyde
Chloroalkyl ethers
p-Chloroaniline
Chlorobenzene
Chlorobenzilate
1-(p-Chlorobenzoyl)-5-methoxy-2-methylindole-3-acetic acid
p-Chloro-m-cresol
1-Chloro-2,3-epoxybutane
2-Chloroethyl vinyl ether
Chloroform
Chloromethane
Chloromethyl methyl ether
2-Chloronaphthalene
2-Chlorophenol
1-(o-Chlorophenyl)thiourea
3-Chloropropionitrile
alpha-Chlorotoluene
Chlorotoluene, N.O.S.
Chromium and compounds, N.O.S.
Chrysene
Citrus red No. 2
Copper cyanide
Creosote
Crotonaldehyde
Cyanides (soluble salts and complexes) N.O.S.
Cyanogen
Cyanogen bromide
Cyanogen chloride
Cycasin
2-Cyclohexyl-4,6-dinitrophenol
Cyclophosphamide
Daunomycin
DDD
DDE
DDT
Diallate
Dibenz[a,h]acridine
Dibenz[a,j]acridine

[1] The abbreviation N.O.S. signifies those members of the general class "not otherwise specified" by name in this listing.

Dibenz[a,h]anthracene(Dibenzo[a,h] anthracene)
7H-Dibenzo[c,g]carbazole
Dibenzo[a,e]pyrene
Dibenzo[a,h]pyrene
Dibenzo[a,i]pyrene
1,2-Dibromo-3-chloropropane
1,2-Dibromoethane
Dibromomethane
Di-n-butyl phthalate
Dichlorobenzene, N.O.S.
3,3'-Dichlorobenzidine
1,1-Dichloroethane
1,2-Dichloroethane
trans-1,2-Dichloroethane
Dichloroethylene, N.O.S.
1,1-Dichloroethylene
Dichloromethane
2,4-Dichlorophenol
2,6-Dichlorophenol
2,4-Dichlorophenoxyacetic acid (2,4-D)
Dichloropropane
Dichlorophenylarsine
1,2-Dichloropropane
Dichloropropanol, N.O.S.
Dichloropropene, N.O.S.
1,3-Dichloropropene
Dieldrin
Diepoxybutane
Diethylarsine
0,0-Diethyl-S-(2-ethylthio)ethyl ester of phosphorothioic acid
1,2-Diethylhydrazine
0,0-Diethyl-S-methylester phosphorodithioic acid
0,0-Diethylphosphoric acid, 0-p-nitrophenyl ester
Diethyl phthalate
0,0-Diethyl-0-(2-pyrazinyl)phosphorothioate
Diethylstilbestrol
Dihydrosafrole
3,4-Dihydroxy-alpha-(methylamino)-methyl benzyl alcohol
Di-isopropylfluorophosphate (DFP)
Dimethoate
3,3'-Dimethoxybenzidine
p-Dimethylaminoazobenzene
7,12-Dimethylbenz[a]anthracene
3,3'-Dimethylbenzidine
Dimethylcarbamoyl chloride
1,1-Dimethylhydrazine
1,2-Dimethylhydrazine
3,3-Dimethyl-1-(methylthio)-2-butanone-0-((methylamino) carbonyl)oxime
Dimethylnitrosoamine
alpha,alpha-Dimethylphenethylamine
2,4-Dimethylphenol
Dimethyl phthalate
Dimethyl sulfate
Dinitrobenzene, N.O.S.
4,6-Dinitro-o-cresol and salts
2,4-Dinitrophenol
2,4-Dinitrotoluene
2,6-Dinitrotoluene Di-n-octyl phthalate
1,4-Dioxane
1,2-Diphenylhydrazine
Di-n-propylnitrosamine
Disulfoton
2,4-Dithiobiuret
Endosulfan
Endrin and metabolites
Epichlorohydrin
Ethyl cyanide
Ethylene diamine
Ethylenebisdithiocarbamate (EBDC)
Ethyleneimine
Ethylene oxide
Ethylenethiourea
Ethyl methanesulfonate
Fluoranthene
Fluorine
2-Fluoroacetamide
Fluoroacetic acid, sodium salt
Formaldehyde
Glycidylaldehyde
Halomethane, N.O.S.
Heptachlor
Heptachlor epoxide (alpha, beta, and gamma isomers)
Hexachlorobenzene
Hexachlorobutadiene
Hexachlorocyclohexane (all isomers)
Hexachlorocyclopentadiene
Hexachloroethane
1,2,3,4,10,10-Hexachloro-1,4,4a,5,8,8a-hexahydro-1,4:5,8-endo,endo-dimethanonaphthalene
Hexachlorophene
Hexachloropropene
Hexaethyl tetraphosphate
Hydrazine
Hydrocyanic acid
Hydrogen sulfide
Indeno(1,2,3-c,d)pyrene
Iodomethane
Isocyanic acid, methyl ester
Isosafrole
Kepone
Lasiocarpine
Lead and compounds, N.O.S.
Lead acetate
Lead phosphate
Lead subacetate
Maleic anhydride
Malononitrile
Melphalan
Mercury and compounds, N.O.S.
Methapyrilene
Methomyl
2-Methylaziridine
3-Methylcholanthrene
4,4'-Methylene-bis-(2-chloroaniline)

Methyl ethyl ketone (MEK)
Methyl hydrazine
2-Methyllactonitrile
Methyl methacrylate
Methyl methanesulfonate
2-Methyl-2-(methylthio)propionaldehyde-o-(methylcarbonyl) oxime
N-Methyl-N'-nitro-N-nitrosoguanidine
Methyl parathion
Methylthiouracil
Mustard gas
Naphthalene
1,4-Naphthoquinone
1-Naphthylamine
2-Naphthylamine
1-Naphthyl-2-thiourea
Nickel and compounds, N.O.S.
Nickel carbonyl
Nickel cyanide
Nicotine and salts
Nitric oxide
p-Nitroaniline
Nitrobenzene
Nitrogen dioxide
Nitrogen mustard and hydrochloride salt
Nitrogen mustard N-oxide and hydrochloride salt
Nitrogen peroxide
Nitrogen tetroxide
Nitroglycerine
4-Nitrophenol
4-Nitroquinoline-1-oxide
Nitrosamine, N.O.S.
N-Nitrosodi-N-butylamine
N-Nitrosodiethanolamine
N-Nitrosodiethylamine
N-Nitrosodimethylamine
N-Nitrosodiphenylamine
N-Nitrosodi-N-propylamine
N-Nitroso-N-ethylurea
N-Nitrosomethylethylamine
N-Nitroso-N-methylurea
N-Nitroso-N-methylurethane
N-Nitrosomethylvinylamine
N-Nitrosomorpholine
N-Nitrosonornicotine
N-Nitrosopiperidine
N-Nitrosopyrrolidine
N-Nitrososarcosine
5-Nitro-o-toluidine
Octamethylpyrophosphoramide
Oleyl alcohol condensed with 2 moles ethylene oxide
Osmium tetroxide
7-Oxabicyclo[2.2.1]heptane-2,3-dicarboxylic acid
Parathion
Pentachlorobenzene
Pentachloroethane
Pentachloronitrobenzene (PCNB)
Pentacholorophenol
Phenacetin
Phenol
Phenyl dichloroarsine
Phenylmercury acetate
N-Phenylthiourea
Phosgene
Phosphine
Phosphorothioic acid, O,O-dimethyl ester, O-ester with N,N-dimethyl benzene sulfonamide
Phthalic acid esters, N.O.S.
Phthalic anhydride
Polychlorinated biphenyl, N.O.S.
Potassium cyanide
Potassium silver cyanide
Pronamide
1,2-Propanediol
1,3-Propane sultone
Propionitrile
Propylthiouracil
2-Propyn-1-ol
Pryidine
Reserpine
Saccharin
Safrole
Selenious acid
Selenium and compounds, N.O.S.
Selenium sulfide
Selenourea
Silver and compounds, N.O.S.
Silver cyanide
Sodium cyanide
Streptozotocin
Strontium sulfide
Strychnine and salts
1,2,4,5-Tetrachlorobenzene
2,3,7,8-Tetrachlorodibenzo-p-dioxin (TCDD)
Tetrachloroethane, N.O.S.
1,1,1,2-Tetrachloroethane
1,1,2,2-Tetrachlorcethane
Tetrachloroethene (Tetrachloroethylene)
Tetrachloromethane
2,3,4,6-Tetrachlorophenol
Tetraethyldithiopyrophosphate
Tetraethyl lead
Tetraethylpyrophosphate
Thallium and compounds, N.O.S.
Thallic oxide
Thallium (I) acetate
Thallium (I) carbonate
Thallium (I) chloride
Thallium (I) nitrate
Thallium selenite
Thallium (I) sulfate
Thioacetamide
Thiosemicarbazide
Thiourea
Thiuram
Toluene
Toluene diamine
o-Toluidine hydrochloride

Tolylene diisocyanate
Toxaphene
Tribromomethane
1,2,4-Trichlorobenzene
1,1,1-Trichloroethane
1,1,2-Trichloroethane
Trichloroethene (Trichloroethylene)
Trichloromethanethiol
2,4,5-Trichlorophenol
2,4,6-Trichlorophenol
2,4,5-Trichlorophenoxyacetic acid (2,4,5-T)
2,4,5-Trichlorophenoxypropionic acid (2,4,5-TP) (Silvex)
Trichloropropane, N.O.S.
1,2,3-Trichloropropane
0,0,0-Triethyl phosphorothioate
Trinitrobenzene
Tris(1-azridinyl)phosphine sulfide
Tris(2,3-dibromopropyl) phosphate
Trypan blue
Uracil mustard
Urethane
Vanadic acid, ammonium salt
Vanadium pentoxide (dust)
Vinyl chloride
Vinylidene chloride
Zinc cyanide
Zinc phosphide

The trial burn is one phase of a four-phase permit process. Permitting provides for a one-month shake-down phase prior to performance of a test burn. The test burn is the second phase of the permitting process. The third phase comprises limited operation while results of the test burn are evaluated. The fourth phase is the final, or permanent, operating phase. The permit requirements, promulgated in 40 CFR 122, 40 CFR 264, and 40 CFR 265 include the following provisions.

1. Identification, by the generator, of the Appendix VIII (Table 9-10) compounds present, from which the POHC(s) will be selected.
2. Operation of incineration equipment to achieve a *destruction and removal efficiency* (DRE) of at least 99.99 percent (*four nines destruction*). The DRE is defined, with W_{in} the POHC mass flow rate into the system and W_{out} the POHC mass flow rate leaving in the incinerator exhaust to atmosphere, as follows:

$$DRE = 100\% * (W_{in} - W_{out}) / W_{in}$$

3. If hydrogen chloride exits the stack at less than four pounds per hour, no HC1 removal is necessary. If the stack emission contains in excess of four pounds per hour of hydrogen chloride, 99 percent of the hydrogen chloride exiting the incinerator must be removed from the exhaust gas stream.
4. Particulate emissions into the atmosphere must not exceed 0.08 grains per dry standard cubic foot when corrected to 50 percent excess air. A correction to 50 percent excess air is calculated by the following formula:

$$P_c = P_m * 14 / (21 - Y),$$

where P_m is the measured particulate, Y is the measured oxygen by volume and P_c is the corrected particulate concentration.

5. Continuous monitoring of combustion temperature, waste feed rate, combustion gas flow rate, and carbon monoxide (CO in the exhaust stream) is required.

   In lieu of a trial burn, which requires EPA (or state) approval prior to its implementation, operating criteria can be established on the basis of published data on disposal of the POHC(s) in question. The published data, however, must have been generated from incinerator equipment similar in type to the proposed incinerator, in a regulated and verified test burn.

POHC DETERMINATION

The permit writer (the EPA or state regulatory agency) identifies the POHC from the waste analysis provided by the generator. A number of methods have been proposed to identify the compound most difficult to incinerate. A ranking based on flash point has been proposed, as well as rankings based on heating value and on concentration of the contaminant in the waste stream.

The best indicator of incinerability appears to be a function of both heating value and concentration, as follows:

$$I = C + (a/H),$$

where
a = dimensional constant, 100 kcal/gram;
I = incinerability index, dimensionless;
C = concentration, percent; and
H = heating value, kcal/gram.

Values for H are listed in Table 2-11, alphabetically by compound, and in Table 2-12, by heating value.

The higher the incinerability index, the more difficult the compound is to incinerate. As an example, a hazardous-waste stream contains five Appendix VIII compounds in concentrations as follows:

Compound	C	H	I
Dichloromethane	1.7	1.70	60.5
Chlordane	0.5	2.71	37.4
Hexachlorobenzene	1.1	1.79	57.0
Tribromomethane	0.1	0.13	769.3
DDT	2.0	4.51	24.2

The likely candidate for the POHC from this waste stream is tribromomethane. Although it has the lowest concentration of any of the Appendix VIII compounds present, its heating value is extremely low. The other compounds have incineration indices at about the same order of magnitude. An alternate, or additional POHC, if required by the permit writer, would likely include all of these other Appendix VIII compounds in the waste.

Table 2-11. Heat of Combustion of Appendix VIII Compounds, Alphabetical Listing.

Hazardous Constituent	Heat of Combustion kcal/gram
Acetonitrile	7.37
Acetophenone	8.26
3-(alpha-Acetonylbenzyl) -4-hydroxycoumarin and salts (Warfarin)	7.00
2-Acetylaminofluorene	7.92
Acetyl chloride	2.77
1-Acetyl-2-thiourea	4.55
Acrolein	6.95
Acrylamide	5.75
Acrylonitrile	7.93
Aflatoxins	5.73
Aldrin	3.75
Allyl alcohol	7.75
4-Aminobiphenyl	9.00
6-Amino-1,1a,2,8,8a,8b-hexahydro-8-(hydroxymethyl) 8a-methoxy 5-methylcarbamate azirino (2',3':3,4)pyrrolo(1,2-a)indole-4,7-dione (ester)(Mitomycin C)	5.41
5-(Aminomethyl)-3-isoxazolol	4.78
4-Aminopyridine	7.37
Amitrole	4.01
Aniline	8.73
Auramine	7.69
Azaserine	3.21
Benz(c)acridine	8.92
Benz(a)anthracene	9.39
Benzene	10.03
Benzenearsonic acid	3.40
Benzenethiol	8.43
Benzidine	9.18
Benzo(b)fluoranthene	9.25
Benzo(j)fluoranthene	9.25
Benzo(a)pyrene	9.25
Benzoqulnone	6.07
Benzotrichloride	3.90*
Benzyl chloride	6.18
Bis(2-chloroethoxy)methane	4.60
Bis(2-chloroethyl)ether	3.38
N,N-Bis(2-chloroethyl)-2-naphthylamine	6.64
Bis(2-chlorolsopropyl)ether	4.93
Bis(chloromethyl)ether	1.97
Bis(2-ethylhexyl)phthalate	8.42
Bromoacetone	2.66

Hazardous Constituent	Heat of Combustion kcal/gram
Bromomethane	1.70
4-Bromophenyl phenyl ether	5.84
Brucine	7.42
2-Butanone peroxide	6.96
Butyl benzyl phthalate	8.29
2-sec-Butyl-4,6-dinitrophenol (DNBP)	5.46
Chloral(Trichloroacetaldehyde)	0.80
Chlorambucil	5.93
Chlordane	2.71
Chlorinated benzenes, N.O.S.	N/A
Chlorinated ethane, N.O.S.	N/A
Chlorinated fluorocarbons	N/A
Chlorinated naphthalene, N.O.S.	N/A
Chlorinated phenol, N.O.S.	N/A
Chloroacetaldehyde	2.92
Chloroalkyl ethers	N/A
p-Chloroaniline	6.14
Chlorobenzene	6.60
Chlorobenzilate	5.50
p-Chloro-m-cresol	5.08
1-Chloro-2,3-epoxybutane	5.19
2-Chloroethyl vinyl ether	5.19
Chloroform	.75
Chloromethane	3.25
Chloromethyl methyl ether	3.48
2-Chloronaphthalene	7.37
2-Chlorophenol	6.89
1-(o-Chlorophenyl)thiourea	5.30
3-Chloropropionitrile	4.50
Chrysene	9.37
Citrus Red No. 2	—
Coal tars	N/A
Creosote	N/A
Cresol	8.18
Cresylic Acid	8.09
Crotonaldehyde	7.73
Cyanogen	6.79
Cyanogen bromide	.81
Cyanogen chloride	1.29
Cycasin	3.92
2-Cyclohexyl-4,6-dinitrophenol	5.74
Cyclophosphamide	3.97
Daunomycin	5.70
DDD	5.14
DDE	5.05
DDT	4.51
Diallate	5.62
2,4-D	3.62
Dibenz(a,h)acridine	9.53

Hazardous Constituent	Heat of Combustion kcal/gram
Dibenz(a,j)acridine	9.53
Dibenz(a,h)anthracene	9.40
(Dibenzo(a,h)anthracene) 7ll-Dibenzo(c,g)carbazole	8.90
Dibenzo(a,c)pyrene	9.33
Dibenzo(a,h)pyrene	9.33
Dibenzo(a,i)pyrene	9.33
1,2-Dibromo-3-chloropropane	1.48
1,2-Dibromoethane	1.43
Dibromomethane	0.50
Di-n-butyl phthalate	7.34
Dichlorobenzene, N.O.S.	4.57
3,3'-Dichlorobenzidine	5.72
1,4-Dichloro -2-butene	4.27
Dichlorodifluoromethane	0.22
1,1-Dichloroethane	3.00
1,2-Dichloroethane	3.00
trans-1,2-Dichloroethene	3.00
Dichloroethylene, N.O.S.	2.70
1,1-Dichloroethylene	2.70
Dichloromethane	1.70
Dichloromethylbenzene	5.09
2,4-Dichlorophenol	3.81
2,6-Dichlorophenol	3.81
Dichloropropane	3.99
Dichlorophenylarsine	2.31
1,2-Dichloropropane	3.99
Dichloropropanol, N.O.S.	2.84
Dichloropropene, N.O.S.	3.44
1,3-Dichloropropene	3.44
Dieldrin	5.56
Diepoxybutane	5.74
Diethylarsine	5.25
1,2-Diethylhydrazine	8.68
Diethyl phthalate	6.39
Dihydrosafrole	7.66
3,4-Dihydroxy-alpha-(methylamino)-methyl benzyl alcohol	6.05
Dimethoate	4.02
3,3'-Dimethoxybenzidine	7.36
p-Dimethylaminoazobenzene	6.97
7,12-Dimethylbenz(a)anthracene	9.61
3,3'-Dimethylbenzidine	8.81
Dimethylcarbamoyl chloride	5.08
1,1-Dimethylhydrazine	7.87
1,2-Dimethylhydrazine	7.87
3,3-Dimethyl-1-(methylthio)-2-butanone-0-(methylamino)carbonyl oxime	5.82
Dimethylnitrosoamine	5.14

Hazardous Constituent	Heat of Combustion kcal/gram
alpha,alpha-Dimethylphenethylamine	9.54
2,4-Dimethylphenol	8.51
Dimethyl phthalate	5.74
Dimethyl sulfate	2.86
Dinitrobenzene, N.O.S.	4.15
4,6-Dinitro-o-cresol and salts	4.06
2,4-Dinitrophenol	3.52
2,4-Dinitrotoluene	4.68
2,6-Dinitrotoluene di-n-octyl phthalate	6.67
1,4-Dioxane	6.41
Diphenylamine	9.09
1,2-Diphenylhydrazine	8.73
Di-n-propylnitrosamine	7.83
Disulfoton	5.73
2,4-Dithiobiuret	2.12
Endosulfan	2.33
Endrin	3.46
Ethyl carbamate	4.73
Ethylenebisdithiocarbamate	5.70
Ethyl cyanide	4.57
Ethyleneimine	7.86
Ethylene oxide	6.86
Ethylenethiourea	5.98
Ethyl methacrylate	7.27
Fluoranthene	9.35
2-Fluoroacetamide	3.24
Formaldehyde	4.47
Formic acid	1.32
Glycidylaldehyde	5.74
Halomethane, N.O.S.	N/A
Heptachlor	2.96
Heptachlor epoxide	2.71
Hexachlorobenzene	1.79
Hexachlorobutadiene	2.12
Hexachlorocyclohexane (all isomers)	1.12
Hexachlorocyclopentadiene	2.10
Hexachloroethane	.46
1,2,3,4,10,10-Hexachloro-1,4,4a,5,8,8a -hexahydro-1,4:5,8-endo, endo-dimethanonaphthalene	3.38
Hexachlorophene	3.82
Hexachloropropene	0.70
Hydrazine	4.44
Indeno(1,2,3-c,d)pyrene	8.52
Iodomethane	1.34
Isocyanic acid, methyl ester	4.69
Isobutyl alcohol	8.62
Isosafrole	7.62
Kepone	2.15

Hazardous Constituent	Heat of Combustion kcal/gram
Lasiocarpine	—
Maleic anhydride	3.40
Maleic hydrozide	4.10
Malononitrile	5.98
Melphalan	5.21
Methacrylonitrile	8.55
Methanethiol	5.91
Methapyrilene	7.93
Methomyl	5.20
Methoxychlor	5.59
2-Methylaziridine	9.09
3-Methylcholanthrene	9.57
4,4'-Methylene-bis-(2-chloroaniline)	4.84
Methyl ethyl ketone (MEK)	8.07
Methyl hydrazine	6.78
2-Methyllactonitrile	6.43
Methyl methacrylate	6.52
Methyl methanesulfonate	3.74
2-Methyl-2-(methylthio)propionaldehyde-o-(methylcarbonyl) oxime	5.34
N-Methyl-N'-nitro-N-nitrosoguanidine	4.06
Methylparathion	4.00
Methylthiouracil	4.79
Mustard gas	4.06
Naphthalene	9.62
1,4-Naphthoquinone	6.97
1-Naphthylamine	8.54
2-Naphthylamine	8.54
1-Naphthyl-2-thiourea	7.50
Nicotine and salts	8.92
p-Nitroaniline	5.50
Nitrobenzene	5.50
Nitrogen mustard and hydrochloride salt	4.28
Nitrogen mustard N-oxide and hydrochloride salt	3.56
Nitroglycerine	3.79
4-Nitrophenol	4.95
4-Nitroquinoline-1-oxide	5.59
5-Nitro-o-toluidine	5.98
Nitrosoamine, N.O.S.	N/A
N-Nitrosodi-N-butylamine	8.46
N-Nitrosodiethanolamine	7.02
N-Nitrosodiethylamine	6.86
N-Nitrosodimethylamine	5.14
N-Nitroso-N-ethylurea	3.92
N-Nitrosomethylethylamine	6.13
N-Nitroso-N-methylurea	2.89
N-Nitroso-N-methylurethane	4.18
N-Nitrosomethylvinylamine	7.91

Hazardous Constituent	Heat of Combustion kcal/gram
N-Nitrosomopholine	5.22
N-Nitrosonornicotine	7.07
N-Nitrosopiperidine	7.04
N-Nitrosopyrrolidine	6.43
N-Nitrososarosine	3.19
7-Oxabicyclo(2.2.1)heptane-2,3-dicarboxylic acid	4.70
Paraldehyde	6.30
Parathion	3.61
Pentachlorobenzene	2.05
Pentachloroethane	0.53
Pentachloronitrobenzene(PCNB)	1.62
Pentachlorophenol	2.09
Phenacetin	7.17
Phenol	7.78
Penylenediamine	7.81
Phenyl dichloroarsine	3.12
Phenylmercury acetate	2.71
N-Phenylthiourea	6.91
Phthalic acid eaters, N.O.S.	N/A
Phthalic anhydride	5.29
2-Picoline	8.72
Polychlorinated biphenyl isomers	
Monochloro	7.75
Dichloro	6.36
Trichloro	5.10
Tetrachloro	4.29
Pentachloro	3.66
Hexachloro	3.28
Heptochloro	2.98
Octachloro	2.72
Nonachloro	2.50
Decachloro	2.31
Pronamide	5.72
1,3-Propane sultona	3.67
n-Propylamine	9.58
Propylthiouracil	6.28
2-Propyn-1-ol	7.43
Pyridine	7.83
Reserpine	6.70
Resorcinol	6.19
Saccharin	4.49
Safrola	7.68
Strychnine and salts	8.03
2,4,5-TP	5.58
2,4,5-T	2.87
1,2,4,5-Tetrachlorobenzene	2.61
TCDD	3.43
Tetrachloroethane, N.O.S.	1.39
1,1,1,2-Tetrachloroethane	1.39

Hazardous Constituent	Heat of Combustion kcal/gram
1,1,2,2-Tetrachloroethane	1.39
Tetrachloroethene (Tetrachloroethylene)	1.19
Tetrachloromethane (Carbon tetrachloride)	0.24
2,3,4,6-Tetrachlorophenol	2.23
Tetraethyl lead	4.04
Tetranitromethane	0.41
Thioacetamide	5.95
Thiosemicarbaxide	4.55
Thiourea	4.55
Thiuram	5.85
Toluene	10.14
Toluene diamine	8.24
o-Toluidine hydrochloride	6.63
Toluene diisocyanate	5.92
Toxephene	2.50
Tribromomethane	0.13
1,2,4-Trichlorobenzene	3.40
1,1,1-Trichloroethane	1.99
1,1,2-Trichloroethane	1.99
Trichloroethane (Trichloroethylene)	1.74
Trichloromethanethiol	0.84
Trichloromonofluoro methane	0.11
2,4,5-Trichlorophenol	2.88
2,4,6-Trichlorophenol	2.88
Trichloropropane, N.O.S.	2.81
1,2,3-Trichloropropane	2.81
Trypan blue	3.84
Uracil mustard	4.00
Vinyl chloride	4.45

Table 2-12. Combustion of Appendix VIII Compounds, by Heat of Combustion.

Hazardous Constituent	Heat of Combustion kcal/gram
Trichloromonofluoromethane	0.11
Tribromomethane	0.13
Dichlorodifluoromethane	0.22
Tetrachloromethane (Carbon tetrachloride)	0.24
Tetranitromethane	0.41
Hexachloroethane	0.46
Dibromomethane	0.50
Pentachloroethane	0.53

Hazardous Constituent	Heat of Combustion kcal/gram
Hexachelloropropane	0.70
Chloroform	0.75
Chloral(trichloroacetaldehyde)	0.80
Cyanogen bromide	0.81
Trichloromethanetiol	0.84
Hexachlorocyclohexane	1.12
Tetrachloroethane (Tetrachloroethylene)	1.19
Cyanogen chloride	1.29
Formic acid	1.32
Iodomethane	1.34
Tetrachloroethane, N.O.S.	1.39
1,1,1,2-Tetrachloroethane	1.39
1,1,2,2-Tetrachloroethane	1.39
1,2-Dibromomethane	1.43
1,2-Dibromo-3-chloropropane	1.48
Pentachloronitrobenzene	1.62
Bromomethane	1.70
Dichloromethane	1.70
Trichloroathene (Trichloroethylene)	1.74
Hexachlorobenzene	1.79
Bis (chloromethyl) ether	1.97
1,1,1-Trichloroethane	1.99
1,1,2-Trichlorethane	1.99
Pentachlorobenzene	2.05
Pentachlorophenol	2.09
Hexachlorocyclopentadiene	2.10
Hexachlorobutadiene	2.12
Kepone	2.15
2,3,4,6-Tetrachlorophenol	2.23
Dichlorophenylarsine	2.31
Decachlorobiphenyl	2.31
Endosulfan	2.33
Nonachlorobiphenyl	2.50
Toxaphena	2.50
1,2,4,5-Tetrachlorobenzene	2.61
Bromoacetone	2.66
Dichloroethylene, N.O.S.	2.70
1,1-Dichloroethylene	2.70
Chlordane	2.71
Heptachlor epoxide	2.71
Phenylmercury acetate	2.71
Octachlorobiphenyl	2.72
Acetyl chloride	2.77
Trichloropropane, N.O.S.	2.81
1,2,3-Trichloropropane	2.81
Dichloropropanol, N.O.S.	2.84
Dimethyl sulfate	2.86
2,4,5-T	2.87
2,4,5-Trichlorophenol	2.88

Hazardous Constituent	Heat of Combustion kcal/gram
2,4,6-Trichlorophenol	2.88
N-Nitroso-N-methylurea	2.89
Heptachlorobiphenyl	2.98
1,1-Dichloroethane	3.00
1,2-Dichloroethane	3.00
trans-1,2-Dichloroethane	3.00
Phenyl dichloroarsine	3.12
N-Nitrosoarcosine	3.19
Azaserine	3.21
2-Fluoroacetamide	3.24
Chloromethane	3.25
Hexachlorobiphenyl	3.28
Bis (2-chloroethyl) ether	3.38
1,2,3,4,10,10-Hexachloro-1,4,4a,5,7,8a-hexahydro-1,4:5,8-endo, endo-dimet hanonaphthalene	3.38
Benzenearsonic acid	3.40
Maleic anhydride	3.40
1,2,4-Trichlorobenzene	3.40
TCDD	3.43
Dichloropropene, N.O.S.	3.44
1,3-Dichloropropene	3.44
Endrin	3.46
Chloromethyl methyl ether	3.48
2,4-Dinitrophenol	3.52
Nitrogen mustard N-oxide and hydrochloride salt	3.56
Parathion	3.61
2,4-D	3.62
Pentachlorobiphenyl	3.66
1,3-Propane sultone	3.67
Methyl methanesulfonate	3.74
Aldrin	3.75
Nitroglycerine	3.79
2,4-Dichlorophenol	3.81
2,6-Dichlorophenol	3.81
Hexachlorophene	3.82
Trypan blue	3.84
Benzotrichloride	3.90
Cycasin	3.92
N-Nitroso-N-ethylurea	3.92
Cyclophosphamide	3.97
Dichloropropane, N.O.S.	3.99
1,2-Dichloropropane	3.99
Methylparathion	4.00
Uracil mustard	4.00
Amitrole	4.01
Dimethoate	4.02

Hazardous Constituent	Heat of Combustion kcal/gram
Tetraethyl lead	4.04
4,6-Dinitro-o-cresol and salts	4.06
N-Methyl-N -nitro-N-nitrosoguanidine	4.06
Mustard gas	4.06
Maleic hydrazide	4.10
Dinitrobenzene, N.O.S.	4.15
N-Nitroso-N-methylurethane	4.18
1,4-Dichloro-2-butene	4.27
Nitrogen mustard and hydrochloride salt	4.28
Tetrachlorobiphenyl	4.29
Hydrazine	4.44
Vinyl chloride	4.45
Formaldehyde	4.47
Saccharin	4.49
3-Chloropropionitrile	4.50
DDT	4.51
Thiourea	4.51
1-Acetyl-2-thiourea	4.55
Thiosemicarbazide	4.55
Dichlorobenzene, N.O.S.	4.57
Ethyl cyanide	4.57
Bis (2-chloroethoxy) methane	4.60
2,4-Dinitrotoluene	4.68
Isocyanic acid, methyl ester	4.69
7-Oxabicyclo (2.2.1) heptane-2,dicarboxylic acid	4.70
Ethyl carbamate	4.73
5-(Aminomethyl)-3-isoxazolol	4.78
Methylthiouracil	4.79
4,4'-Methylene-bis-(2-chloroaniline)	4.84
Bis (2-chloroisopropyl) ether	4.93
4-Nitrophenol	4.95
DDE	5.05
Dimethylcarbamoyl chloride	5.08
p-Chloro-m-cresol	5.08
Dichloromethylbenzene	5.09
Trichlorobiphenyl	5.10
DDD	5.14
Dimethylnitrosoamine	5.14
N-Nitrosodimethylamine	5.14
Diethylarsine	5.25
Phthalic anhydride	5.29
1-(o-chlorophenyl) thiourea	5.30
2-Methyl-2-(methylthio) propionaldehyde-o-(methylcarbonyl) oxime	5.34
2-sec-Butyl-4,6 dinitrophenol (DNBP)	5.46
p-Nitroaniline	5.50
Chlorobenzilate	5.50

Hazardous Constituent	Heat of Combustion kcal/gram
Dieldrin	5.56
2,4,5-TP	5.58
Methoxychlor	5.59
4-Nitroquinoline-1-oxide	5.59
Diallate	5.62
Daunomycin	5.70
Ethylenebisdithiocarbonate	5.70
3,3'-Dichlorobenzidine	5.72
Pronamide	5.72
Aflatoxins	5.73
Disulfoton	5.73
4,6-Dinitrophenol	5.74
Diepoxybutane	5.74
Dimethyl phthalate	5.74
Glycidylaldehyde	5.74
Acrylamide	5.75
3,3-Dimethyl-1-(methylthio)-2-butanone-0-(methylamino)carbonyl oxime	5.82
4-bromophenyl phenyl ether	5.84
thiuram	5.85
Methanethiol	5.91
Tolylene diisocyanate	5.92
Chlorambucil	5.93
Thioacetamide	5.95
Ethylenethiourea	5.98
Malononitrile	5.98
5-Nitro-o-toluidine	5.98
Nitrobenzene	6.01
3,4-Dihydroxy-alpha-(methylamino)methyl benzyl alcohol	6.05
Benzoquinone	6.07
N-Nitrosomethylethylamine	6.13
p-Chloroaniline	6.14
Benzyl chloride	6.18
Resorcinol	6.19
Propylthiouracil	6.28
Paraldehyde	6.30
Dichlorobiphenyl	6.36
Diethyl phthalate	6.39
Dioxane	6.41
2-Methyllactonitrile	6.43
N-Nitrosopyrrolidone	6.43
Methyl methacrylate	6.52
Chlorobenzene	6.60
o-Toluidine hydrochloride	6.63
N, N-Bis (2-chloroethyl)-2-naphthylamine	6.64
2,6-Dinitrotoluene di-n-octyl phthalate	6.67
Reserpine	6.70
Methyl hydrazine	6.78

Hazardous Constituent	Heat of Combustion kcal/gram
Cyanogen	6.79
Ethylene oxide	6.86
N-Nitrosodiethylamine	6.86
2-Chlorophenol	6.89
N-Phenylthiourea	6.93
Acrolein	6.96
2-Butanone peroxide	6.96
p-Dimethylaminoazobenzene	6.97
1,4-Naphthoquinone	6.97
3-(alpha-Acetonylbenzyl)-4-hydroxycoumarin and salts (Warfarin)	7.00
N-Nitrosodiethanolamine	7.02
N-Nitrosopiperidine	7.04
N-Nitrosonornicotine	7.07
Phenacetin	7.17
Ethyl methacrylate	7.27
Di-n-butyl phthalate	7.34
3,3'-Dimethoxybenzidine	7.36
Acetonitrile	7.37
4-Aminopyridine	7.37
2-Chloronaphthalene	7.37
2 Propyn-l-ol	7.43
1-Naphthyl-2-thiourea	7.50
Isosafrole	7.62
Dihydrosafrole	7.66
Safrole	7.68
Auramine	7.69
Crotonaldehyde	7.73
Allyl alcohol	7.75
Monochlorobiphenyl	7.75
Phenol	7.78
Phenylenediamine	7.81
Di-n-propylnitrosoamine	7.83
Pyridine	7.83
Ethyleneimine	7.86
1,1-Dimethylhydrazine	7.87
1,2-Dimethylhydrazine	7.87
N-Nitrosomethylvinylamine	7.91
2-Acetylaminofluorine	7.82
Acrylonitrile	7.93
Methapyrilene	7.93
Strychnine and salts	8.03
Methyl ethyl ketone (MEK)	8.07
Cresylic acid	8.09
Cresol	8.18
Toluene diamine	8.24
Acetophenone	8.26
Butyl benzyl phthalate	8.29

Hazardous Constituent	Heat of Combustion kcal/gram
Ethyl cyanide	8.32
Bis (2-ethylhexyl) phthalate	8.42
Benzenethiol	8.43
N-Nitrosodi-N-butylamine	8.46
2,4-Dimethylphenol	8.51
Indenol (1,2,3-c,d) pyrene	8.52
Diethylstilbestrol	8.54
1-Naphthylamine	8.54
2-Naphthylamine	8.54
Methacrylonitrile	8.55
Isobutyl alcohol	8.62
1,2-Diethylhydrazine	8.68
2-Picoline	8.72
Aniline	8.73
1,2-Diphenylhydrazine	8.73
3,3'-Dimethoxybenzidine	8.81
7H-Dibenzo (c,g) carbazole	8.90
Benz (c) acridine	8.92
Nicotine and salts	8.92
4-Amino biphenyl	9.00
Diphenylamine	9.09
2-Methylaziridine	9.09
Benzidine	9.18
Benzo (b) fluoranthene	9.25
Benzo (j) fluoranthene	9.25
Benzo (a) pyrene	9.25
Dibenzo (a,e) pyrene	9.33
Dibenzo (a,h) pyrene	9.33
Dibenzo (a,i) pyrene	9.33
Fluoranthene	9.35
Benz (a) authracene	9.39
Dibenz (a,h) anthracene (Dibenzo (a,b) anthracene)	9.40
Dibenz (a,h) acridine	9.53
Dibenz (a,j) acirdine	9.53
alpha, alpha-Dimethylphenethylamine	9.54
3-Methylcholanthrene	9.57
n-Propylamine	9.58
7,12-Dimethylbenz (a) anthracene	9.61
Naphthalene	9.62
Benzene	10.03
Toluene	10.14

TRIAL-BURN PLAN

Unless a waiver, based on the use of published data on a trial burn previously run on similar waste in a similar incinerator, is granted, a trial burn of the incinerator is required to demonstrate compliance of the unit with the RCRA incinerator operating standards. The trial-burn operating conditions will then be written into the permit if the incinerator complies with these standards.

The trial burn is conducted in accordance with a detailed trial-burn plan prepared by the applicant and submitted with the other elements of the application. The trial burn is a significant part of the RCRA Part B submittal and is unique to incineration facilities. It documents all important design and operating features of the unit and establishes the basis for future facility operation.

The trial-burn plan consists of five principal elements:

- Waste analysis. A thorough waste analysis is required as part of the trial-burn plan. Information regarding the presence and concentration of all Appendix VIII constituents, as well as routine variations in these constituents and in over-all waste composition, must be included in this characterization. These data will be used to place limits on the composition of the waste that can be fed to the incinerator; therefore, they must accurately reflect the range of waste feed composition that the facility is likely to process. In addition, from these data the permit writer (the EPA or state agency) will select the POHC(s) for analysis.
  The ash (or non-combustible) content of the waste must be determined to specify permit conditions for allowable variations in the waste feed. Measurement of carbon, hydrogen, sulfur, nitrogen, phosphorus, and oxygen concentrations and the water content of the waste feed is needed to calculate air requirements and to evaluate the proposed excess air rates. Measurement of organically bound chlorides is necessary to determine potential emissions of gaseous hydrogen chloride and to establish permit conditions for allowable variations in waste constituent quantities.
- Engineering description of the unit. The trial-burn plan must include an engineering description of the incinerator system. Items such as incinerator component dimensions, design rates of air and waste feed, auxiliary fuel systems, and descriptions of continuous monitoring devices should also be submitted in this section.
- Sampling and analysis procedures. The sampling and analysis procedures, including those for monitoring process operations, waste and supplemental fuel feed rates, and stack gas composition with respect to POHCs, particulate, and HCl, should be noted. These sampling and analysis methods must be in accordance with published EPA procedures and must be sufficient to allow calculation of the fate of POHCs in the system.
- Test schedule and protocol. A trial-burn schedule and protocol must document the dates and duration of the trial burn, the quantity of

waste feed to be burned and the planned operating conditions for each performance test.

- Control information. Procedures for stopping waste feed, shutting down the incinerator, and controlling emissions in the event of an equipment malfunction or other emergency are to be included in the plan. The *set points* for each operating parameter used with this system must be specified. These are the levels that would activate feed cut-off. These operating limits will be written into the permit and must be sufficient to allow the operator some flexibility in unit operation.

DIOXIN-CONTAINING WASTES

The *Federal Register*, January 14, 1985 (Parts 261, 264, 256 and 270) addressed the identification, treatment, and disposal of wastes containing dioxins and related compounds. Major features of this regulation include the following:

1. All isomers of chlorodibenzo-p-dioxin (CDDS) are classified as equally toxic, and all chlorinated dibenzofurans (CDFs) are classified as equally toxic.
2. Wastes containing CDDS or CDFs are classified under RCRA as "acutely hazardous wastes".
3. An incinerator must be fully permitted under RCRA and it must have demonstrated its ability to successfully destroy POHCs that are more difficult to destroy than CDDs and CDFs before CDD and CDF containing material can be fired.
4. The DRE for CDDs and CDFs must be a minimum of six nines (99.9999 percent).

WHEN CRITERIA ARE NOT AVAILABLE

For purposes of initial design, where test data are not immediately available for a particular waste, a reasonable estimate of combustion criteria can be assumed, as follows:

1. For a non-halogenated hazardous waste provide 1,832° F (1,000° C) in the combustion chamber, with a gas residence time of 2 seconds and 2 percent oxygen in the exhaust (dry volume basis).
2. For a halogenated waste (one containing at least 0.5 percent chlorine) provide a temperature of 2,192° F (1,200° C) in the combustion chamber with a gas residence time of 2 seconds, and 3 percent (dry volume basis) in the exhaust.

Note that these criteria are not stated within RCRA or other federal statutory requirements. They are only suggested here as criteria to be used as a guide to equipment design where no other data are available. They are not to be substituted for information required from a test burn.

PCBs AND TSCA

Incineration of substances containing PCBs (polychlorinated biphenyls) is covered by the Toxic Substances Control Act (TSCA), 44 CFR 106 Paragraph 761.41. Permitting under TSCA is reserved for EPA. States do not have the authority to issue TSCA permits.

Under TSCA a material containing less than 50 ppm PCBs is not controlled. Over 50 ppm PCBs the material must be handled, stored, transported, treqated, and disposed of under the aegis of TSCA, with a TSCA permit. A material to be disposed of with a PCB content in excess of 500 ppm must be disposed of by thermal means (incineration). Between 50 and 500 ppm PCBs thermal treatment or other methods of disposal may be applied.

PCB INCINERATION

The incineration of PCBs is subject to the following criteria, excerpted from TSCA:

1. Combustion at 2,192° F (1,200° C), with a 2 second retention time and 3 percent oxygen in the exhaust gas or 2,192° F (1,600° C), with a 1½ second retention time and 2 percent oxygen in the exhaust.
2. Combustion efficiency (CE) of 99.9 percent. With $[CO_2]$ and [CO] being, respectively, the concentrations of CO_2 and CO in the exhaust gas, the CE is calculated as follows:

$$CE = 100\% \times [CO_2] \,/\, [(CO_2] + [CO])$$

3. The PCB charging rate and total feed must be monitored at least once every 15 minutes.
4. The combustion temperature must be monitored on a continuous basis.
5. Upon a drop in temperature below 2,912° F/1,200° C (or 2,192° F/1,600° C) the flow of PCBs shall automatically cease. PCB flow shall also cease if there is a failure in any of the monitoring operations or if measured oxygen falls below the minimum required.
6. When an incinerator is initially used for the disposal of PCBs the following stack emissions must be monitored:

   - Oxygen (O_2)
   - carbon monoxide (CO)

- Carbon dioxide (CO_2)
- Oxides of nitrogen (NO_x)
- Hydrogen chloride (HCl)
- Total chlorinated organic constituents (RCl)
- PCBs
- Total particulate matter

7. During normal operation of the incinerator the CO_2 concentration in the exhaust gas shall be monitored on a periodic basis. The O_2 and CO fractions of the exhaust gas shall be monitored on a continuous basis.
8. Water scrubbers or equivalent gas-cleaning equipment shall be used to control the HCl emission in the exhaust gas. Spent scrubber water must be monitored and shall be in compliance with applicable effluent standards.
9. If the PCBs to be incinerated are non-liquid, in addition to the above requirements, the emissions discharge shall not be greater than one pound PCB per million pounds of PCBs (six nines, or 99.9999 percent) charged into the furnace.

COMMENTARY ON TSCA INCINERATOR REGULATIONS

The EPA intends to include PCB regulations in the RCRA statutes at a later date, removing these regulations from TSCA. In this transfer of authority changes may be made to these regulations.

REFERENCE

1. *Guidance Manual for Hazardous Waste Permits,* United States Environmental Protection Agency Office of Solid Waste and Emergency Response, PB84-100577, July 1983.

3

Rotary-Kiln Incineration Systems

The rotary-kiln incinerator can handle solid, liquid, gaseous, and sludge wastes. Throughout the world, more rotary-kiln incinerators are used for the destruction of non-liquid hazardous wastes than any other incinerator. The types of rotary kilns and the differences among them will be presented in this chapter.

KILN SYSTEM

A rotary-kiln system used for the incineration of hazardous waste is shown in Fig. 3-1. It includes provisions for waste feed, air injection, the kiln itself, an afterburner, and an ash collection system. The gas discharge from the afterburner is directed to an air emissions control system. An *induced-draft* (ID) fan is provided within the control system to draw gases from the kiln through the equipment line and discharges through a stack to the atmosphere.

As shown in Fig. 3-1, there are a number of areas within the system where leakage can occur. The feeding ports cannot be completely sealed; the kiln seals are also areas of potential leakage. The ash system is normally provided with a water seal, but with dry ash collection there will always be some leakage. To assure that the leakage is into the system, and that no hot, dirty gases leak out of the kiln to the surrounding area, the kiln is maintained with a negative draft. The ID fan is sized to maintain

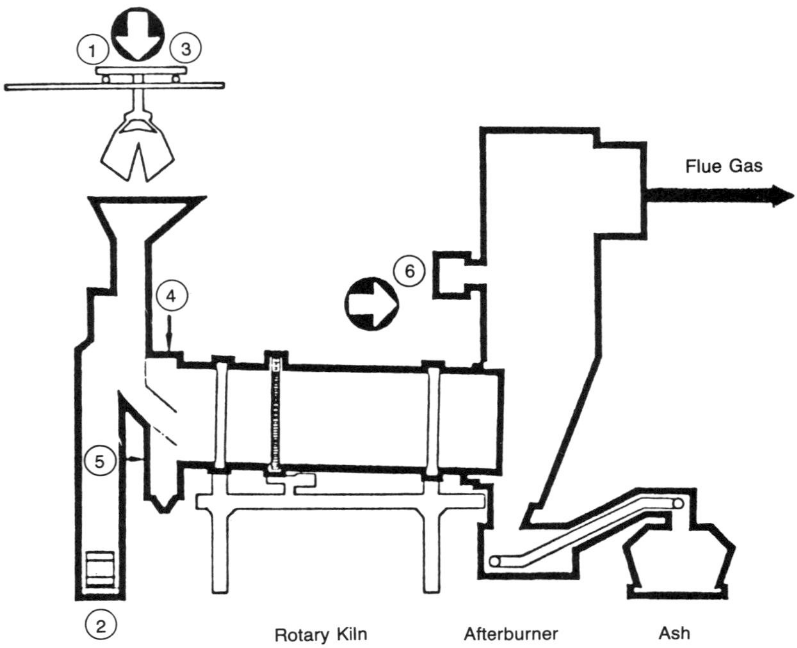

FEED

(1) SOLID WASTE

(2) BARREL CHARGING

(3) SLUDGE

(4) EXHAUST AIR FROM PRODUCTION ROOMS AND TANKS

(5) FUEL OIL

Fig. 3-1. Rotary-kiln system.

a negative pressure throughout the system so that leakage is always into, not out of, the kiln system.

KILN APPLICATION

The rotary kiln can incinerate a wide variety of wastes; however, its application has limitations. Advantages and disadvantages in the use of a rotary kiln as an incinerator can be summarized as follows:

Advantages:

- Ability to incinerate a variety of waste streams
- Minimal waste pre-processing
- Direct disposal of wastes in metal drums

- Ability to incinerate varied types of wastes (solids, liquids, sludges, etc.) at the same time
- Availability of many types of feed mechanisms (ram feeder, screw, direct injection, etc.)
- Readily controlled residence time of waste in kiln
- High turbulence and effective contact with air within kiln

Disadvantages:

- Relatively high particulate carryover to the gas stream
- Separate afterburner normally required for destruction of volatiles
- Conditions along kiln length impossible to control
- Relatively high amount of excess air, nominally 100 percent of stoichiometric, required
- Effective kiln seal difficult to obtain
- Significant amount of heat is lost in the ash discharge
- Operation in a slagging mode to process inorganic wastes or metal drums increases kiln maintenance requirements.

THE ROTARY KILN

The conventional rotary kiln, as shown in Fig. 3-2, is a horizontal cylinder, lined with refractory, which turns about its longitudinal axis. Waste is deposited in the kiln at one end and the waste burns out to an ash by the time it reaches the other end. Kiln rotation speed varies in the range of ¾ to 4 revolutions per minute.

Fig. 3-2. The rotary kiln. Copyright © C. R. Brunner, 1979.

Most kiln designs utilize smooth refractory on the kiln interior. Some designs, particularly those for the processing of granular material (dirt or powders) may have internal vanes or paddles to encourage motion along the kiln length and to promote turbulence of the feed. See Fig. 3-3. Care must be taken in the provision of internal baffles of any kind. With certain material consistencies, such as soil of from 10 to 20 percent moisture content, baffles may tend to retard the movement of material through the kiln.

Fig. 3-3. Paddle system. Source: R. Miller, IT Corp.

The kiln is supported by at least two trunnions. One or more sets of trunnion rollers are idlers. Kiln rotation can be by a set of powered trunnion rollers, as shown in Fig. 3-4, by a gear drive around the kiln periphery, or through a chain driving a large sprocket around the body of the kiln, as shown in Fig. 3-5.

The kiln trunnion supports are adjustable in the vertical direction. The kiln is normally supported at an angle to the horizontal, or *rake*. The rake will normally vary from 2 to 4 percent (¼ inch to ½ inches per foot of length), with the higher end at the feed end of the kiln. Other kiln designs have a zero or slightly negative rake, with lips at the input and discharge ends. These kilns are operated in the slagging mode, with the internal kiln geometry designed to maintain a pool of molten slag between the kiln lips.

A source of heat is required to bring the kiln up to operating temperature, and to maintain its temperature during incineration of the waste feed. Supplemental fuel is normally injected into the kiln through a conventional burner or a ring burner, when gas fuel is used.

Fig. 3-4. Trunnion drive. Copyright © C. R. Brunner, 1981.

There are a number of variations in kiln design, including the following:

- Parallel or counterflow
- Slagging or non-slagging
- Refractory or bare-wall.

The most commonly used kiln design, referred to as the conventional kiln, is a parallel-flow, non-slagging, refractory-lined system.

KILN EXHAUST-GAS FLOW

When gas flow through the kiln is in the same direction as the waste flow, the kiln is said to have parallel or co-current flow, as indicated in the kiln in Fig. 3-1. With counter-current flow, the gas flows opposite the flow of waste. The burner(s) is placed at the front of the kiln, the face of the kiln from which the air-gas mixture originates.

Generally, a counter-current kiln is used when an aqueous waste (a waste with at least 60 percent water content) is to be incinerated. Waste is introduced at the end of the kiln far from the burner. The gases exiting the kiln will dry the aqueous waste, and its temperature will drop. If aqueous waste were dropped into a kiln with co-current flow, water would be evaporated at the feed end of the kiln. The feed end would be the end of the kiln at the lowest temperature, and a much longer kiln would be required for burn-out of the waste.

Fig. 3-5. Chain drive. Copyright © C. R. Brunner, 1982.

Wastes containing a light volatile fraction (containing greases, for instance) should utilize a kiln with co-current flow. These volatiles will likely be released from the feed immediately upon entering the kiln. The co-current kiln provides residence time that a counter-current kiln would not for the burn-out of these volatiles.

SLAGGING MODE

At temperatures in the range of 2000 to 2200° F, ash will start to deform and as the temperature increases, the ash will melt. The actual temperatures of initial deformation and subsequent physical changes to the ash is a function of the chemical constituents present in the waste residual. It is also a function of the presence of oxygen in the furnace. The ash deformation temperatures will vary with reducing versus oxidizing atmospheres. Eutectic properties can be controlled by the use of additives to the molten material.

A kiln can be designed to generate and maintain molten ash during operation. Operation in a slagging mode provides a number of advantages over non-slagging operation. When operation in a non-slagging mode is intended; however, and slagging occurs, it is undesirable and must be eliminated.

The differences in slagging versus non-slagging kilns are outlined in Table 3-1. As noted, the construction of a slagging kiln is more complex than that of a non-slagging kiln, requiring a lip at the kiln exit to contain the molten material. A non-slagging kiln will have a smooth transition with no impediments to the smooth discharge of ash.

Table 3-1. Slagging versus Non-slagging Kiln.

FACTOR	EFFECT
Construction	More complex with slagging kiln
Duty	Slagging kiln can accept drums, salt laden wastes, non-slagging kiln is limited
Temperature	Higher with slagging kiln
Retention time	Greater residence required in non-slagging kiln
Process control	Thermal inertia / forgiveness in slagging kiln
Emissions	Less particulate, greater NO_x in slagging kiln
Slag	Slagging kiln may require CaO, $A1_2O_3$, SiO_2 additives, dissolves drums, salts
Ash	Wet, less leachable with slagging, wet or dry with non-slagging kiln
Maintenance	Higher with slagging kiln
Refractory	More critical with slagging kiln

Slagging kilns have been designed and operated with a negative rake; i.e., the outer surface of the kiln at the feed end is lower than the kiln surface at the discharge end. This permits the accumulation of more slag in the kiln than zero or positive rake. The kiln internal surface must be designed for this operating mode. For instance, as noted previously, an internal refractory lip is required on the kiln feed end.

The slagging kiln can accept metal drums. The ash eutectic properties at the molten slag temperatures will tend to promote the melting of a metal drum placed into the kiln. While depositing drums containing waste in a kiln may be undesirable from a safety and maintenance standpoint (even with the tops of the drums removed, localized heating of the drum surface may occur, causing an explosion, and the impact of a dropping drum will eventually damage kiln refractory), if drums are to be placed in a kiln, slagging kilns are able to absorb the drum into a homogeneous residue discharge. The non-slagging kiln can only move the drum through the unit and must include specialty equipment for handling the drum body as it exits the kiln.

Salt-laden wastes will tend to melt in the range of 1300 to 1600° F and can produce severe caking, or deposits, in a non-slagging kiln. Often, salt-bearing wastes are prohibited from kilns because they will produce an unacceptable build-up on the kiln surface, which can eventually choke off the kiln. A slagging kiln, however, is maintained at temperatures high enough to keep the salts in a molten state. The salts combine with the molten ash in the pool at the bottom of the kiln and are maintained in their molten state until quenched.

The temperature in a slagging kiln must be sufficiently high to maintain the ash as a molten slag. Temperatures as high as 2600 to 2800° F are not uncommon. A non-slagging kiln will normally operate at temperatures below 2000° F.

The destruction of organic compounds is achieved by a combination of high temperature and residence time. Generally, the higher the temperature, the shorter residence time required for destruction. Conversely, the higher the residence time, the lower the required temperature. The use of higher temperatures in the slagging kiln reduce the residence time requirements for the off-gas. The afterburner associated with a slagging kiln can often be much smaller than that required for a non-slagging kiln.

The molten slag can weigh hundreds or thousands of pounds. As a concentrated material, a liquid, it represents a significant thermal inertia within the kiln. The molten slag tends to act as a heat sink which provides thermal stability to the system. The slagging kiln is much less subject to temperature extremes than the non-slagging kiln because of the presence of this massive melt. It will maintain a relatively constant temperature profile under rapid changes in kiln loading. This stability leads to a more predictable system behavior. Safety factors employed in the design and operation of downstream equipment (such as an exhaust gas scrubber or the induced draft fan) can be reduced when a slagging kiln is used.

The tumbling action of a rotating kiln encourages the release of particles to the gas stream. From 5 to 25 percent of the non-volatile solids in a feed stream may become airborne with the use of a conventional non-slagging kiln. The molten slag in a slagging kiln acts as does the fluid ash in a PC (pulverized coal) burner. The slag will absorb particulate matter from the gas stream and can reduce particulate emissions from the kiln to 25 to 75 percent of the emissions from a non-slagging kiln. On the other hand, emissions of NO_x are greater with a slagging kiln than with a non-slagging kiln. The generation of NO_x is practically insignificant until the process temperature exceeds 2000° F. Above this temperature the formation of NO_x will increase substantially. At 2600° F the generation of NO_x is almost ten times as great as its generation at 1800° F.

A danger in slagging kiln operation is that the slag will solidify. When this happens the kiln will be off-balance. With an eccentric, turning kiln, if rotation of the kiln is not stopped, damage to kiln supports and to the kiln drive may occur. The incineration process will degrade during a slag freeze. Operating stability will be lost and demands on downstream equipment (the gas scrubbing system, for instance) may be too severe. One reason for the slag freezing besides a drop in temperature, is a change in the feed quality. To assure the maintenance of adequate eutectic parameters, additives may have to be employed. These additives may include CaO, Al_2O_3, SiO_2 or other compounds, depending upon the nature of the waste. Additives will help maintain the eutectic point, to assure that the slag will remain molten.

The molten slag from a slagging kiln is dropped into a wet sump. The slag immediately hardens into a granular material (*frit*), with the appearance of a dark glass. The ash from a non-slagging kiln can be collected wet or dry. There is a chance that the bath water from a slagging kiln may be considered hazardous, but the frit is much less leachable than ash from a non-slagging kiln. Ash from a non-slagging kiln can be more difficult to de-list.

Refractory for slagging kiln service will experience more severe duty than for non-slagging kiln service. The higher operating temperatures will directly affect refractory life, as will the corrosive effect of the melt. In addition, if steel drums are dropped into the kiln, the physical impact of the drum on the kiln surface will be damaging. The molten slag will absorb the steel and ferrous metals, as well as other metals, which are highly corrosive to refractory. The refractory must resist this corrosive attack, high temperatures, and impact loading. The resulting refractory system will be expensive and will require high maintenance.

OPERATION

The waste retention time in a kiln can vary. It is a function of kiln geometry and kiln speed, represented in the following equation:

$$t = 2.28 \times (L/D) \,/\, (S \times N)$$

where
- t = mean residence time, minutes
- L/D = internal length to diameter ratio
- N = rotational speed, revolutions per minute
- S = kiln rake, inches per foot of length

For a given L/D ratio and rake, the solids residence time within the unit is inversely proportional to the kiln speed. By doubling the speed, one halves the residence time.

KILN SEAL

Sealing a kiln is a difficult task. Efficient kiln operation requires that, to control the infiltration of unwanted air flow into the system, kiln seals be provided and maintained. With too much air, fuel use increases and process control deteriorates.

The kiln turns between two stationary yokes. The kiln's diameter can vary from 6 to 20 feet; its perimeter varies from 18 to 60 feet. At one foot per minute velocity, the kiln surface is moving at a rate of up to 60 feet per minute. A seal must close the gap between the yoke and the kiln surface while the kiln is moving at this surface velocity. The kiln surface is not a machined surface and will have variations in texture and dimension, making the task of sealing very difficult. An additional problem is that the kiln interior is normally at relatively high temperatures, which tends to encourage wear of the kiln surface.

Figure 3-6 illustrates a compression-type kiln seal. Rings of heat resistant, flexible material are held in a frame bolted to the yoke. As the edge of this seal wears, it is adjusted forward, to maintain a seal against the kiln surface.

A friction seal, Fig. 3-7, comprises a series of flexible fingers, holding heat-resistant strips to the kiln surface. As the kiln turns, variations in the surface dimension of the kiln are absorbed by the flexibility and resilience of the seal.

LIQUID WASTE INJECTION

Liquid waste will usually be either aqueous (at least 60 percent water content) or organic. This waste stream can be injected into or outside of the flame envelope.

Fig. 3-6. Compression seal. Copyright © C. R. Brunner, 1981.

Fig. 3-7. Friction seal. Source: R. Miller, IT Corp.

Supplemental fuel that will be heating the kiln should be allowed to burn completely. An aqueous waste stream should be injected downstream of the supplemental fuel flame envelope. Were it injected within this envelope it would quench the flame and the flame would not develop its full temperature.

Liquid waste with a significant heat content, at least 6,000 BTU/lb, would supplement the heat content of the fuel, and can be injected into the flame envelope.

SYSTEM SELECTION

The kiln will burn out solids and will volatilize organics. All the organics will generally not be incinerated in the kiln and will require that a high temperature be maintained at a specific residence time for destruction. This is the purpose of a secondary combustion chamber, or afterburner. An afterburner is normally placed immediately downstream of the kiln. It is a stationary piece of equipment that is designed to maintain the temperature of the gas stream exiting the kiln at a pre-selected temperature of destruction for a specific period of time. The released volatiles will exit the kiln in the flue gas and will enter an afterburner, as shown in Fig. 3-3. The afterburner will normally contain at least one burner to provide the supplemental fuel required for burn-out of the organics in the gas stream.

There are kiln system designs in which the volatiles released from the kiln have a high enough heating value that they require no external source of supplemental fuel for complete combustion. In these instances, the afterburner does not require supplemental fuel and it acts as an extension of the kiln, providing the residence time for burn-out of the organics. This secondary combustion chamber would be initially heated by hot gases from the kiln.

4

Liquid-Waste Incineration

The majority of wastes generated by industry, and the majority of hazardous wastes, are liquid. Land disposal of liquid wastes is severely limited, and hazardous liquid wastes are banned from land disposal. Not all liquids are candidates for incineration, but many are, and a discussion of liquid waste incineration criteria and equipment is presented in this chapter.

LIQUID-WASTE CHARACTERISTICS

Liquid wastes may be aqueous or non-aqueous. An aqueous waste is one with at least a 60 percent water content. Non-aqueous wastes may or may not burn *autogenously* (without the need for supplemental fuel) at their temperature of destruction, but aqueous wastes will always need a source of supplemental fuel for effective burning.

Viscosity is another quality important to a discussion of liquid waste incineration. The higher the viscosity of a material, the most difficult it is to pump, to inject through a nozzle, and to burn.

The line between liquid and non-liquid is not always well-defined. A material is considered a liquid if it can be pumped to a burner and atomized, i.e., fired in suspension. In general, a material can be pumped if its viscosity is less that 10,000 Saybolt-seconds universal (SSU). Atomization is a function of nozzle type as well as liquid properties. For conventional nozzles the stream viscosity should be no greater than 750 SSU for proper atomization to occur. Specialty nozzles have been developed that can atomize a liquid stream with up to 5,000 SSU viscosity.

ATOMIZATION

Atomization is the breaking up of a liquid stream into very small particles in which the ratio of surface area to volume is very large (the area to volume ratio of a spherical particle is 3/r, where r is the particle radius). By maximizing this ratio, the evaporation of water is encouraged and the volatilization and burning of organics within the stream will be rapid and efficient. The process of atomization produces a fine mist or fog.

Almost every liquid stream nozzle will include a means of atomization. Mechanical nozzles utilize a plate or other surface as a target for the liquid stream. The liquid stream is pumped at a relatively high pressure, and the high velocity flow created by this high pressure will strike the target and form small, atomized droplets.

Air atomization utilizes an external air stream to break up the waste stream. Approximately 100 standard cubic feet (scf) of air is required per gallon of waste for atomization. The atomizing air stream is normally provided at from 50 to 75 psig above the pressure of the liquid stream.

Steam is often used for atomization. Besides providing atomization of the liquid stream, steam will heat the flow and will tend to reduce the viscosity of most liquids. As the viscosity decreases, the efficiency of atomization generally increases. Approximately one pound of steam is required to atomize each pound of liquid, or seven pounds of steam per gallon of flow. The steam pressure will normally be set at 75 to 100 psig above the pressure of the liquid stream.

AIR REQUIREMENTS

Whenever an organic material is to be incinerated, a supply of oxygen is necessary for combustion. Usually, oxygen is provided through a supply of air. Air is required to fire liquid wastes and for other purposes, as noted:

- Atomization, to encourage evaporation and efficient burning
- Primary air supply, to provide air for combustion of the organic content of the waste stream
- Secondary air supply, injected external to the burner to shape the flame exiting the burner, to keep the flame away from the walls, and to reduce the generated temperatures as required.

NOZZLE TYPES

The nozzle is a major element in the design of the incinerator and in the selection of ancillary equipment. A nozzle will have a distribution pattern that affects the size of a furnace chamber. It may require a relatively high pressure liquid stream, which must be taken into account in the selection of pumps and piping. If the nozzle is sensitive to solids within the waste stream, strainers or solids filters may be required in the supply piping.

Types of nozzles in common use on waste incineration systems are described as follows:

Mechanical Atomizing Nozzles

Mechanical atomizing nozzles are the most common types of burner nozzles in current use. They require no external source of atomization, such as steam or compressed air. Typical mechanical atomizing nozzles are illustrated in Fig. 4-1 and Fig. 4-2. The liquid stream is pumped into the nozzle at a pressure of 75 to 150 psig through a small fixed-orifice discharge. The stream is given a strong *cyclonic*, or whirling, velocity before it is released through the orifice. Primary combustion air is provided around the periphery of the conical liquid spray generated by the nozzle. The combination of primary combustion air introduced into the burner and the action of the swirling liquid produce effective atomization. Normal turndown ratios are in the range of 2.5:1 to 3.5:1. By utilizing a return flow line for liquid flow the turndown ratio can be increased to as high as 10:1. Typical burner capacities are in the range of 10 to 100 gallons per hour. A major disadvantage of this type of burner/nozzle is its susceptibility to erosion and plugging from solids components in the liquid stream. Flames

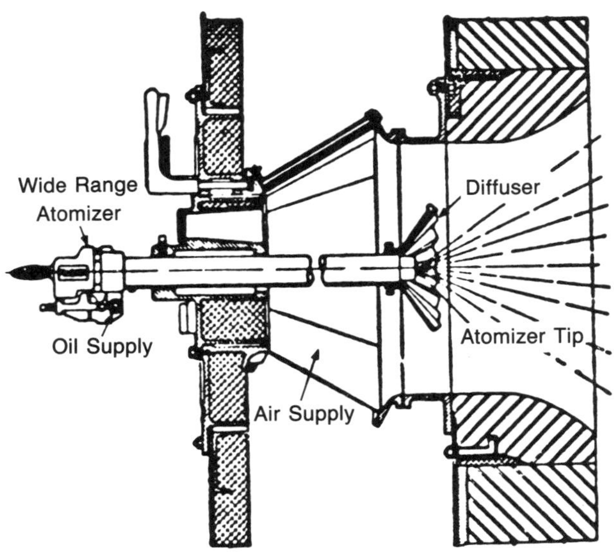

Fig. 4-1. Mechanical atomizing burner. Source: Reference 1.

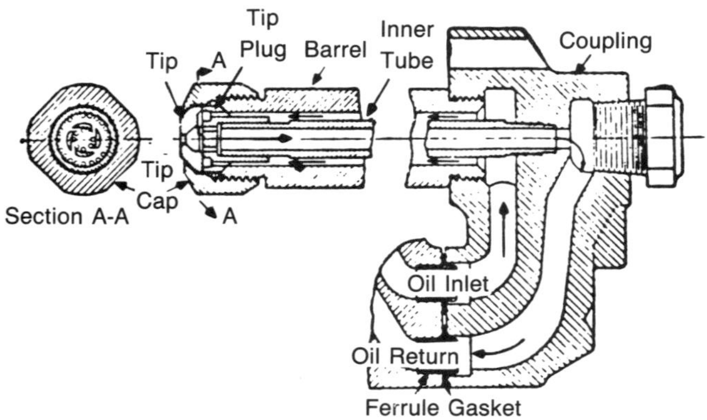

Fig. 4-2. Mechanical atomizing burner. Source: Reference 1.

tend to be short, bush, with a low velocity, and this results in slower combustion requiring relatively large combustion-chamber volumes. The mechanical atomizing burner is applicable for fluids with relatively low viscosity, under 100 SSU.

Rotary-Cup Burners

As shown in Fig. 4-3, atomization is provided by discharging the liquid stream centrifugally, from a rotating cup or plate. Liquid is thrown from the lip of the cup in the form of conical sheets which break up into droplets to minimize surface energy. No air is mixed with the liquid feed prior to atomization. Instead, it is introduced through an annular space around the rotating cup. Normally a common motor drives the liquid-feed pump, the rotating cup, and the combustion-air blower. The liquid pressure required for this burner is relatively low, because atomization is a

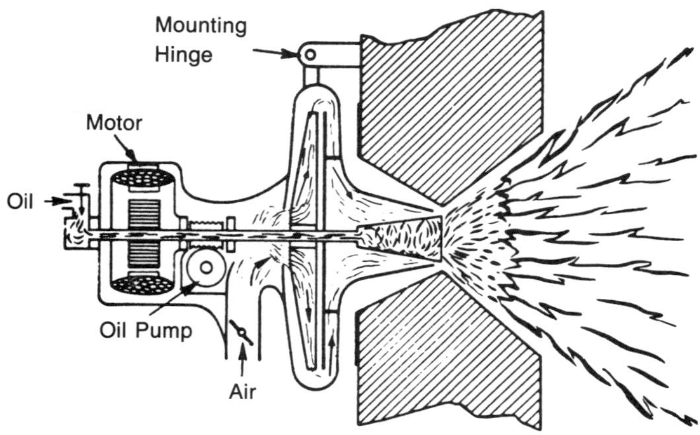

Fig. 4-3. Rotary-cup burner. Source: Reference 1.

function of cup rotation and combustion-air discharge, not liquid-feed pressure. This low pressure requirement and the relatively large openings along the burner liquid path allow the passage of liquids with relatively high solids content, as high as 20 percent by weight. Burner capacities range from low flows (under 10 gph) to over 250 gph. Rotary cup burners have a turndown ratio of approximately 5:1 and can fire liquids with viscosity up to 300 SSU. Rotary cup burners are sensitive to combustion air flow adjustment; insufficient air flow will result in liquid impingement on furnace walls, while excessive combustion air will cause a flame-out.

External Low-Pressure, Air-Atomizing Burner

The major portion of the combustion air required for this type burner is provided, at a pressure of 1 to 5 psig, at the burner tip. Air is injected externally to the nozzle and is directed to the liquid stream, producing high turbulence and effective atomization. The liquid pressure for operation need not exceed that required for positive delivery, normally less than 1½ psig. The quantity of atomizing air required decreases with increasing atomization pressure and may range from 400 to 1000 SCF per gallon of liquid feed. Secondary combustion air is provided around the periphery of the atomized liquid mixture, to shape the flame as well as to allow complete combustion of the liquid. The flame is relatively short because of the high amount of atomization and secondary combustion air provided at the burner. The short flame permits smaller combustion chambers. These burners normally operate with liquids in the range of 200 to 1500 SSU viscosity and can handle solids concentrations of up to 30 percent. Figure 4-4 illustrates a low pressure air atomizing burner. A small quantity of the air passes around the liquid discharge to optimize the liquid flow pattern.

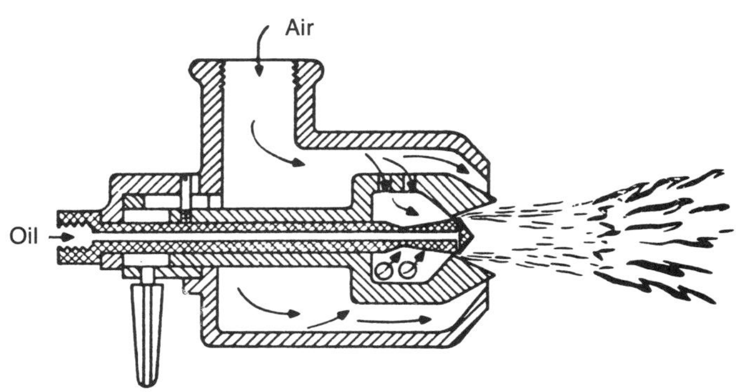

Fig. 4-4. Low-pressure, air-atomizing burner. Source: Reference 1.

External High-Pressure, Two-Flow Burner

The atomizing fluid, air, or steam (or nitrogen or other gas) impinges the liquid stream at high velocity, generating small particles that vaporize quickly. A typical burner is shown in Fig. 4-5. The steam requirement is 2 to 5 pounds per gallon of liquid feed; when air is used for atomization, 20 to 200 scf is required per gallon. The required atomization pressure varies from 30 to 150 psig. Turndown is in the range of 3:1 to 4:1. The flame produced is relatively long, requiring appropriately designed combustion chambers. The liquid viscosity normally handled by these burners ranges from 150 to 5000 SSU for both compressed air and steam atomization. A solids content of up to 70 percent can be accommodated by these burners.

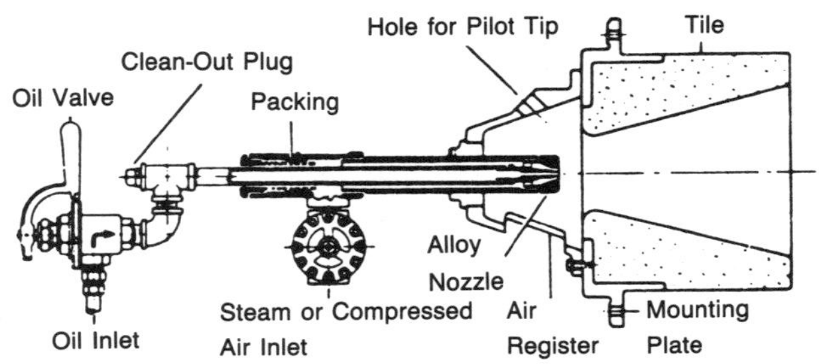

Fig. 4-5. External high-pressure, two-flow burner. Source: Reference 1.

Internal Mix Nozzles

Air or steam is introduced within the nozzle, as shown in Fig. 4-6, to provide impingement of atomization steam or air on the liquid stream prior to discharge. The turndown ratio for this type of burner is from 3:1 to 4:1. These nozzles cannot tolerate a significant solids content and can only handle viscosities under 100 SSU; this burner is used for clean, low viscosity liquids. Its advantage is its low cost compared to other burners.

Sonic Nozzles

These nozzles utilize a compressed gas, such as air or steam, to create high frequency sound waves, which are directed at the liquid stream.

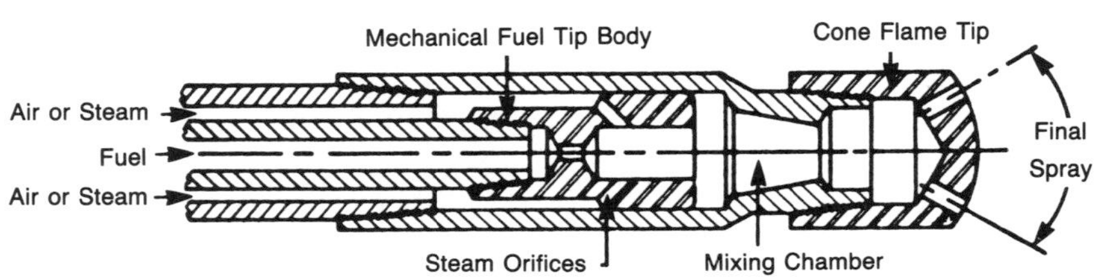

Fig. 4-6. Internal mix nozzle. Source: Reference 1.

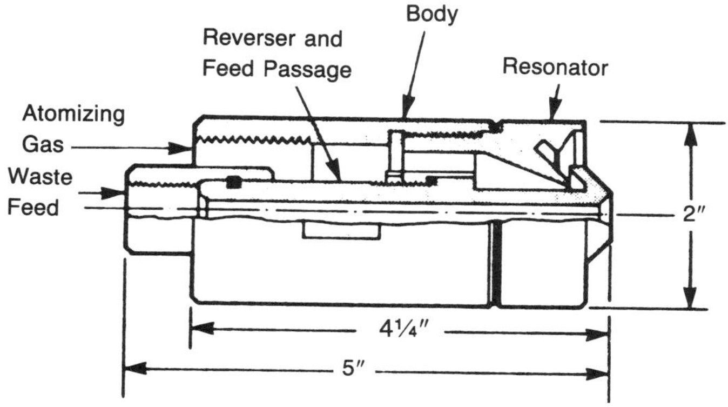

Fig. 4-7. Sonic nozzle. Source: Reference 1.

This acoustic energy is transferred to the liquid stream and creates an atomizing force, breaking the stream into minute particles. The feed nozzle diameter is relatively large, allowing passage of streams with high solids content, such as slurries and sludges. Little liquid stream pressurization is required. The spray pattern is amorphous, with finely atomized, uniformly distributed droplets travelling at low velocities. These nozzles are difficult to adjust, have low turndown, and generate an extremely high noise level during operation. A typical sonic nozzle is shown in Fig. 4-17.

Vortex Burners

High efficiency burner systems have been developed which utilize relatively high-pressure combustion air to generate a vigorous swirl or vortex exiting the burner. The burner discharges tangentially into a burner well. A supply of secondary combustion air at relatively high pressure is introduced within the well to increase the turbulence of the gases exiting the burner. The vortex generated has a high swirl velocity and completely fills the well with flame. This type of burner is illustrated in Fig. 4-8. Figure 4-9 shows a vortex burner in operation. Note that the flame completely fills the burner well, providing uniform and complete burning of the fuel and waste.

LIQUID-WASTE INCINERATOR

The type, size, and shape of a furnace is a function of waste characteristics, burner design, air distribution, and furnace wall design. The furnace can be as simple in design as a vertical, refractory-lined chamber, shown in Fig. 4-10, or it can be relatively complex, utilizing combustion air preheat and firing multi-fuel streams, as shown in Fig. 4-11.

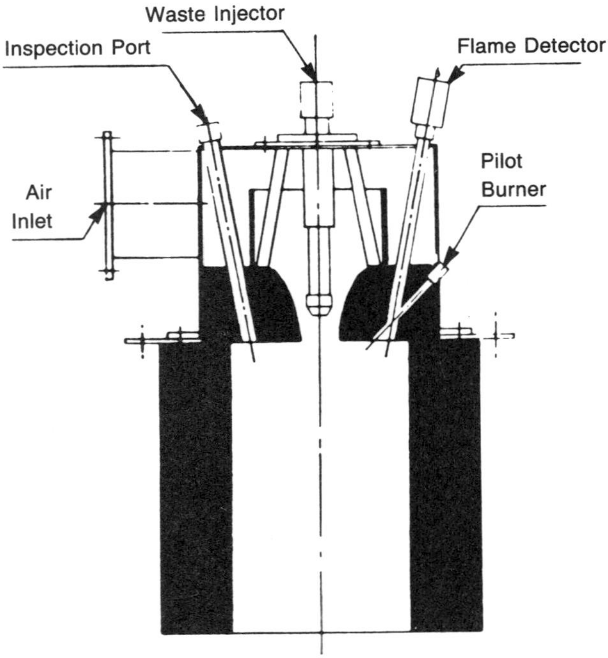

Fig. 4-8. Vortex burner. Source: Reference 1.

Fig. 4-9. Vortex burner operation. Courtesy of J. Santoleri, Four Nines, Inc.

Fig. 4-10. Liquid-waste incinerator. Courtesy of J. Santoleri, Four Nines, Inc.

Except in furnace designs utilizing vortex burners, flame impingement on a furnace wall is undesirable, in that it creates the potential for refractory corrosion and results in lost energy. A furnace must be designed to avoid impingement. Impingement is a function of liquid atomization and

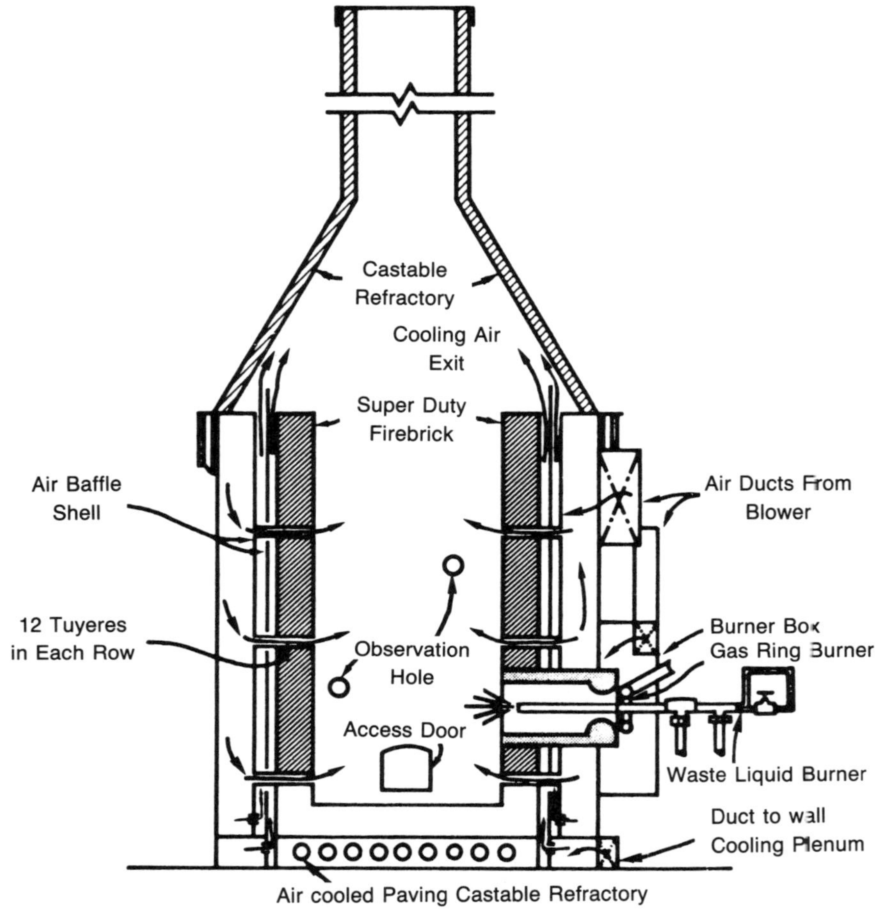

Fig. 4-11. Dual-fuel liquid-waste incinerator.

vaporization, which in turn is dependent on nozzle design, velocity of fluid exiting the burner, air distribution within the furnace, and furnace temperatures.

Primary combustion air is that air flow supplied at the fuel burner to burn the primary fuel within the furnace. It is normally distributed through a *burner register*, an open, fan-shaped component normally surrounding the burner nozzle which imparts a circular motion to the air flow. The register is either fixed or adjustable, with adjustment arms located immediately outside the burner mounting on the furnace exterior. *Secondary air* is that air flow necessary for combustion of the waste feed and is normally introduced into the furnace downstream of the main flame front. In liquid injection furnaces the secondary air supply is often used not just as combustion air, but to create turbulence within the furnace and to provide a relatively cool flow on the furnace refractory surfaces, keeping the temperature of the refractory cooler than that at the center of the furnace. The primary and secondary air flows are also introduced in a manner that aids fuel atomization and helps prevent any unburned material from impinging on the furnace wall. Extraneous impingement often creates *sparklers*, luminous burnout of volatile particles on the furnace wall.

Liquid destruction furnaces require from 5 to 30 percent *excess air* (a total of both the primary and secondary combustion air supplies) to insure adequate combustion. Another furnace parameter is *heat release*. Most liquid burners have a heat release rate of 20,000 to 30,000 Btu per cubic foot per hour. Vortex burners will release 500,000 to 1,000,000 Btu per cubic foot per hour.

SUBMERGED QUENCH REACTORS

Many industrial hazardous waste streams have inorganic components, including metal salts. Many of these salts have a relatively low melting point, in the range of 1400° F to 1600° F. The temperature required for destruction of the organic component of the waste will often be higher than the melting point of the salt constituent. The salt will melt, and will solidify at any lower temperature surface that it comes in contact with, such as air injection ports, feed injections ports, access doors, or other furnace components where there is a leakage of air into the furnace. The result of this activity is the formation of a slag or cake, as illustrated in Fig. 4-12.

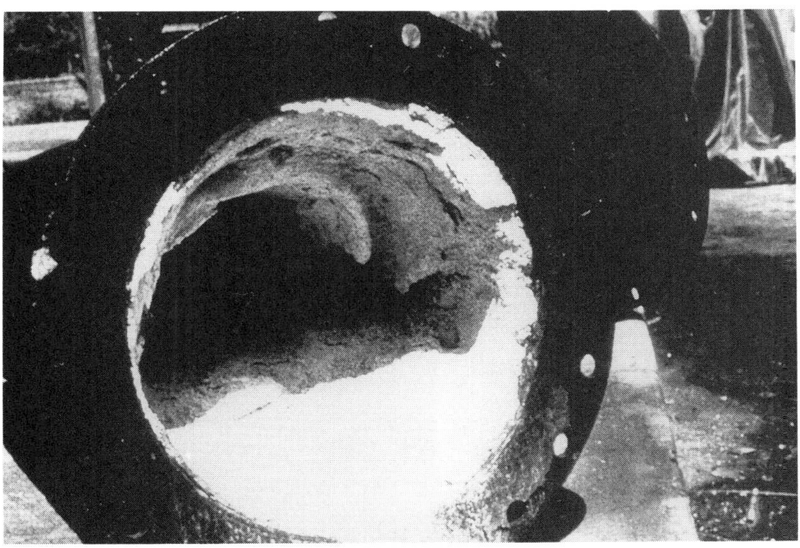

Fig. 4-12. Salt cake accumulation. Copyright © C. R. Brunner.

Slag creates a number of problems. It constricts openings and passages and can react with refractory materials to promote corrosion within the furnace. Slag will normally be extremely hard, and will adhere strongly to refractory surfaces. It must be removed.

The slagging of low melting point wastes is a serious issue. This problem often prevents the use of conventional incineration in the disposal of salt-laden wastes. Submerged combustion systems have been developed to dispose of these troublesome wastes.

Originally designed for the reclamation of brine, these systems have been adapted for use as incinerators. As shown in Fig. 4-13, the incinerator fires into a water tank. The incinerator itself, Fig. 4-14, utilizes a vortex burner which creates a high swirl at the top of the vertical reactor. A temperature of 1800° F is maintained in the reactor. At this temperature, salts in the waste stream will be molten. The reactor provides 2 to 5 seconds retention time at this temperature, which will destroy most organics.

The molten salt is directed against the inside walls of the reactor by the centrifugal action of the gasses coming from the vortex section. It remains molten because the temperature of the chamber is 1800° F, well above the melting point of the salt. The salt falls, by gravity, along the refractory furnace through a distribution section into the water quench tank. Quenched gas passes first through a de-mister to remove entrained moisture, and then through a scrubber.

Larger particulate matter (in excess of 50 μ mean particle size) will probably remain in the quench. Smaller particulate will be airborne and

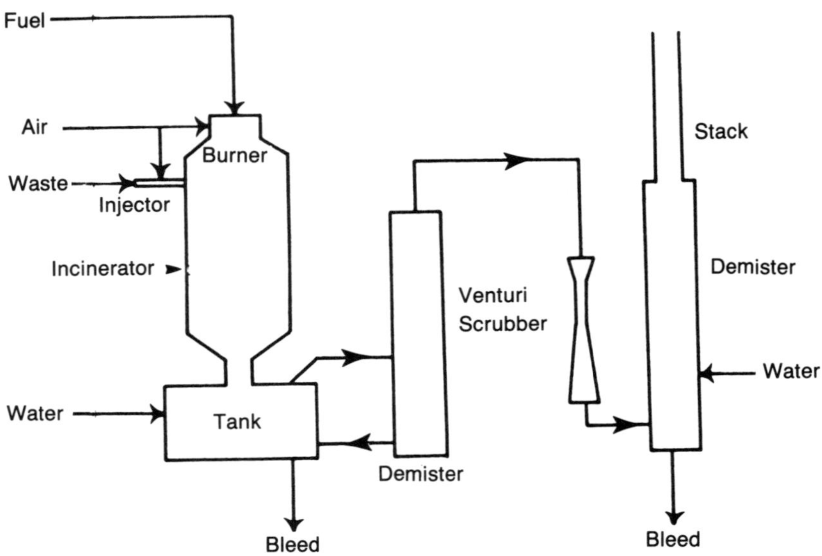

Fig. 4-13. Submerged-quench system. Source: Reference 2.

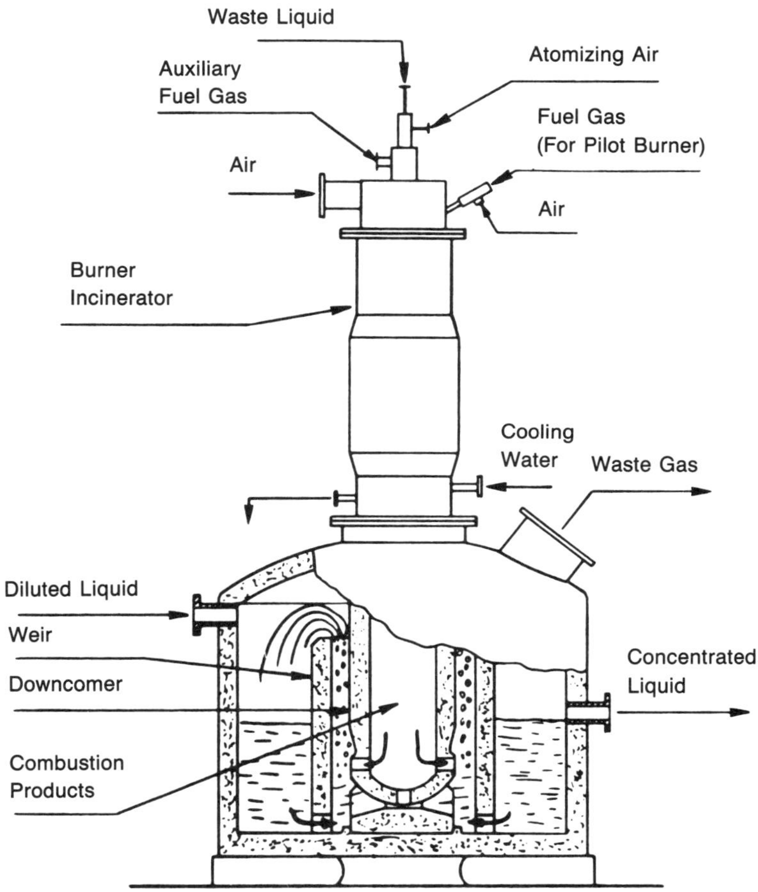

Fig. 4-14. Submerged-quench reactor. Source: Reference 3.

a venturi is normally provided to remove the majority of the particulate matter exiting the quench. Not shown in this illustration, an induced draft fan is normally provided between the de-mister and the stack to generate a partial vacuum for gases exiting the quench tank. The second de-mister is used to remove entrained moisture from the venturi discharge before it exits the stack.

REFERENCES

1. C. R. Brunner, *Incineration Systems: Selection & Design* (Reston, VA: Incinerator Consultants Incorporated, 1988).

2. Y. Kiang, "Incineration of Hazardous Organic Wastes," *Proceedings of the 1980 National Waste Processing Conference*, American Society of Mechanical Engineers, 1980.

3. J. Santoleri, "Chlorinated Hydrocarbon Waste Disposal and Recovery Systems," *Chemical Engineering Progress*, January 1973.

5

Waste-Sludge Incineration

There are several types of furnaces that are used primarily for the incineration of sludge waste. *Sludges* are those materials with semi-liquid consistency that comprise small particles within a liquid matrix, as opposed to slurries, which are large particles in a liquid medium (usually water).

INCINERATOR TYPES

The following types of incineration equipment will be discussed in this chapter: multiple-hearth furnaces, fluid-bed furnaces, and infrared systems.

Other furnace systems, such as the rotary kiln, may be applicable to the incineration of sludge waste; however, they have not been developed specifically for sludge.

MULTIPLE-HEARTH FURNACE

The multiple-hearth furnace is the most prevalent furnace used for sludge incineration this country. It was originally developed for ore roasting in the early 1900's. In the 1930's, it was adapted for sludge incineration. There are over 400 of these furnaces in operation today, the majority of which are used for sludge incineration at wastewater treatment plants. They are also used for carbon regeneration and lime recalcining.

Figure 5-1 is an outdoor installation of a multiple-hearth furnace system. It is continuously fed by a conveyor system entering diagonally on

Fig. 5-1. Multiple-hearth outdoor installation. Source: P. Karr, City of Atlanta.

the left. Ash from the process is discharged from the ash silo to a truck.

The multiple-hearth furnace itself, shown in Fig. 5-2, is a vertical structure, a steel shell, lined internally with refractory. Sludge cake is fed by gravity at the top of the furnace. Sludge can also be fed to the furnace from the side, through a screw-type feeder.

The furnace interior is composed of a series of circular refractory hearths, one above the other. The hearths are self-supporting, each acting as an arch supported from the refractory lining the furnace shell. The hearths are designated with number 1 as the top hearth, number 2, next to the top, etc. There are from 5 to 9 hearths in a typical furnace, and they are normally built in diameters of from 10 feet to 25.

A vertical shaft is positioned in the center of the furnace. Rabble arms are attached to the center shaft above each hearth. The center shaft rotates and drives the rabble arms. A series of rabble teeth on each rabble arm wipes sludge across each hearth. The shaft is driven at a speed of ¾ to 2 revolutions per minute by a variable speed drive at the bottom of the furnace. A bull-and-pinion gear system is normally used to drive the center shaft.

Alternating hearths have a large annular opening between the hearth and the center shaft. These are termed *in-hearths*. The teeth on the rabble arms will rabble sludge to the center of these hearths, "in" towards the center of the furnace, where sludge will drop off the edge of the refractory, landing on the hearth below, an *out-hearth*. An in-hearth, center shaft, and rabble arms are shown in Fig. 5-3.

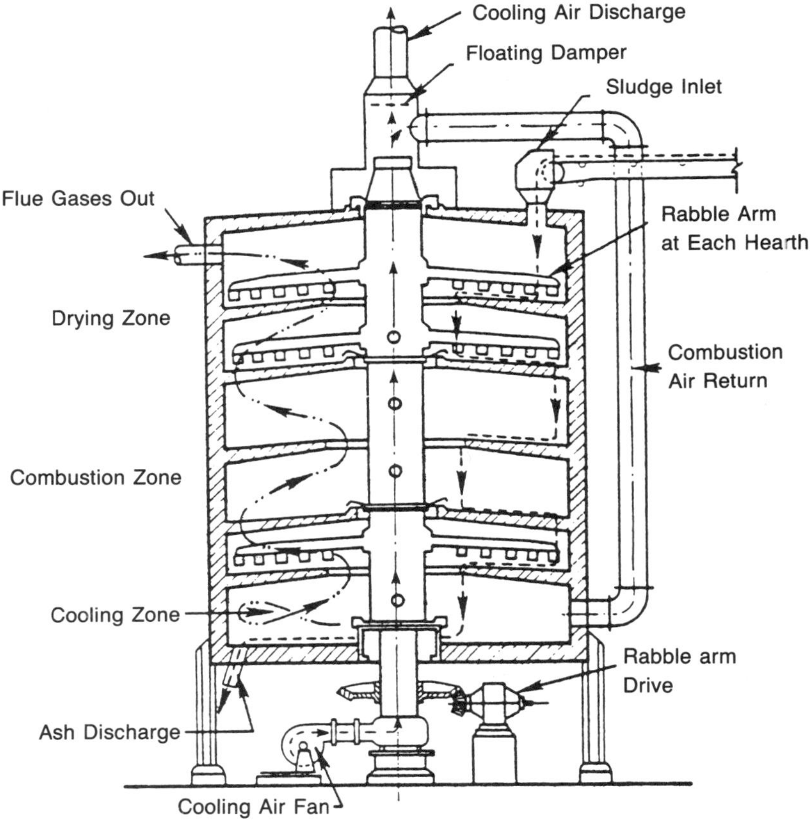

Fig. 5-2. Multiple-hearth furnace. Source: Reference 1.

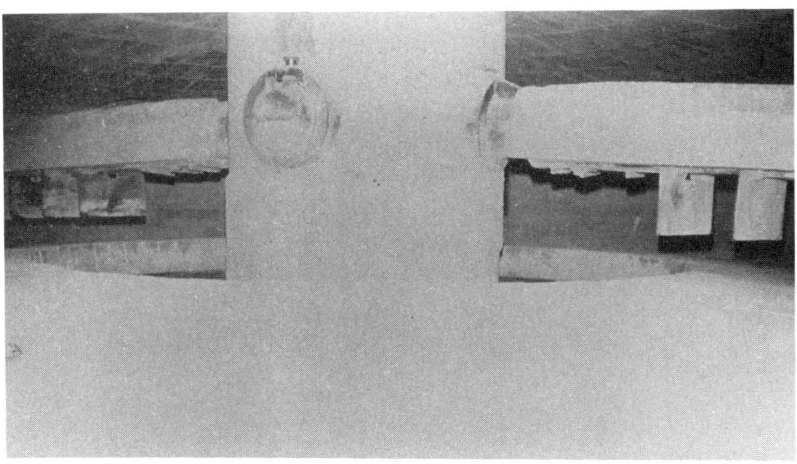

Fig. 5-3. In-hearth. Copyright © C. R. Brunner, 1983.

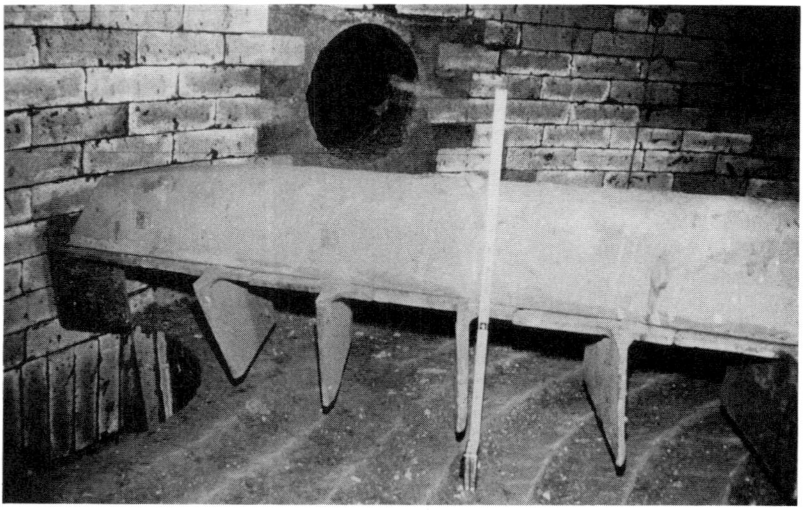

Fig. 5-4. Out-hearth drop holes. Copyright © C. R. Brunner, 1983.

The out-hearths allow sludge to rabble "out", away from the center of the furnace. A series of drop holes, shown in Fig. 5-4, are placed equidistant around the periphery of the out-hearth to allow sludge to drop down to an in-hearth before it reaches the vertical furnace wall.

Sludge combustion air is introduced at the bottom of the furnace. It rises through the furnace, passing over each hearth, picking up the products of combustion and elutriated ash particles with it. Additional sludge combustion air is often supplied above additional hearths within the furnace. Generally, center shaft cooling air is recycled as sludge combustion air in the furnace. The temperature of this air stream ranges from 250° F to 450° F. Its use in the furnace represents waste heat energy reclamation.

Off-gas exits from the top of the furnace. From 10 to 20 percent of the non-combustible (ash) component of the sludge will be airborne and will exit with the off-gas.

The rabble teeth above each hearth agitate the sludge, exposing new surfaces to the gas flow within the furnace. The fall from one hearth to another also exposes new surfaces to the hot gas.

The upper hearths of the furnace comprise the drying zone, where the sludge cake gives up moisture while cooling the hot flue gas. Approximately 10 pounds of moisture are released per square foot of hearth area per hour. When the moisture content of sludge within the furnace is reduced to 30 percent, the sludge begins to burn. The burning rate is approximately 10 pounds of combustibles per square foot of hearth area per hour. The furnace heat release rate is approximately 10,000 Btu/ft^3 per hour, applied above the burning hearths.

A set of multiple-hearth furnace sizing curves were developed from the above criteria. These curves, Fig. 5-5, provide a rapid means of determining the required size of a furnace for burning an aqueous sludge.

In a typical, six-hearth furnace, the upper two hearths will normally be the drying hearths, where sludge loses most of its moisture. The two middle hearths are the burning hearths, where both the air and other gases passing through the furnace and the sludge are heated to combustion temperature. The sludge residual (ash) will burn out to a sterile ash on the two lowest hearths. The ash cools, heating the air passing over it.

Treatment plant sludge requires a minimum temperature of 1400° F and a retention time of at least 0.5 seconds to destroy the majority of organic material present. In a multiple-hearth furnace, to insure that this temperature and retention time are maintained, a temperature of 1600° F should be kept above the two burning hearths. If the sludge is too dry (in excess of 25 percent solids content), or if the solids grease content is too high (in excess of 5 percent), an afterburner would probably be required. The purpose of the afterburner is to maintain the entire off-gas stream at a controlled temperature for a specific retention time. With a high solids or grease content, combustibles would otherwise volatilize above the burning zone and exit the incinerator system unburned, causing smoke and odor emissions.

All sludge off-gas must pass through the afterburner. The most effective afterburner is one that is separate from the incinerator, immediately downstream of the incinerator exit, with one or more burners to provide the required temperature. It should be sized for a well-defined retention time. A less efficient design makes the afterburner part of the multiple-hearth furnace itself. Often, the top hearth of the furnace is expanded in height. Sludge is fed sidewise to hearth number 2, or is dropped to hearth number 2 through an opening placed in the top hearth.

Sludge that can be processed in the multiple-hearth furnace is limited in consistency to the range of approximately 15 to 50 percent solids. With a solids content below 15 percent, the sludge is more liquid than solid, and will not move properly within the furnace. Such sludge tends to flow along the hearth, and the rabble action will not be effective. Above 50 percent solids, the sludge is excessively viscous, and tends to plug the rabble teeth, climb up and build up on the rabble arms, and clog drop holes on out-hearths.

Supplemental fuel is required for incinerator heat-up, and to maintain incinerator operating temperatures when burning sludge. Burners are mounted on the shell of the multiple-hearth furnace. Two to four burners are normally provided at alternate hearths. A typical six-hearth furnace has two burners on each of hearths 1, 3, and 5, a total of six burners. The fuel source can be either fuel oil (normally #2 grade fuel oil) or gas (either natural gas or a lower heating value gas). Attempts to use solid

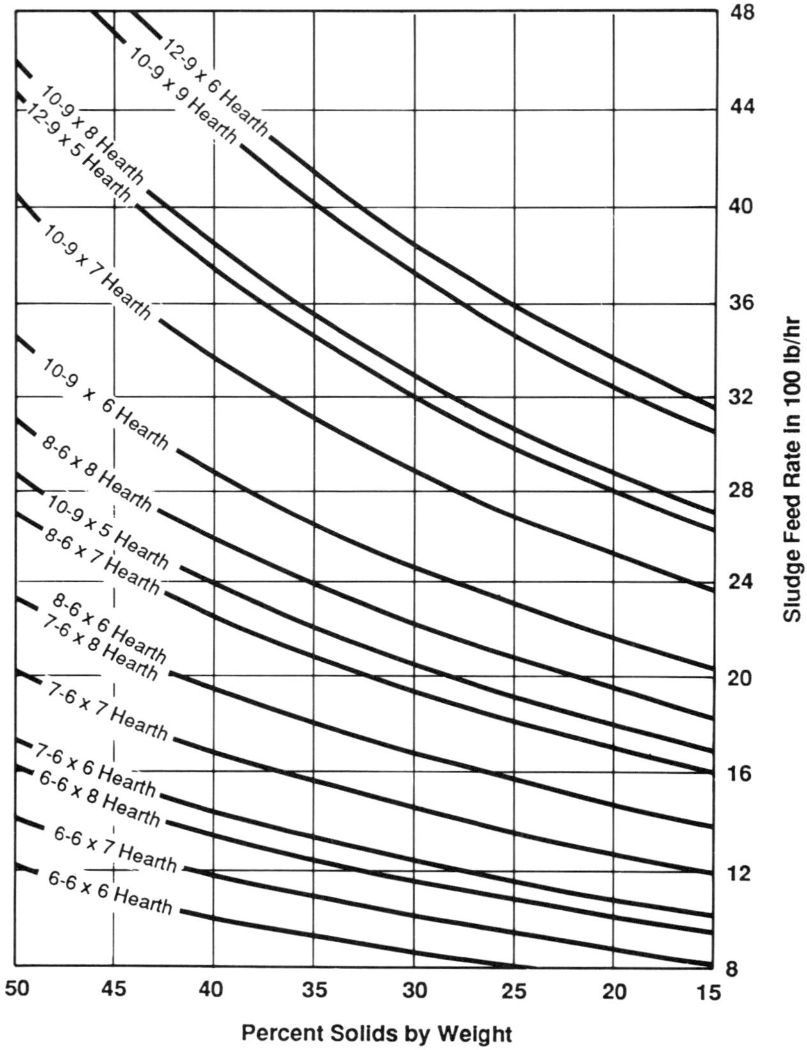

Fig. 5-5. Multiple-hearth furnace sizing. Source: Reference 1.

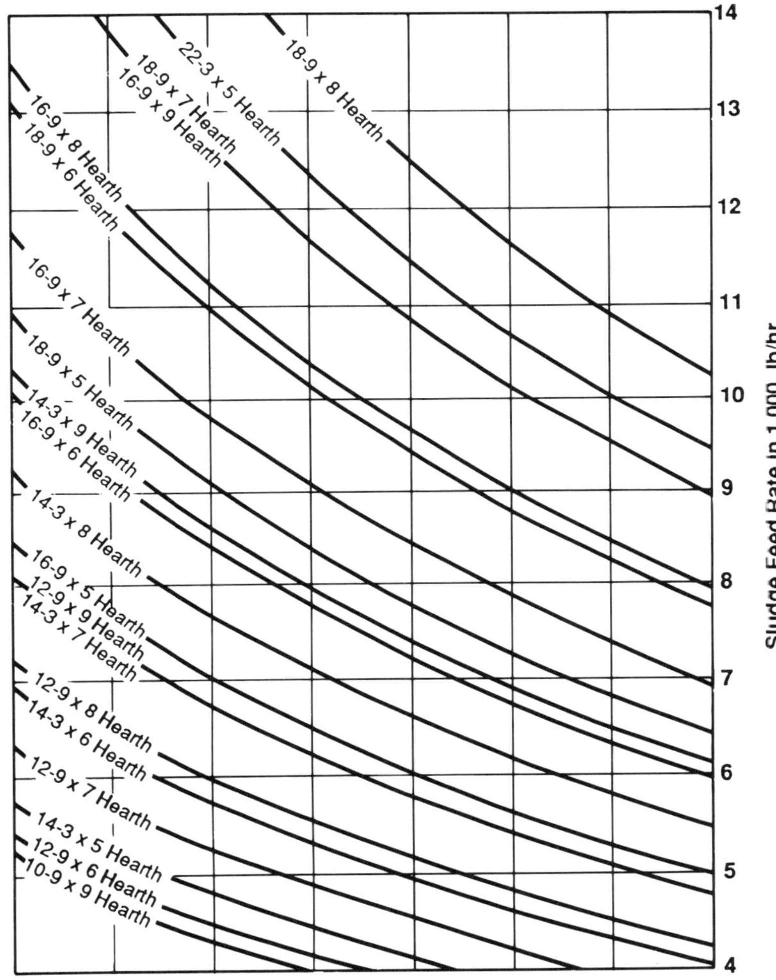

SELECTION OF INCINERATOR

(continued)

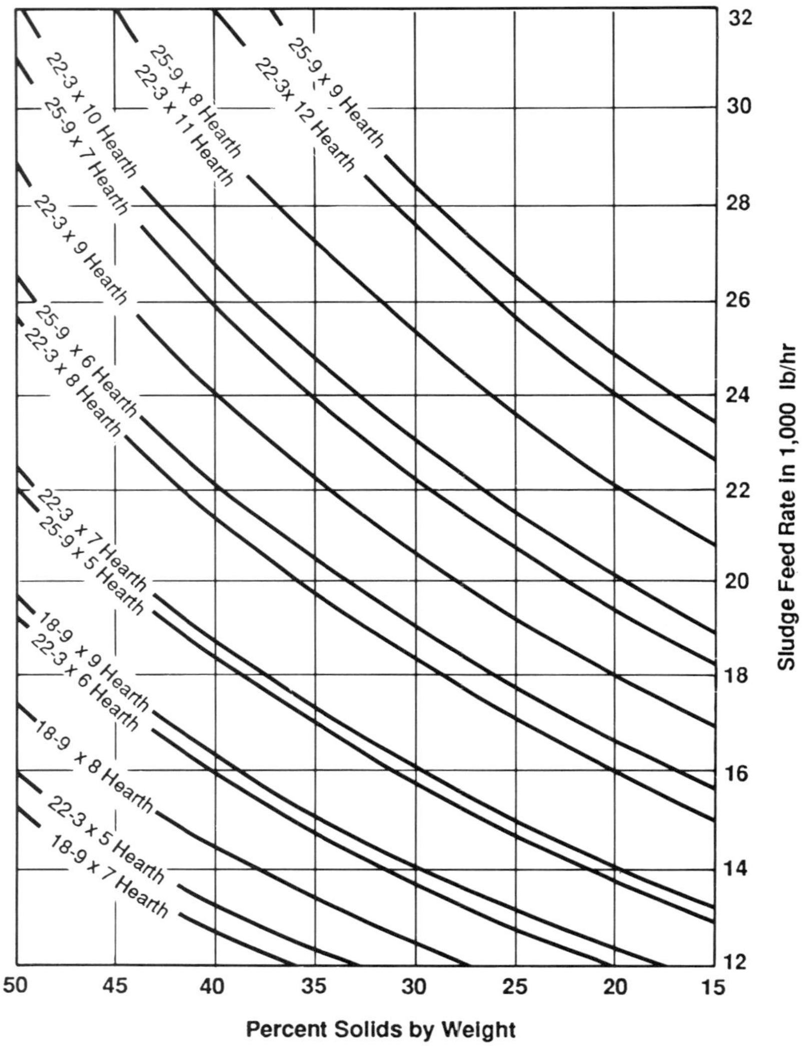

SELECTION OF INCINERATOR

(continued)

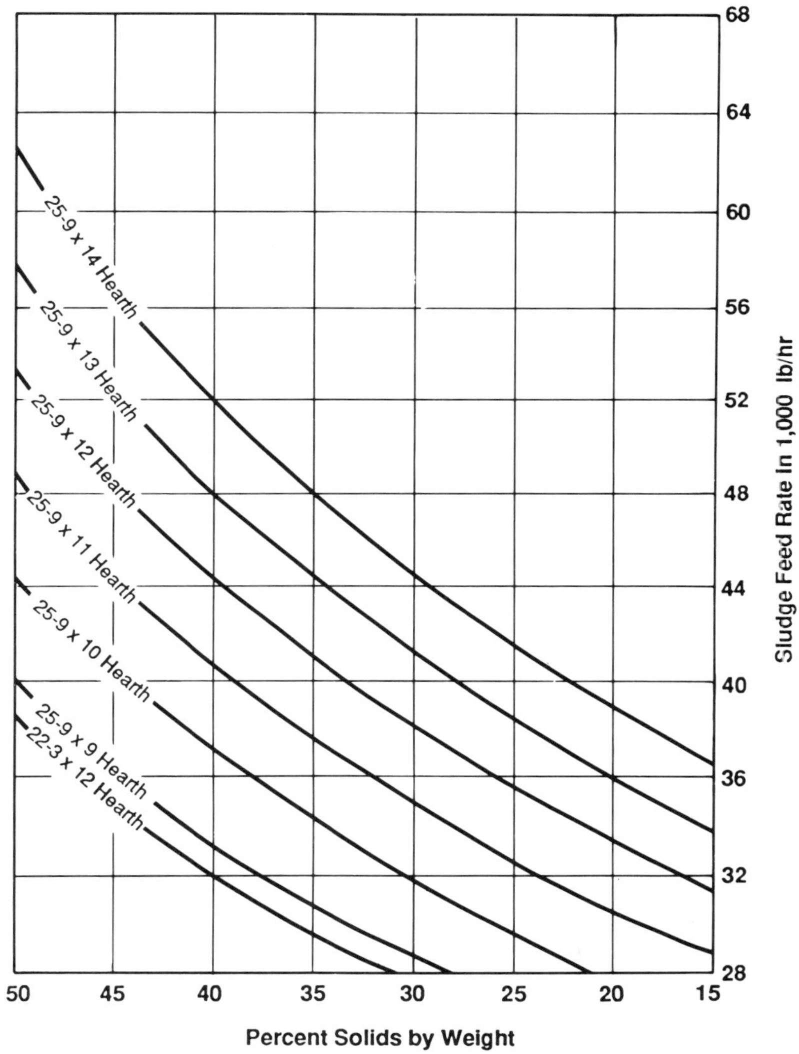

SELECTION OF INCINERATOR
(continued)

fuel such as coal or wood waste as a source of supplemental heat have not met with success. Solid fuels tend to burn high in the furnace because of their relatively low moisture content, and they cannot provide the temperature profile required for effective burning and burn-out of sludge.

The multiple-hearth furnace is maintained at a negative pressure, or draft. This furnace has numerous leakage points, such as at access/observation doors (normally two per hearth), the sludge inlet and outlet mechanisms, and the top and bottom center shaft seals. Although most furnace systems are designed to minimize leakage from these sources, the multiple-hearth furnace is normally not air-tight, and some leakage will always occur. If draft is maintained, leakage will be into the furnace; hot gases will not leak out of the furnace. The most important operating problem occurs when a multiple-hearth furnace goes positive, i.e., loses draft. When this happens, hot, dirty gases find their way through furnace leak points; this uncontrolled discharge is a personnel hazard and can cause serious equipment damage.

A typical multiple-hearth furnace system is shown in Fig. 5-6. An induced-draft fan (ID), immediately before the stack, draws gas through the system and creates furnace draft. The damper ahead of the ID fan modulates system pressure and is normally automatically controlled by furnace draft.

This example is of a multiple-hearth furnace that was designed for duty as a sludge incinerator as well as a lime recalciner. In the lime-recalcining mode, the lime sludge must be maintained at a temperature of at least 1400° F. This sludge was expected to have an organic constituent which would be released from the sludge at a temperature of 1200° F to 1300° F. Sludge is introduced into the furnace at hearth 2; the top hearth is used as an afterburner hearth. It is maintained at a temperature in the range of 1400° F to 1600° F to assure the destruction of the trace organic component of the sludge.

Lime sludge has negligible heat content. Heat must be provided to the sludge to maintain calcination. Forty burners are required to provide sufficient heat within the furnace. Four of these burners are located above the top hearth, to perform an afterburner function. The other 36 burners are distributed four burners per hearth, except for the bottom hearth, which contains no burners. For burning organic sludge, which has a significant heat content, no more than eight burners, inclusive of the afterburner (top) hearth burners, are necessary to maintain the required temperature level.

A cyclonic separator is placed at the exit of the incinerator to capture larger particles of lime when the furnace is operated in the recalcining mode. This lime residual is reinjected into the furnace below the burning zone. Generally, a cyclonic separator is not used for organic sludges. The

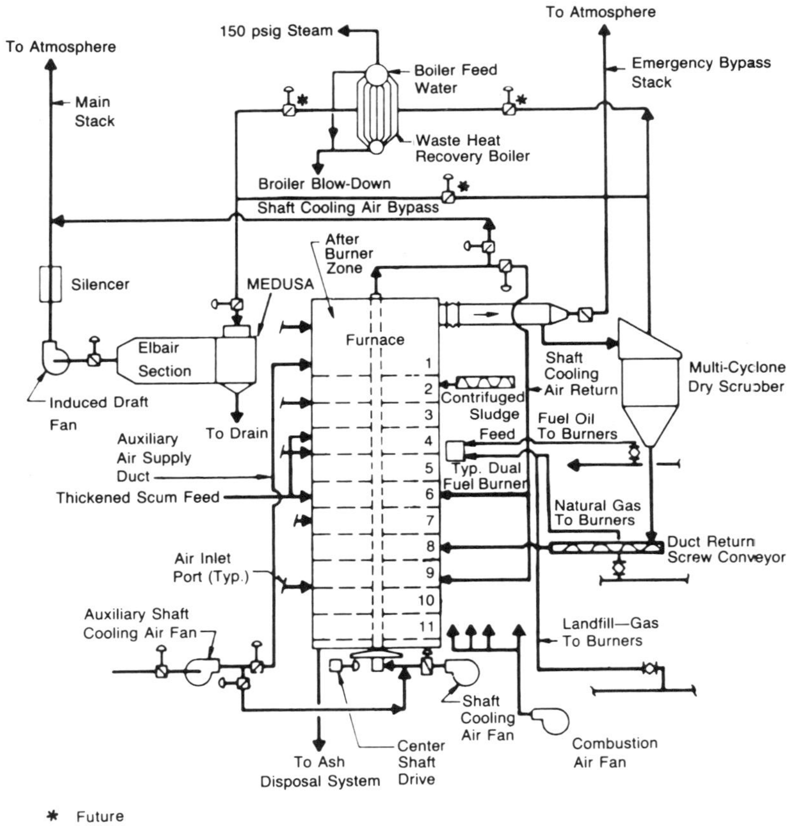

Fig. 5-6. Multiple-hearth incinerator system.

efficiency of these types of collectors is poor for the relatively small elutriated particulate matter derived from organic sludges.

A waste-heat boiler extracts heat from the incinerator exhaust gas to generate process steam. From the boiler, the gas stream, which has been reduced in temperature to within 100° F of the generated steam, is drawn through a wet scrubber. The scrubber removes the majority of particulate matter remaining in the gas stream and discharges it in a spent water stream. The gas stream is cooled to approximately 120° F. It passes through an induced-draft fan, and is directed through a noise silencer to an exhaust stack for discharge to the atmosphere.

FLUID-BED INCINERATION

As illustrated in Fig. 5-7, the fluid-bed furnace is a cylindrical, refractory-lined shell with a supporting structure above its bottom sur-

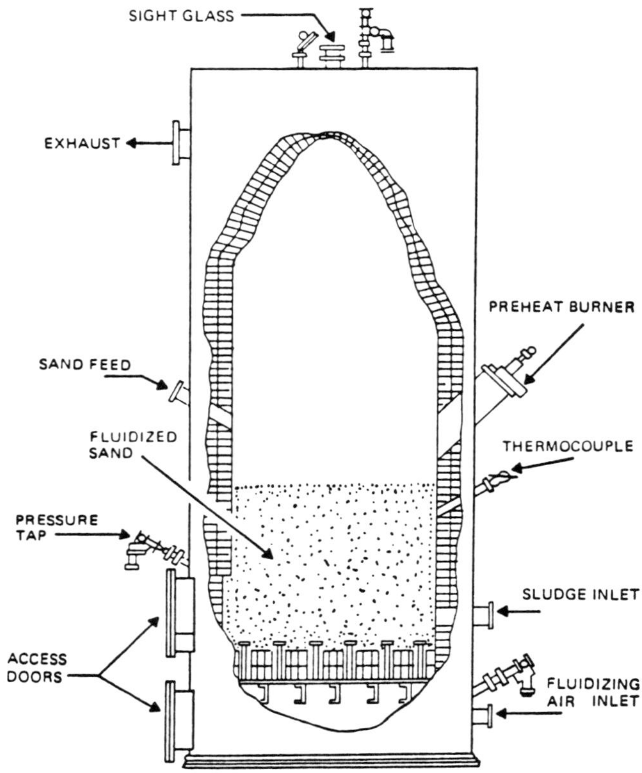

Fig. 5-7. Fluid-bed furnace. Source: Reference 1.

face to hold. The structure has a series of tuyeres, which allow the passage of air upward into the bed and tend to prevent the passage of sand. This structure, a *tuyere plate*, can be either refractory or constructed of steel alloy.

Air is introduced through the fluidizing-air inlet at pressures in the range of 3.5 to 5 psig. The air passes through the windbox, up through the tuyere plate, and into the sand bed. This air creates a high degree of turbulence in the sand bed. The top of the bed undulates and has the appearance of a fluid.

Air can be introduced into the windbox cold (slightly above ambient temperature) or, as is usually the case, preheated by the exiting flue gas. The simple system illustrated in Fig. 5-8 includes an air preheater.

When fluidized with air at a velocity of 1½ to 3 feet per second, the sand bed expands 40 to 60 percent in volume. (This velocity is the equivalent or superficial space velocity determined by dividing the actual volumetric flow rate by the internal bed cross-sectional area.) The bed material is most commonly silica sand, but may also be limestone, alumina, or ce-

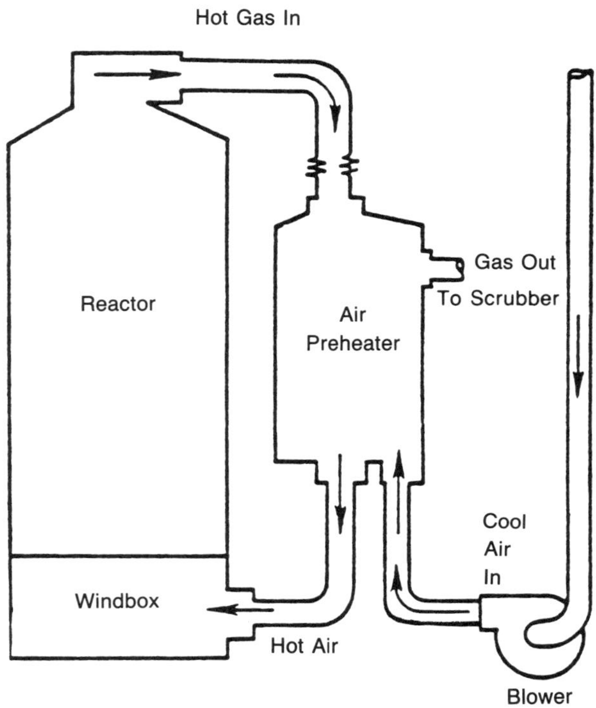

Fig. 5-8. Conventional fluid-bed incineration system.

ramic material.

Sludge cake is normally introduced within, or just above, the sand bed. The size of the furnace is a function of the moisture in the feed. The greater the moisture content, the larger the bed surface. For this reason, when sludge has a relatively high moisture content (above 80 percent), it is often introduced from the center of the furnace ceiling. By the time sludge reaches the sand bed, it has lost much of its moisture. The bed can be sized smaller than if the sludge were injected directly within the bed.

The drying process is practically instantaneous; moisture flashes into steam upon entering the hot bed. Fluidization provides maximum contact of air with the sludge surface, maximizing the efficiency of the burning process.

The furnace itself is extremely simple, with few moving parts. Figure 5-9 illustrates a typical system, showing the reactor and supplemental fuel piping. The major equipment external to the furnace is the fluidizing-air blower. The blower is sized to provide the air required to fluidize the sand bed and for combustion of the sludge organics.

Some furnaces, particularly those of European design, also utilize an induced draft fan, which is sized for drawing flue gas from the freeboard,

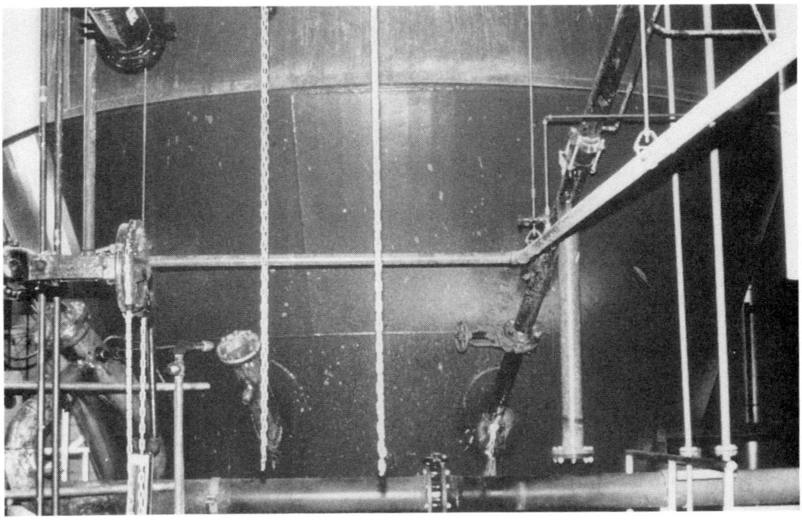

Fig. 5-9. Fluid-bed system. Copyright © C. R. Brunner, 1982.

through the scrubber, and out the stack to the atmosphere. A typical dual-fan design schematic is shown in Fig. 5-10. Table 5-1 lists advantages and disadvantages of these two system configurations, positive pressure and negative pressure. A typical European-type fluid-bed furnace is illustrated in Fig. 5-11.

Ash and some sand become airborne and exit the furnace within the flue gas stream. Generally, sand has to be made up (added to the bed) at the rate of approximately 5 percent of the bed volume every 100 hours of operation.

Maintenance of the bed integrity is a function of the material being incinerated. The waste non-combustible content (ash) will either remain in the bed, or it will be airborne and will exit with the flue gas. Salts, for instance, will generally remain in the bed. As the bed volume increases with accumulated residual, it must be tapped. If ash is airborne (most organic sludges containing little or no metals fit into this category), all of the ash will exit in the flue gas, and the gas cleaning system must be sized for this relatively high particulate loading. Typical fluid bed systems operating with airborne ash discharge require from 40 to 60 inches water column (WC) across a venturi scrubber for effective ash removal.

Bed agglomeration can be a problem with fluid bed furnaces. Some waste materials, particularly those containing metal salts, may cause the bed particles to grow in size and, eventually, to combine, or agglomerate. Clinkers may be formed, and these clinkers will reduce the fluidization of the bed, leading to bed seizure. Agglomeration will be apparent through an increase in the differential pressure across the bed. An agglomeration can be a temporary occurrence, because as residence time increases in

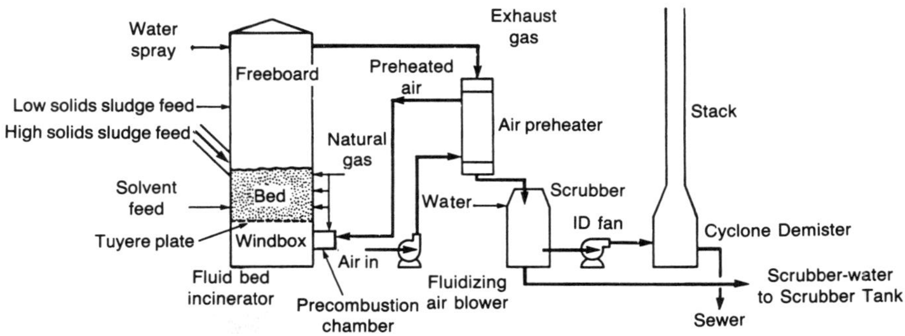

Fig. 5-10. Dual-fan fluid-bed incineration system.

Table 5-1. Fluid-Bed Furnace: Pressurized versus Unpressurized Operation.

Feature	Positive Pressure	Negative Pressure
Fan	Fluidized air blower required	Induced draft fan plus fluidized air blower required
Control	Complex, single fan controls bed and freeboard	Blower controls bed, ID fan controls freeboard
Feeding	Critical, positive feed required	Not critical, gravity feed is acceptable
Construction	Airtight	Not necessarily airtight with fugitive emissions possible
Operation	Simple, single fan	More complex with dual fan system

a bed, particles that are agglomerated may be worn down through continual contact with adjacent particles in the bed. Test burns on materials that have not previously been fired in a fluid-bed furnace must usually be performed to determine bed reaction, agglomerating or non-agglomerating properties, and seizure potential. If it is found that bed seizure may occur, bed additives may be available that would discourage such action.

Fuel is used for start-up, reheat, and, depending on the net heating value of the sludge, maintenance of temperature during incineration. It

Fig. 5-11. European fluid-bed incineration system. Copyright © C. R. Brunner, 1986.

can be injected within the bed, sprayed on top of the bed, or fired beneath the bed within the windbox. Supplemental fuel for start-up is often fired in the windbox, and after start-up is complete, the windbox burner is turned off, and fuel is fired through lances within the sand bed. Solid fuel (coke or pulverized coal) or liquid fuel (fuel oil) can be used. Generally, fuel gas will not burn properly in a fluid bed; it tends not to pass vertically through the bed and creates local hot spots. Gas can be fired in the windbox without affecting bed operation.

Because of thorough mixing of air and sludge in the fluid bed, excess air requirements are low, from 40 to 60 percent. The bed is maintained in the range of 1300° F to 1500° F, depending on the nature of the sludge. The temperature of the open volume above the bed (termed the *freeboard*) is usually no more than 100° F above the bed temperature. The residence time of gases in the freeboard is normally in the range of 3 to 6 seconds.

The fluid-bed furnace system is compact, airtight and, in most designs, maintained as a positive pressure system. Because the entire system is under positive pressure, the furnace must be airtight. This feature is useful in applications where the furnace is required to operate on a non-continuous basis. If the furnace is to operate five days a week, for instance, its airtight construction allows it to be sealed effectively. Dampers on its inlet and outlet will hold in its heat. The internal refractory and the mass of sand within the bed provide thermal inertia to maintain an effective heat sink. A fluid-bed furnace, after a shutdown, will lose as little as 10° F per hour. It can be shut down on Friday evening and will require only a few hours of heat up on Monday morning to be available for waste feeding.

A major area of maintenance in this system is the *recuperator*, or air pre-heater. In domestic designs, this equipment is normally provided as a gas-to-air tube-in-shell heat exchanger. Its tubes are usually constructed of high-temperature-resistant steel alloy. At temperatures in excess of 1700° F (or at sustained operation in excess of 1600° F), severe corrosion will occur, reducing the heat exchanger tube wall dimension to the point of weakness and subsequent failure. The design of the heat exchanger should insure against the development of these high temperatures. Parallel, as opposed to counterflow, heat exchangers help reduce the possibility of reaching these high temperatures within the recuperator.

Corrosion is encouraged by oxygen in the flue gas. Oxygen will decarburize steel at temperatures in excess of 1650° F. De-carburization results in tube *wastage*, the creation of a friable tube surface adjacent to the hot gas stream. Other heat exchanger designs, such as insertion-type units used by a number of European manufacturers, utilize an intermediate heat-exchange medium, such as heated oil, hot water, or steam, to provide process heat. These units are less costly than tube-in-shell exchangers and can be replaced more quickly and less costly.

An internal water-spray system is often employed in fluid-bed systems to help protect the heat exchanger from high-temperature excursions. Water is automatically injected into the freeboard when the heat exchanger inlet temperature is above a pre-set figure, generally no higher than 1650° F. The location and design of these sprays are critical to prevent their damage from high incinerator temperatures. When not in use, cooling air should be supplied to them to prevent damage from furnace heat.

A fluid-bed furnace requires a minimum amount of air to maintain bed fluidization, regardless of loading rate. This requirement is based on the furnace rating at design load. When the furnace is operating at a feed quantity below the design point, approximately the same air quantity is required to maintain fluidization as was required at the design load rate. Fuel consumption is directly related to the quantity of fluidization air required, and, while the furnace may have good fuel efficiency at its design point, its efficiency is poor at lower loads.

A fluid-bed furnace's feed system requires particular attention. Except for some European-manufactured fluid-bed furnaces, in which negative pressure is maintained in the freeboard and waste can be dropped into the furnace by gravity, fluid-bed furnaces are under positive pressure. Waste must be forced into the furnace; gravity feed cannot be utilized. Positive displacement pumps are normally used for feeding fluid-bed furnaces under positive pressure.

The relatively high residence time within the fluid bed incinerator may allow lower temperatures for destruction of organic compounds. If an organic compound that requires 2200° F at a residence time of 1 second for "four nines" destruction is subject to a residence time of, for instance, 5 seconds, only 1600° F may be required for the same level of destruction. This provides some degree of flexibility in the use of a fluid-bed furnace for destruction of organics. On the other hand, if a temperature of destruction in excess of 1600° F is required, a fluid-bed furnace is probably not applicable. An afterburner is normally not provided for fluid-bed incineration systems.

The common mode of heat reclamation in a fluid-bed furnace is internal, through the recycling of heat in the recuperator. Fluid-bed furnaces have been used in the power industry for the generation of steam, but this has not generally been applied to fluid-bed furnaces used as incinerators.

Fluid-bed furnace size is a function of the amount of water in the sludge feed. Size criteria for typical furnaces operating on organic sludge is given in Table 5-2. The nominal size of the furnace is the maximum outside diameter of the furnace. Fluid bed facility arrangement is shown in Fig. 5-12, which is a typical single-fan (conventional) fluid-bed furnace system.

THE INFRARED FURNACE

The infrared (or electric or radiant heat) furnace is a conveyor-belt system passing through a long, refractory-lined chamber, as shown in Fig. 5-13. An induced-draft (ID) fan maintains a negative pressure throughout the system.

Table 5-2. Fluid-Bed Furnace Sizing.

Nominal Size (feet)	Water in Sludge Cake (1b/hr) Hot Windbox	Cold Windbox
9	2,800	2,240
10	3,470	2,770
11	4,200	3,350
12	4,950	3,940
13	5,800	4,620
14	6,750	5,530
15	7,750	6,150
16	8,760	7,000
17	9,900	7,900
18	11,100	8,700
19	12,350	9,750
20	13,600	10,800
21	15,100	11,900
22	16,550	13,100
23	18,100	14,300
24	19,700	15,500
25	21,400	16,800

Combustion air is introduced at the discharge end of the belt, shown as the viewport. Air picks up heat from the hot, burned sludge as sludge and air travel counter to each other. Supplemental heat is provided by electric infrared heating elements within the furnace above the belt. Cooling air is injected into the incinerator chamber to prevent local hot spots in the immediate vicinity of the heaters and is used as secondary combustion air within the furnace. The furnace is designed to provide and maintain a temperature of 1600° F in the space above the travelling conveyor.

The conveyor belt is continuous, woven-wire, high-temperature alloy steel mesh that will withstand the 1300° F to 1600° F temperatures encountered within the furnace. The refractory is not brick, but ceramic felt. It does not have a high capacity for holding heat (*thermal inertia*) and can therefore be started up relatively quickly, from 1 to 2 hours. The furnace cross-section is illustrated in Fig. 5-14.

Sludge cake is fed by gravity to the belt, and is immediately leveled to a depth of approximately one inch. There is no other sludge-contact or sludge-handling mechanism on most of these units. The belt speed and travel is selected to provide burn-out of the sludge without agitation. This feature results in a relatively low level of particulate emissions.

An excess air rate of 20 to 30 percent is claimed for this furnace. With this low excess air and without agitation of the sludge the scrubber

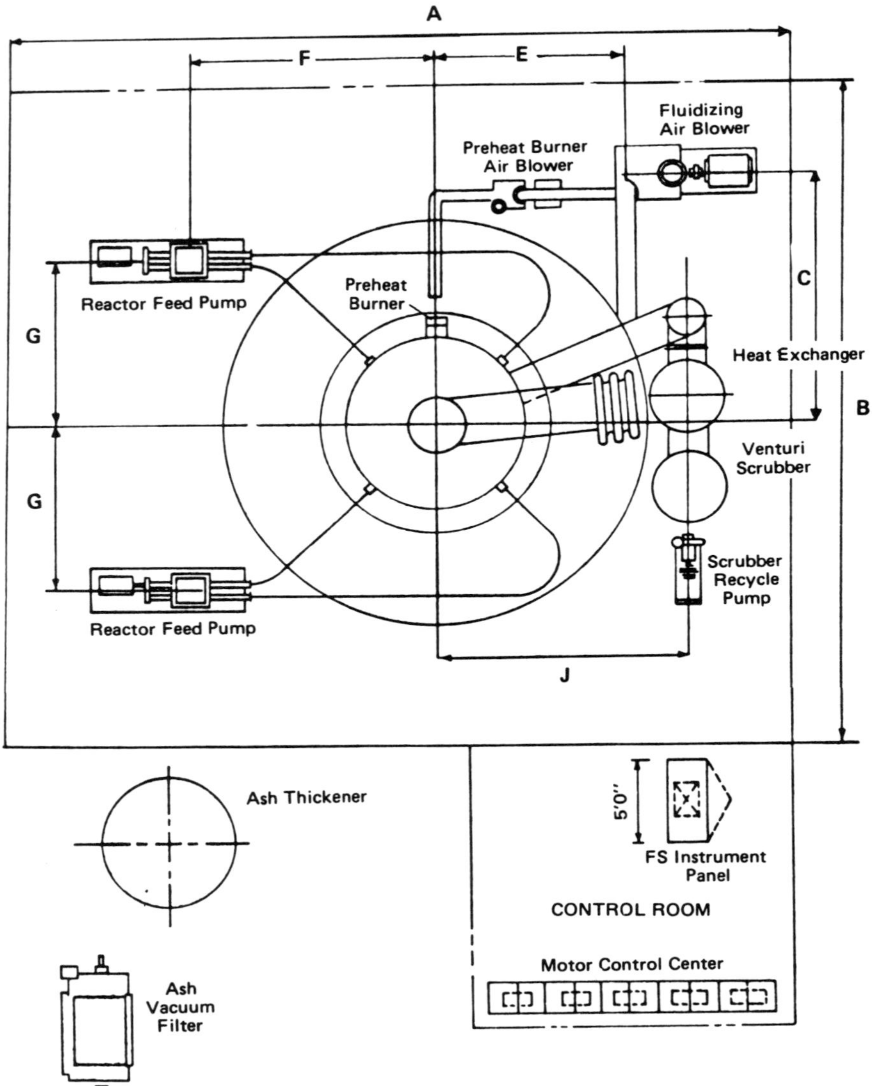

Fig. 5-12. Fluid-bed furnace system configuration. Source: Dorr-Oliver, Inc.

differential-pressure requirement is relatively low, below 15 inches WC.

Supplemental fuel (electric power) is required for start-up. Electric power is also used to provide the heat required to maintain combustion temperatures. Electric energy costs approximately four times more than fossil fuel on a heat value basis. Unless the sludge will burn autogenously, an electric furnace may not be cost-effective. Even with autogenously burning sludge; however, the cost of installed kilowatts (the demand charge)

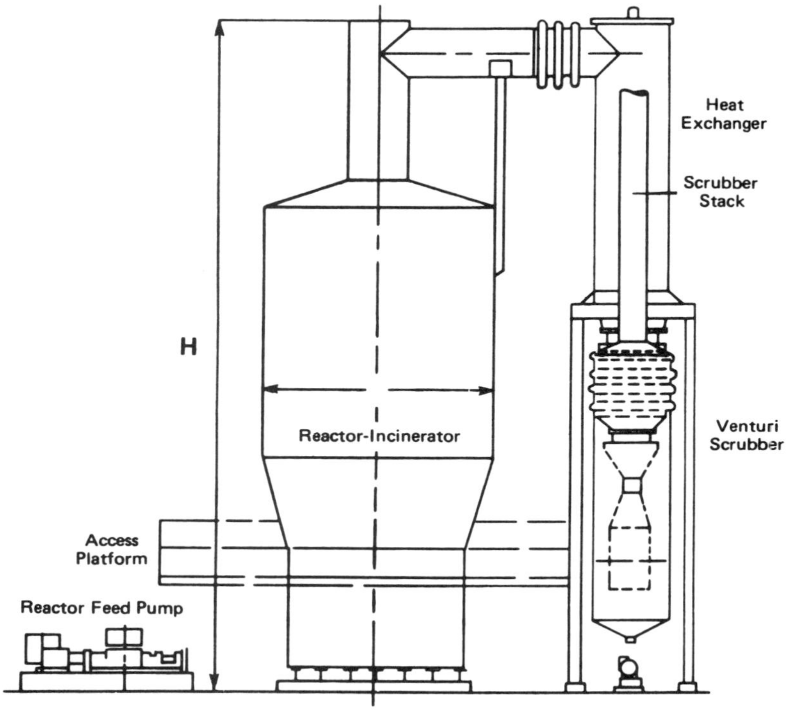

NOM SIZE REACTOR	A	B	C	D	E	F	G	H	J
9′-0″ ϕ	52′-0″	42′-0″	15′-0″	11′-4″	11′-0″	18′-0″	—	45′-0″	15′-0″
10′-0″ ϕ	52′-0″	42′-0″	15′-0″	12′-4″	11′-0″	18′-0″	10′-0″	45′-0″	15′-0″
11′-0″ ϕ	54′-0″	42′-0″	16′-0″	13′-4″	12′-0″	18′-0″	10′-0″	47′-0″	16′-0″
12′-0″ ϕ	54′-0″	42′-0″	16′-0″	14′-4″	13′-0″	18′-0″	8′-0″	47′-0″	17′-0″
13′-0″ ϕ	58′-0″	45′-0″	16′-0″	15′-4″	13′-0″	18′-0″	8′-0″	48′-0″	17′-0″
14′-0″ ϕ	60′-0″	45′-0″	17′-0″	16′-4″	14′-0″	19′-0″	12′-0″	48′-0″	18′-0″
15′-0″ ϕ	62′-0″	45′-0″	17′-0″	17′-4″	14′-0″	19′-0″	12′-0″	50′-0″	18′-0″
16′-0″ ϕ	62′-0″	54′-0″	18′-0″	18′-4″	15′-0″	20′-0″	12′-0″	45′-0″	19′-0″
17′-0″ ϕ	63′-0″	54′-0″	20′-0″	19′-4″	15′-0″	20′-0″	12′-0″	47′-0″	19′-0″
18′-0″ ϕ	63′-0″	54′-0″	21′-0″	20′-4″	16′-0″	20′-0″	12′-0″	48′-0″	19′-0″
19′-0″ ϕ	64′-0″	54′-0″	22′-0″	21′-4″	16′-0″	20′-0″	12′-0″	49′-0″	20′-0″
20′-0″ ϕ	64′-0″	55′-0″	24′-0″	22′-4″	17′-0″	20′-0″	12′-0″	50′-0″	20′-0″
21′-0″ ϕ	65′-0″	55′-0″	25′-0″	23′-4″	17′-0″	20′-0″	14′-0″	52′-0″	20′-0″
22′-0″ ϕ	65′-0″	55′-0″	25′-0″	24′-4″	18′-0″	22′-0″	14′-0″	52′-0″	22′-0″
23′-0″ ϕ	65′-0″	56′-0″	26′-0″	25′-4″	18′-0″	22′-0″	14′-0″	53′-0″	22′-0″
24′-0″ ϕ	66′-0″	56′-0″	26′-0″	26′-4″	19′-0″	22′-0″	14′-0″	54′-0″	24′-0″
25′-0″ ϕ	67′-0″	58′-0″	27′-0″	27′-4″	19′-0″	22′-0″	14′-0″	55′-0″	26′-0″

LARGER SIZES AVAILABLE

Fig. 5-5. (continued)

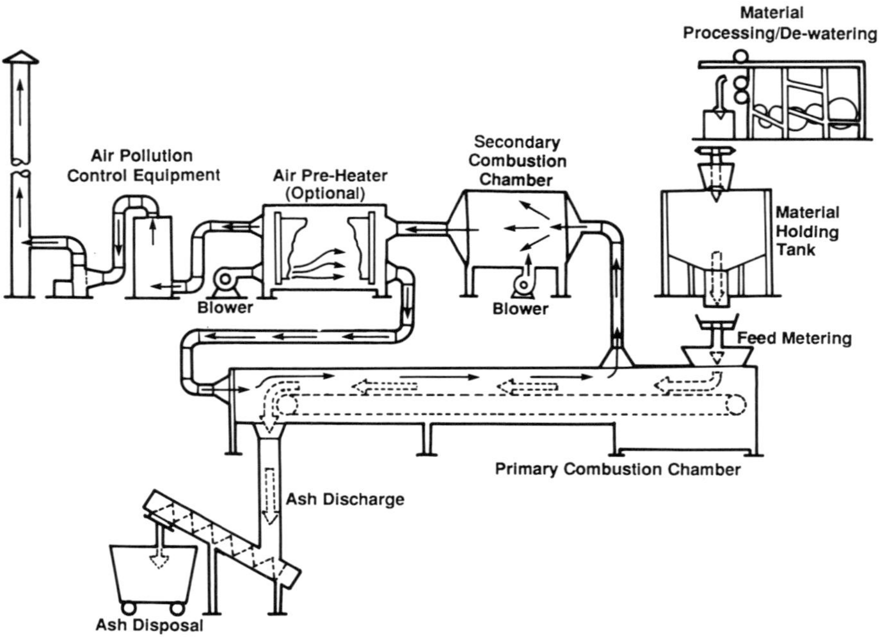

Fig. 5-13. Infrared incineration system.

may be prohibitive. In some parts of the country, the electrical demand charge may be as large as the power charge.

There are a number of advantages to this system. The lack of agitation of the sludge on the conveyor belt results in a smaller fraction of ash carryover to the gas stream; air emissions control requirements are easier to meet. Ceramic-fiber insulation, which has very low thermal inertia, provides the ability to heat up the furnace to operating temperature in under two hours. This feature allows the use of infrared furnaces for intermittent or infrequent loading. Another important advantage is that the heat generated by the electric elements does not produce additional flue gas, as does the burning of fossil fuel.

This furnace system has been adapted to the clean up of soils contaminated with trace organics. The soils are heated in the basic unit to release their organic contamination. The organics are directed, with the exiting gas stream, to an external afterburner, where they are fired at sufficient temperature and for a specific residence time for destruction. This system is illustrated in Fig. 5-15. Fuel oil is added to the soil to heat it above 1200° F, to assure the release of organics. The use of fuel oil as the supplemental fuel reduces operating costs compared to the use of electric energy. To encourage the firing of the fuel-oil-laden soil, an agita-

Fig. 5-14. Infrared furnace cross-section. Source: P. Daily, Shirco Inc.

tor bar is installed above the conveyor, near the start of its travel down the incinerator chamber.

These modifications to the standard sludge-incineration unit for the incineration of soils tend to reduce some of the advantages of the infrared

Fig. 5-15. Infrared furnace with afterburner. Source: P. Daily, Shirco Inc.

Table 5-3. Infrared Furnace Sizing.

Size (w × l, ft)	Belt Size (w × l, ft)	Wet Feed (lb/hr)	Dry Solids (lb/hr)	Loading (lb wet feed/ft2)
6 × 32	4.5 × 21	1,070	430	11.3
8.5 × 72	7 × 54.5	4,520	900	11.8
8.5 × 88	7 × 68.5	6,330	2,530	13.2
9.5 × 88	8 × 70	8,000	3,200	14.3
		Average		12.6

system. The use of fuel oil generates off-gas from the burning of the fuel oil itself. An all electric-powered unit does not generate this additional flue gas. Additionally, fuel oil will volatilize within the furnace and, without the presence of an open flame, there is a potential explosion problem. By burning within the sludge bed on the conveyor and with an external agitator, the particulate carryover by the flue gas is increased.

Table 5-3 is a sizing guide for selecting an infrared furnace firing an organic sludge feed. If additional capacity is required, multiple furnaces are normally used.

REFERENCES

1. C.R. Brunner, *Design of Sewage Sludge Incineration Systems* (Park Ridge, New Jersey: NOYES, 1980).

2. C.R. Brunner, *Incineration Systems: Selection & Design* (Reston, VA: Incinerator Consultants Incorporated, 1988).

3. C.R. Brunner, *Hazardous Air Emissions from Incineration*, Second Printing (New York: Chapman & Hall, 1986).

6

Site Clean-Up

There are thousands of sites in the United States that have been found to be contaminated with toxic or hazardous materials. Many of these sites represent potential sources of danger to human, animal, or plant life. Over the past five years they have been catalogued, isolated, studied, and fenced, like the site in Southern California shown in Fig. 6-1. In this chapter, the nature of these sites will be discussed, and thermal destruction techniques for the clean-up of these sites will be described.

THE NATIONAL PRIORITIES LIST

In 1980 Congress voted into law the Comprehensive Environmental Response, Compensation, and Liability Act (CERCLA), commonly known as Superfund. Superfund established a program to identify sites from which releases of toxic and hazardous substances into the environment might occur or may have already occurred. The provision of this legislation include procedures to ensure that either the "principal responsible party" (PRP) or the government cleans up these sites and that restitution be made to both the environment and the affected parties.

CERCLA requires that the EPA inventory and investigate hazardous-waste sites and prioritize them for clean-up. The most critical of these sites are listed on the National Priority List (NPL). Criticality is based

Fig. 6-1. Hazardous waste site. Copyright © C. R. Brunner, 1986.

on the pathways to exposure at the site, namely: ground water, surface water, air, direct contact, and fire and explosion.

Generally, the first three pathways to environmental damage require remediation, which can include neutralization, containment, or treatment, as well as removal. When danger is found from direct contact or from fire and explosion removal is necessary.

As of June 10, 1986, there were 703 sites on the NPL. Table 6-1 lists these sites, by rank. The lower the number, the more dangerous the site in terms of its effect on the public health. The sites are grouped by fifties, as indicated in the NPL. This list is regularly upgraded with removals as sites are cleaned up, and additions as new sites are found or existing ones re-evaluated. The EPA region referred to in the NPL is the region corresponding to the state location, in accordance with the map in Fig. 6-2.

CONTAMINANTS

No sites have been found to be truly homogeneous. Some of them have a scattering of buried or exposed drums contaminated with toxic or hazardous organic or inorganic components. Many sites include ponds with significant hazardous constituents, but no drums or other containers, and some sites have no visible contamination, but have underground wastes.

Sites with contaminated soils often contain other materials, which may also be contaminated. For instance, trees, tree stumps, rocks, and vegetation will likely be tainted and must be included in a remediation plan.

Table 6-1. National Priorities List.

NPL RANK	EPA RG	ST	SITE NAME *	CITY/COUNTY	RESPONSE CATEGORY#				
			GROUP 1						
1	02	NJ	Lipari Landfill	Pitman		R	F		
2	03	DE	Tybouts Corner Landfill *	New Castle County	V	R	F		
3	03	PA	Bruin Lagoon	Bruin Borough		R			
4	02	NJ	Helen Kramer Landfill	Mantua Township		R		S	
5	01	MA	Industri-Plex	Woburn	V	R			
6	02	NJ	Price Landfill *	Pleasantville		R	F		
7	02	NY	Pollution Abatement Services *	Oswego		R			
8	07	IA	LaBounty Site	Charles City	V		F		
9	03	DE	Army Creek Landfill	New Castle County	V		F		
10	02	NJ	CPS/Madison Industries	Old Bridge Township					D
11	01	MA	Nyanza Chemical Waste Dump	Ashland		R			
12	02	NJ	Gems Landfill	Gloucester Township		R		S	
13	05	MI	Berlin & Farro	Swartz Creek	V	R	F	S	
14	01	MA	Baird & McGuire	Holbrook		R	F		
15	02	NJ	Lone Pine Landfill	Freehold Township	V	R	F		
16	01	NH	Somersworth Sanitary Landfill	Somersworth		R			
17	05	MN	FMC Corp. (Fridley Plant)	Fridley	V				
18	06	AR	Vertac, Inc.	Jacksonville	V		F		
19	01	NH	Keefe Environmental Services	Epping		R		S	
20	08	SD	Whitewood Creek *	Whitewood	V				
21	08	MT	Silver Bow Creek	Sil Bow/Deer Lodge		R			
22	06	TX	French, Ltd.	Crosby	V	R	F		
23	01	NH	Sylvester *	Nashua		R		S	
24	05	MI	Liquid Disposal, Inc.	Utica		R			
25	03	PA	Tysons Dump	Upper Merion Twp		R	F		
26	03	PA	McAdoo Associates *	McAdoo Borough		R			
27	06	TX	Motco Inc. *	La Marque		R	F		
28	05	OH	Arcanum Iron & Metal	Darke County		R	F		
29	08	MT	East Helena Site	East Helena	V		F		
30	06	TX	Sikes Disposal Pits	Crosby		R			
31	04	AL	Triana/Tennessee River	Limestone/Morgan	V	R	F		
32	09	CA	Stringfellow *	Glen Avon Heights		R	F		
33	01	ME	McKin Co.	Gray		R	F	S	
34	06	TX	Crystal Chemical Co.	Houston		R			
35	02	NJ	Bridgeport Rental & Oil Services	Bridgeport		R			
36	08	CO	Sand Creek Industrial	Commerce City		R	F		
37	06	TX	Geneva Industries/Fuhrmann Energy	Houston		R	F		
38	01	MA	W. R. Grace & Co. (Acton Plant)	Acton	V		F	S	
39	05	MN	Reilly Tar (St. Louis Park Plant)	St. Louis Park		R	F	S	
40	02	NJ	Burnt Fly Bog	Marlboro Township		R		S	
41	02	NJ	Vineland Chemical Co., Inc.	Vineland	V		F		

*: STATES' DESIGNATED TOP PRIORITY SITES
#: V = VOLUNTARY OR NEGOTIATED RESPONSE;
R = FEDERAL AND STATE RESPONSE;
F = FEDERAL ENFORCEMENT;
S = STATE ENFORCEMENT;
D = ACTIONS TO BE DETERMINED.

NPL RANK	EPA RG	ST	SITE NAME *	CITY/COUNTY	RESPONSE CATEGORY#				
			GROUP 1 (CON'T)						
42	04	FL	Schuylkill Metals Corp.	Plant City					D
43	05	MN	New Brighton/Arden Hills	New Brighton	V	R			
44	02	NY	Old Bethpage Landfill	Oyster Bay	V			S	
45	02	NJ	Shieldalloy Corp.	Newfield Borough	V			S	
46	04	FL	Reeves SE Galvanizing Corp.	Tampa					D
47	08	MT	Anaconda Co. Smelter	Anaconda	V		F		
48	10	WA	Western Processing Co., Inc.	Kent	V	R	F	S	
49	05	WI	Omega Hills North Landfill	Germantown				S	
50	04	FL	American Creosote (Pensacola)	Pensacola		R	F		
			GROUP 2						
51	02	NJ	Caldwell Trucking Co.	Fairfield		R		S	
52	02	NY	GE Moreau	South Glen Falls	V		F	S	
53	05	IN	Seymour Recycling Corp. *	Seymour	V	R	F		
54	04	FL	Peak Oil Co./Bay Drum Co.	Tampa		R			
55	05	OH	United Scrap Lead Co., Inc.	Troy		R			
56	06	OK	Tar Creek (Ottawa County)	Ottawa County		R			
57	07	KS	Cherokee County	Cherokee County		R			
58	02	NJ	Brick Township Landfill	Brick Township	V			S	
59	05	MI	Northernaire Plating	Cadillac		R			
60	05	WI	Janesville Old Landfill	Janesville			F		
61	10	WA	Frontier Hard Chrome, Inc.	Vancouver		R			
62	04	SC	Independent Nail Co.	Beaufort		R			
63	04	SC	Kalama Specialty Chemicals	Beaufort				S	
64	05	WI	Janesville Ash Beds	Janesville			F		
65	04	FL	Davie Landfill	Davie					D
66	05	OH	Miami County Incinerator	Troy			F		
67	04	FL	Gold Coast Oil Corp.	Miami					D
68	05	IN	International Minerals (E. Plant)	Terre Haute					D
69	05	WI	Wheeler Pit	La Prairie Township				S	
70	09	AZ	Tucson Intl Airport Area	Tucson		R			
71	09	CA	Operating Industries, Inc. Lndfll	Monterey Park			F		
72	02	NY	Wide Beach Development	Brant		R			
73	09	CA	Iron Mountain Mine	Redding		R			
74	02	NJ	Scientific Chemical Processing	Carlstadt	V		F	S	
75	08	CO	California Gulch	Leadville			F		
76	02	NJ	D'Imperio Property	Hamilton Township		R			
77	05	MN	Oakdale Dump	Oakdale	V				
78	05	MI	Gratiot County Landfill *	St. Louis	V	R	F	S	
79	01	RI	Picillo Farm *	Coventry		R	F	S	
80	01	MA	New Bedford Site *	New Bedford	V	R	F	S	
81	06	LA	Old Inger Oil Refinery *	Darrow		R			
82	05	OH	Chem-Dyne *	Hamilton	V	R	F	S	

*: STATES' DESIGNATED TOP PRIORITY SITES
#: V = VOLUNTARY OR NEGOTIATED RESPONSE;
R = FEDERAL AND STATE RESPONSE;
F = FEDERAL ENFORCEMENT;
S = STATE ENFORCEMENT;
D = ACTIONS TO BE DETERMINED.

NPL RANK	EPA RG	ST	SITE NAME *	CITY/COUNTY	RESPONSE CATEGORY#				
			GROUP 2 (CON'T)						
83	04	SC	SCRDI Bluff Road *	Columbia	V	R	F		
84	01	CT	Laurel Park, Inc. *	Naugatuck Borough	V			S	
85	08	CO	Marshall Landfill *	Boulder County			F		
86	05	IL	Outboard Marine Corp. *	Waukegan		R	F		
87	06	NM	South Valley *	Albuquerque	V	R	F		
88	01	VT	Pine Street Canal *	Burlington					D
89	03	WV	West Virginia Ordnance *	Point Pleasant			F		
90	07	MO	Elliisville Site *	Ellisville		R	F	S	
91	08	ND	Arsenic Trioxide Site *	Southeastern N.D.		R			
92	03	VA	Matthews Electroplating *	Roanoke County		R			
93	07	IA	Aidex Corp. *	Council Bluffs		R			
94	09	AZ	Mountain View Mobile Home Estates	Globe		R	F		
95	04	TN	North Hollywood Dump *	Memphis	V	R		S	
96	04	KY	A.L. Taylor (Valley of Drums) *	Brooks		R	F		
97	09	GU	Ordot Landfill *	Guam		R			
98	04	MS	Flowood Site *	Flowood	V				
99	08	UT	Rose Park Sludge Pit *	Salt Lake City	V				
100	07	KS	Arkansas City Dump *	Arkansas City		R			
			GROUP 3						
101	05	IL	A & F Material Reclaiming, Inc.	Greenup			F		
102	03	PA	Douglassville Disposal	Douglassville		R			
103	02	NJ	Krysowaty Farm	Hillsborough		R			
104	05	MN	Koppers Coke	St. Paul	V			S	
105	01	MA	Plymouth Harbor/Cannon Engnrng	Plymouth	V	R		S	
106	10	ID	Bunker Hill Mining & Metallurg	Smelterville			F	S	
107	02	NY	Hudson River PCBs	Hudson River		R			
108	02	NJ	Universal Oil Products (Chem Div)	East Rutherford	V			S	
109	09	CA	Aerojet General Corp.	Rancho Cordova			F		
110	10	WA	Com Bay, South Tacoma Channel	Tocoma	V	R	F	S	
111	03	PA	Osborne Landfill	Grove City	V			S	
112	08	UT	Portland Cement (Kiln Dust 2 & 3)	Salt Lake City	V			S	
113	01	CT	Old Southington Landfill	Southington				S	
114	02	NY	Syosset Landfill	Oyster Bay					D
115	09	AZ	Nineteenth Avenue Landfill	Phoenix				S	
116	10	OR	Teledyne Wah Chang	Albany					D
117	10	WA	Midway Landfill	Kent		R			
118	02	NY	Sinclair Refinery	Wellsville		R			
119	04	AL	Mowbray Engineering Co.	Greenville		R			
120	05	MI	Spiegelberg Landfill	Green Oak Township		R			
121	04	FL	Miami Drum Services	Miami		R			
122	02	NJ	Reich Farms	Pleasant Plains		R			
123	10	ID	Union Pacific Railroad Co.	Pocatello					D
124	02	NJ	South Brunswick Landfill	South Brunswick	V		F		

*: STATES' DESIGNATED TOP PRIORITY SITES
#: V = VOLUNTARY OR NEGOTIATED RESPONSE;
R = FEDERAL AND STATE RESPONSE;
F = FEDERAL ENFORCEMENT;
S = STATE ENFORCEMENT;
D = ACTIONS TO BE DETERMINED.

NPL RANK	EPA RG	ST	SITE NAME *	CITY/COUNTY	RESPONSE CATEGORY#				
			GROUP 3 (CON'T)						
125	04	AL	Ciba-Geigy Corp. (McIntosh Plant)	McIntosh					D
126	04	FL	Kassauf-Kimerling Battery	Tampa	V	R	F		
127	05	IL	Wauconda Sand & Gravel	Wauconda		R			
128	06	TX	Bailey Waste Disposal	Bridge City		R			
129	01	NH	Ottati & Goss/Kingston Steel Drum	Kingston	V	R	F	S	
130	05	MI	Ott/Story/Cordova	Dalton Township		R	F		
131	05	MI	Thermo-Chem, Inc.	Muskegon					D
132	02	NJ	NL Industries	Pedricktown					D
133	05	MN	St. Regis Paper Co.	Cass Lake				S	
134	02	NJ	Ringwood Mines/Landfill	Ringwood Borough	V		F		
135	04	FL	Whitehouse Oil Pits	Whitehouse		R			
136	04	GA	Hercules 009 Landfill	Brunswick					D
137	05	MI	Velsicol Chemical (Michigan)	St. Louis	V			S	
138	05	OH	Summit National	Deerfield Township		R			
139	02	NY	Love Canal	Niagara Falls		R	F	S	
140	05	MN	Pine Bend Sanitary Landfill	Dakota County				S	
141	07	IA	Lawrence Todtz Farm	Camanche					D
142	05	IN	Fisher-Calo	LaPorte			F		
143	04	FL	Pioneer Sand Co.	Warrington		R		S	
144	05	MI	Springfield Township Dump	Davisburg		R			
145	03	PA	Hranica Landfill	Buffalo Township					D
146	04	NC	Martin Marietta, Sodyeco, Inc.	Charlotte	V				
147	04	FL	Zellwood Ground Water Contam	Zellwood			F		
148	05	MI	Packaging Corp. of America	Filer City	V		F		
149	05	WI	Muskego Sanitary Landfill	Muskego			F		
150	02	NY	Hooker (S Area)	Niagara Falls			F	S	
			GROUP 4						
151	03	PA	Lindane Dump	Harrison Township					D
152	08	CO	Central City-Clear Creek	Idaho Springs		R			
153	02	NJ	Ventron/Velsicol	Wood Ridge Borough	V	R			S
154	04	FL	Taylor Road Landfill	Seffner	V		F		
155	01	RI	Western Sand & Gravel	Burrillville		R		S	
156	04	SC	Koppers Co., Inc (Florence Plant)	Florence				S	
157	02	NJ	Maywood Chemical Co.	Maywood/Rochelle Pk		R			
158	02	NJ	Nascolite Corp.	Millville		R			
159	05	OH	Industrial Excess Landfill	Uniontown		R		S	
160	06	OK	Hardage/Criner	Criner			F		
161	05	MI	Rose Township Dump	Rose Township		R			
162	05	MN	Waste Disposal Engineering	Andover	V	R	F	S	
163	02	NY	Liberty Industrial Finishing	Farmingdale	V			S	
164	02	NJ	Kin-Buc Landfill	Edison Township	V	R	F		
165	05	OH	Bowers Landfill	Circleville	V		F		

*: STATES' DESIGNATED TOP PRIORITY SITES
#: V = VOLUNTARY OR NEGOTIATED RESPONSE;
R = FEDERAL AND STATE RESPONSE;
F = FEDERAL ENFORCEMENT;
S = STATE ENFORCEMENT;
D = ACTIONS TO BE DETERMINED.

NPL RANK	EPA RG	ST	SITE NAME *	CITY/COUNTY	RESPONSE CATEGORY#				
			GROUP 4 (CON'T)						
166	02	NJ	Ciba-Geigy Corp.	Toms River	V		F		
167	05	MI	Butterworth #2 Landfill	Grand Rapids			F		
168	02	NJ	American Cyanamid Co.	Bound Brook	V			S	
169	03	PA	Heleva Landfill	North Whitehall Twp	V	R	F		
170	02	NJ	Ewan Property	Shamong Township		R			
171	02	NY	Batavia Landfill	Batavia	V		F		
172	05	MN	Boise Cascade/Onan/Medtronics	Fridley				S	
173	01	RI	L&RR, Inc.	North Smithfield				S	
174	04	FL	NW 58th Street Landfill	Hialeah		R			
175	02	NJ	Delilah Road	Egg Harbor Township		R			
176	03	PA	Mill Creek Dump	Erie		R			
177	02	NJ	Glen Ridge Radium Site	Glen Ridge		R			
178	02	NJ	Montclair/West Orange Radium Site	Montclair/W Orange		R			
179	04	FL	Sixty-Second Street Dump	Tampa		R			
180	05	MI	G&H Landfill	Utica		R			
181	04	NC	Celanese(Shelby Fiber Operations)	Shelby					D
182	02	NJ	Metaltec/Aerosystems	Franklin Borough		R			
183	05	WI	Schmalz Dump	Harrison		R			
184	05	MI	Motor Wheel, Inc.	Lansing					D
185	02	NJ	Lang Property	Pemberton Township			F		
186	06	TX	Stewco, Inc.	Waskom		R	F		
187	02	NJ	Sharkey Landfill	Parsippany Troy Hls		R			
188	09	CA	Selma Treating Co.	Selma			F		
189	06	LA	Cleve Reber	Sorrento	V	R			
190	05	IL	Velsicol Chemical (Illinois)	Marshall		R			
191	05	MI	Tar Lake	Mancelona Township			F		
192	02	NY	Johnstown City Landfill	Town of Johnstown					D
193	04	NC	NC State U (Lot 86, Farm Unit #1)	Raleigh					D
194	08	CO	Lowry Landfill	Arapahoe County	V	R			
195	05	MN	MacGillis & Gibbs/Bell Lumber	New Brighton		R		S	
196	03	PA	Hunterstown Road	Straban Township		R	F		
197	02	NJ	Combe Fill North Landfill	Mount Olive Twp		R			
198	01	MA	Re-Solve, Inc.	Dartmouth		R	F		
199	02	NJ	Goose Farm	Plumstead Township	V	R	F	S	
200	04	TN	Velsicol Chem (Hardeman County)	Teone					D
			GROUP 5						
201	02	NY	York Oil Co.	Moira		R	F		
202	04	FL	Sapp Battery Salvage	Cottondale		R			
203	04	SC	Wamchem, Inc.	Burton					D
204	02	NJ	Chemical Leaman Tank Lines, Inc.	Bridgeport	V		F		
205	05	WI	Master Disposal Service Landfill	Brookfield		R			

*: STATES' DESIGNATED TOP PRIORITY SITES
#: V = VOLUNTARY OR NEGOTIATED RESPONSE;
R = FEDERAL AND STATE RESPONSE;
F = FEDERAL ENFORCEMENT;
S = STATE ENFORCEMENT;
D = ACTIONS TO BE DETERMINED.

NPL RANK	EPA RG	ST	SITE NAME *	CITY/COUNTY	RESPONSE CATEGORY#				
			GROUP 5 (CON'T)						
206	07	KS	Doepke Disposal Site (Holliday)	Johnson County		R			
207	02	NJ	Florence Land Recontouring LF	Florence Township		R			
208	01	RI	Davis Liquid Waste	Smithfield		R		S	
209	01	MA	Charles-George Reclamation Lf	Tyngsborough		R	F		
210	02	NJ	King of Prussia	Winslow Township	V		F		
211	03	VA	Chisman Creek	York County		R			
212	05	OH	Nease Chemical	Salem	V			S	
213	08	CO	Eagle Mine	Minturn/Redcliff		R		S	
214	02	NJ	W. R. Grace & Co. (Wayne Plant)	Wayne Township		R			
215	02	NJ	Chemical Control	Elizabeth		R		S	
216	04	SC	Leonard Chemical Co., Inc.	Rock Hill				S	
217	05	OH	Allied Chemical & Ironton Coke	Ironton		R	F		
218	05	MI	Verona Well Field	Battle Creek		R			
219	07	MO	Lee Chemical	Liberty					D
220	01	CT	Beacon Heights Landfill	Beacon Falls		R			
221	04	AL	Stauffer Chem (Cold Creek Plant)	Bucks	V				
222	05	MN	Burlington Northern (Brainerd)	Brainerd/Baxter	V				
223	05	MI	Torch Lake	Houghton County					D
224	01	RI	Central Landfill	Johnston	V		F	S	
225	03	PA	Malvern TCE	Malvern					D
226	02	NY	Facet Enterprises, Inc.	Elmira	V		F		
227	03	DE	Delaware Sand & Gravel Landfill	New Castle County		R			
228	03	PA	MW Manufacturing	Valley Township				S	
229	04	TN	Murray-Ohio Dump	Lawrenceburg	V			S	
230	05	IN	Envirochem Corp.	Zionsville	V	R	F		
231	05	IN	MIDCO I	Gary			F		
232	05	OH	South Point Plant	South Point			F		
233	03	PA	Whitmoyer Laboratories	Jackson Township					D
234	04	FL	Coleman-Evans Wood Preserving Co.	Whitehouse		R	F	S	
235	03	PA	Shriver's Corner	Straban Township		R	F		
236	03	PA	Dorney Road Landfill	Upper Macungie Twp		F			
237	05	IN	Northside Sanitary Landfill, Inc.	Zionsville			F	S	
238	04	FL	Florida Steel Corp.	Indiantown	V				
239	05	IL	Pagel's Pit	Rockford					D
240	05	MN	U of Minnesota Rosemount Res Cent	Rosemount				S	
241	05	MN	Freeway Sanitary Landfill	Burnsville					D
242	09	AZ	Litchfield Airport Area	Goodyear/Avondale			F		
243	02	NJ	Spence Farm	Plumstead Township	V	R		S	
244	06	AR	Mid-South Wood Products	Mena			F		
245	04	MS	Newsom Brothers/Old Reichhold	Columbia		R			
246	09	CA	Atlas Asbestos Mine	Fresno County		R			
247	09	CA	Coalinga Asbestos Mine	Coalinga		R			
248	04	FL	Brown Wood Preserving	Live Oak	V		F		
249	02	NY	Port Washington Landfill	Port Washington		R			
250	05	IN	Columbus Old Municipal Lndfll #1	Columbus					D

*: STATES' DESIGNATED TOP PRIORITY SITES
#: V = VOLUNTARY OR NEGOTIATED RESPONSE;
R = FEDERAL AND STATE RESPONSE;
F = FEDERAL ENFORCEMENT;
S = STATE ENFORCEMENT;
D = ACTIONS TO BE DETERMINED.

GROUP 6

NPL RANK	EPA RG	ST	SITE NAME *	CITY/COUNTY	RESPONSE CATEGORY#				
251	02	NJ	Combe Fill South Landfill	Chester Township		R			
252	02	NJ	JIS Landfill	Jamesburg/S. Brnswck				S	
253	02	NY	Tronic Plating Co., Inc.	Farmingdale					D
254	03	PA	Centre County Kepone	State College Boro				S	
255	05	OH	Fields Brook	Ashtabula		R			
256	01	CT	Solvents Recovery Service	Southington			F		
257	08	CO	Woodbury Chemical Co.	Commerce City		R			
258	02	NJ	Waldick Aerospace Devices, Inc.	Wall Township				S	
259	01	MA	Hocomonco Pond	Westborough		R			
260	04	KY	Distler Brickyard	West Point		R	F		
261	02	NY	Ramapo Landfill	Ramapo	V			S	
262	09	CA	Coast Wood Preserving	Ukiah				S	
263	09	CA	South Bay Asbestos Area	Alviso		R			
264	02	NY	Mercury Refining, Inc.	Colonie	V			S	
265	04	FL	Hollingsworth Solderless Terminal	Fort Lauderdale		R			
266	02	NY	Olean Well Field	Olean	V	R	F		
267	04	FL	Varsol Spill	Miami		R			
268	05	MN	Joslyn Manufacturing & Supply Co.	Brooklyn Center	V			S	
269	08	CO	Denver Radium Site	Denver		R			
270	04	FL	Tower Chemical Co.	Clermont		R	F		
271	07	MO	Syntex Facility	Verona	V		F		
272	08	MT	Milltown Reservoir Sediments	Milltown		R			
273	05	MN	Arrowhead Refinery Co.	Hermantown		R			
274	10	OR	Martin-Marietta Aluminum Co.	The Dalles	V				
275	08	CO	Uravan Uranium (Union Carbide)	Uravan					D
276	02	NJ	Pijak Farm	Plumstead Township	V	R		S	
277	02	NJ	Syncon Resins	South Kearny		R			
278	05	MN	Oak Grove Sanitary Landfill	Oak Grove Township		R			
279	09	CA	Liquid Gold Oil Corp.	Richmond				S	
280	09	CA	Purity Oil Sales, Inc.	Malaga		R			
281	01	NH	Tinkham Garage	Londonderry			F	S	
282	04	FL	Alpha Chemical Corp.	Galloway	V				
283	02	NJ	Bog Creek Farm	Howell Township		R			
284	01	ME	Saco Tannery Waste Pits	Saco		R			
285	02	PR	Frontera Creek	Rio Abajo			F		
286	04	FL	Pickettville Road Landfill	Jacksonville	V		F		
287	05	OH	Alsco Anaconda	Gnadenhutten				S	
288	01	MA	Iron Horse Park	Billerica		R			
289	03	PA	Palmerton Zinc Pile	Palmerton	V		F		
290	05	IN	Neal's Landfill (Bloomington)	Bloomington	V		F		
291	05	WI	Kohler Co. Landfill	Kohler	V				
292	04	AL	Interstate Lead Co. (ILCO)	Leeds	V	R	F	S	
293	01	MA	Silresim Chemical Corp.	Lowell		R		S	
294	01	MA	Wells G&H	Woburn	V		F		

*: STATES' DESIGNATED TOP PRIORITY SITES
#: V = VOLUNTARY OR NEGOTIATED RESPONSE;
R = FEDERAL AND STATE RESPONSE;
F = FEDERAL ENFORCEMENT;
S = STATE ENFORCEMENT;
D = ACTIONS TO BE DETERMINED.

NPL RANK	EPA RG	ST	SITE NAME *	CITY/COUNTY	RESPONSE CATEGORY#				
			(GROUP 6 (CON'T)						
295	02	NJ	Chemsol, Inc.	Piscataway	V			S	
296	05	WI	Lauer I Sanitary Landfill	Menomonee Falls				S	
297	05	MI	Petoskey Municipal Well Field	Petoskey			F		
298	05	MN	Union Scrap	Minneapolis				S	
299	02	NJ	Radiation Technology, Inc.	Rockaway Township	V			S	
300	02	NJ	Fair Lawn Well Field	Fair Lawn	V			S	
			GROUP 7						
301	05	IN	Main Street Well Field	Elkhart		R			
302	05	MN	Lehillier/Mankato Site	Lehillier/Mankato		R			
303	10	WA	Lakewood Site	Lakewood		R			
304	03	PA	Industrial Lane	Williams Township			F		
305	05	IN	Fort Wayne Reduction Dump	Fort Wayne		R			
306	05	WI	Onalaska Municipal Landfill	Onalaska		R			
307	05	WI	National Presto Industries, Inc.	Eau Claire					D
308	02	NJ	Monroe Township Landfill	Monroe Township	V			S	
309	02	NJ	Rockaway Borough Well Field	Rockaway Township		R			
310	05	IN	Wayne Waste Oil	Columbia City		R	F		
311	03	MD	Mid-Atlantic Wood Preservers, Inc.	Harmans					D
312	10	ID	Pacific Hide & Fur Recycling Co.	Pocatello			F		
313	07	IA	Des Moines TCE	Des Moines		R			
314	02	NJ	Beachwood/Berkley Wells	Berkley Township		R			
315	02	NY	Vestal Water Supply Well 4-2	Vestal	V			S	
316	02	PR	Vega Alta Public Supply Wells	Vega Alta			F		
317	05	MI	Sturgis Municipal Wells	Sturgis		R			
318	05	MN	Washington County Landfill	Lake Elmo				S	
319	06	TX	Odessa Chromium #1	Odessa		R			
320	06	TX	Odessa Chromium #2 (Andrews Hgwy)	Odessa		R			
321	07	NE	Hastings Ground Water Contamin	Hastings		R			
322	09	AZ	Indian Bend Wash Area	Scottsdale/Tempe	V		F		
323	09	CA	San Gabriel Valley (Area 1)	El Monte		R			
324	09	CA	San Gabriel Valley (Area 2)	Baldwin Park Area		R			
325	09	CA	San Fernando Valley (Area 1)	Los Angeles					D

*: STATES' DESIGNATED TOP PRIORITY SITES
#: V = VOLUNTARY OR NEGOTIATED RESPONSE;
R = FEDERAL AND STATE RESPONSE;
F = FEDERAL ENFORCEMENT;
S = STATE ENFORCEMENT;
D = ACTIONS TO BE DETERMINED.

NPL RANK	EPA RG	ST	SITE NAME *	CITY/COUNTY	RESPONSE CATEGORY#				
			GROUP 7 (CON'T)						
326	09	CA	San Fernando Valley (Area 2)	Los Angeles/Glendale					D
327	09	CA	San Fernando Valley (Area 3)	Glendale					D
328	09	CA	T.H. Agriculture & Nutrition Co.	Fresno					D
329	10	WA	Com Bay, Near Shore/Tide Flats	Pierce County		R	F	S	
330	05	IL	LaSalle Electric Utilities	LaSalle		R			
331	05	IL	Cross Brothers Pail (Pembroke)	Pembroke Township		R			
332	04	NC	Jadco-Hughes Facility	Belmont					D
333	02	NJ	Monitor Devices/Intercircuits Inc.	Wall Township					D
334	02	PR	Upjohn Facility	Barceloneta					D
335	09	CA	McColl	Fullerton		R	F		
336	03	PA	Henderson Road	Upper Merion Twp	V		F		
337	02	NY	Hooker Chemical/Ruco Polymer Corp.	Hicksville					D
338	10	WA	Colbert Landfill	Colbert		R			
339	06	LA	Petro-Processors	Scotlandville	V		F		
340	02	NY	Applied Environmental Services	Glenwood Landing				S	
341	02	PR	Barceloneta Landfill	Florida Afuera					D
342	01	NH	Tibbets Road	Barrington		R			
343	03	MD	Sand, Gravel & Stone	Elkton	V	R	F		
344	05	MI	Spartan Chemical Co.	Wyomng	V			S	
345	02	NJ	Roebling Steel Co.	Florence		R			
346	03	PA	East Mount Zion	Springettsbury Twp		R			
347	04	TN	Amnicola Dump	Chattanooga		R			
348	02	NJ	Vineland State School	Vineland	V			S	
349	01	MA	Groveland Wells	Groveland	V	R		S	
350	02	NY	General Motors (Cent Foundry Div)	Massena	V		F		
			GROUP 8						
351	04	SC	SCRDI Dixiana	Cayce		R	F	S	
352	05	MI	Roto-Finish Co., Inc.	Kalamazoo					D
353	05	MN	Olmsted County Sanitary Landfill	Oronoco					D
354	07	MO	Quality Plating	Sikeston					D
355	07	MO	Fulbright Landfill	Springfield					D
356	03	PA	Presque Isle	Erie		R			
357	02	NJ	Williams Property	Swainton		R			
358	02	NJ	Renora, Inc.	Edison Township	V		F		
359	02	NJ	Denzer & Schafer X-Ray Co.	Bayville	V			S	
360	02	NJ	Hercules, Inc. (Gibbstown Plant)	Gibbstown					D
361	05	IN	Ninth Avenue Dump	Gary		R			
362	10	WA	Toftdahl Drums	Brush Prairie		R			
363	06	TX	Texarkana Wood Preserving Co.	Texarkana					D
364	06	AR	Gurley Pit	Edmondson			F		
365	01	RI	Peterson/Puritan, Inc.	Lincoln/Cumberland					D
366	07	MO	Times Beach Site	Times Beach		R			
367	05	MI	Wash King Laundry	Pleasant Plains Twp		R			
368	05	MN	Whittaker Corp.	Minneapolis				S	
369	05	MN	NL Industries/Taracorp/Golden	St. Louis Park				S	
370	09	CA	Westinghouse (Sunnyvale Plant)	Sunnyvale					D
371	01	CT	Kellogg-Deering Well Field	Norwalk		R			

*: STATES' DESIGNATED TOP PRIORITY SITES
#: V = VOLUNTARY OR NEGOTIATED RESPONSE;
R = FEDERAL AND STATE RESPONSE;
F = FEDERAL ENFORCEMENT;
S = STATE ENFORCEMENT;
D = ACTIONS TO BE DETERMINED.

NPL RANK	EPA RG	ST	SITE NAME *	CITY/COUNTY	RESPONSE CATEGORY#				
			GROUP 8 (CON'T)						
372	01	MA	Cannon Engineering Corp. (CEC)	Bridgewater		R		S	
373	05	MI	H. Brown Co., Inc.	Grand Rapids					D
374	02	NY	Nepera Chemical Co., Inc.	Maybrook	V				
375	02	NY	Niagara County Refuse	Wheatfield					D
376	04	FL	Sherwood Medical Industries	Deland					D
377	04	AL	Olin Corp. (McIntosh Plant)	McIntosh					D
378	05	MI	Southwest Ottawa County Landfill	Park Township	V			S	
379	02	NY	Kentucky Avenue Well Field	Horseheads		R			
380	02	NY	Pasley Solvents & Chemicals, Inc.	Hempstead					D
381	02	NJ	Asbestos Dump	Millington	V		F		
382	04	KY	Lee's Lane Landfill	Louisville	V		F		
383	06	AR	Frit Industries	Walnut Ridge	V		F		
384	05	OH	Fultz Landfill	Jackson Township		R			
385	04	FL	Tri-City Oil Conservationist, Inc	Tampa		R	F		
386	05	OH	Coshocton Landfill	Franklin Township			F		
387	01	RI	Davis (GSR) Landfill	Glocester					D
388	03	PA	Lord-Shope Landfill	Girard Township	V			S	
389	10	WA	FMC Corp. (Yakima Pit)	Yakima				S	
390	05	WI	Northern Engraving Co.	Sparta	V		F		
391	06	TX	South Cavalcade Street	Houston	V		F		
392	01	MA	PSC Resources	Palmer				S	
393	05	MI	Forest Waste Products	Otisville		R	F		
394	03	PA	Drake Chemical	Lock Haven		R			
395	01	NH	Kearsarge Metallurgical Corp.	Conway				S	
396	04	SC	Palmetto Wood Preserving	Dixianna		R			
397	05	IL	Petersen Sand & Gravel	Libertyville		R			
398	05	MI	Clare Water Supply	Clare		R	F		
399	03	PA	Havertown PCP	Haverford			F		
400	03	DE	New Castle Spill	New Castle County					D
			GROUP 9						
401	08	MT	Idaho Pole Co.	Bozeman					D
402	05	IN	Lake Sandy Jo (M&M Landfill)	Gary		R			
403	05	IL	Johns-Manville Corp.	Waukegan			F		
404	05	MI	Chem Central	Wyoming Township				S	
405	05	MI	Novaco Industries	Temperance		R			
406	05	MN	Windom Dump	Windom					D
407	02	NJ	Jackson Township Landfill	Jackson Township					D
408	05	IL	NL Industries/Taracorp Lead Smelt	Granite City	V		F	S	
409	05	MI	K&L Avenue Landfill	Oshtemo Township			F		
410	10	WA	Kaiser Aluminum Mead Works	Mead	V				
411	05	MN	Perham Arsenic Site	Perham		R			
412	05	MI	Charlevoix Municipal Well	Charlevoix		R			
413	02	NJ	Montgomery Township Housing Dev	Montgomery Township		R			

*: STATES' DESIGNATED TOP PRIORITY SITES
#: V = VOLUNTARY OR NEGOTIATED RESPONSE;
R = FEDERAL AND STATE RESPONSE;
F = FEDERAL ENFORCEMENT;
S = STATE ENFORCEMENT;
D = ACTIONS TO BE DETERMINED.

GROUP 9 (CON'T)

NPL RANK	EPA RG	ST	SITE NAME *	CITY/COUNTY	RESPONSE CATEGORY#				
414	02	NJ	Rocky Hill Municipal Well	Rocky Hill Borough		R			
415	02	NJ	Cinnaminson Ground Water Contamin	Cinnaminson Township		R			
416	02	NY	Brewster Well Field	Putnam County		R			
417	02	NY	Vestal Water Supply Well 1-1	Vestal		R			
418	04	NC	Bypass 601 Ground Water Contamin	Concord					D
419	07	MO	Solid State Circuits, Inc.	Republic		R	F	S	
420	07	NE	Waverly Ground Water Contamin	Waverly		R			
421	09	CA	Advanced Micro Devices, Inc.	Sunnyvale					D
422	05	MN	Nutting Truck & Caster Co.	Faribault				S	
423	02	NJ	U.S. Radium Corp.	Orange		R			
424	06	TX	Highlands Acid Pit	Highlands		R			
425	03	PA	Resin Disposal	Jefferson Borough					D
426	08	MT	Libby Ground Water Contamination	Libby			F		
427	04	KY	Newport Dump	Newport		R			
428	03	PA	Moyers Landfill	Eagleville		R			
429	04	FL	Parramore Surplus	Mount Pleasant					D
430	01	NH	Savage Municipal Water Supply	Milford			F		
431	05	IN	Poer Farm	Hancock County		R	F		
432	03	PA	Brown's Battery Breaking	Shoemakersville		R	F		
433	02	NY	SMS Instruments, Inc.	Deer Park					D
434	05	MI	Hedblum Industries	Oscoda			F		
435	06	TX	United Creosoting Co.	Conroe		R	F		
436	02	NY	Byron Barrel & Drum	Byron		R	F		
437	08	WY	Baxter/Union Pacific Tie Treating	Laramie	V		F	S	
438	02	NY	Anchor Chemicals	Hicksville					D
439	05	MI	Waste Management-Mich (Holland)	Holland					D
440	06	TX	North Cavalcade Street	Houston		R			
441	02	NJ	Sayreville Landfill	Sayreville					D
442	01	NH	Dover Municipal Landfill	Dover		R			
443	02	NY	Ludlow Sand & Gravel	Clayville	V			S	
444	05	WI	City Disposal Corp. Landfill	Dunn			F	S	
445	02	NJ	Tabernacle Drum Dump	Tabernacle Township	V	R	F		
446	02	NJ	Cooper Road	Voorhees Township	V			S	
447	07	MO	Minker/Stout/Romaine Creek	Imperial		R			
448	01	CT	Yaworski Waste Lagoon	Canterbury		R		S	
449	03	WV	Leetown Pesticide	Leetown		R			
450	04	FL	Cabot/Koppers	Gainesville		R	S		

GROUP 10

NPL RANK	EPA RG	ST	SITE NAME *	CITY/COUNTY	RESPONSE CATEGORY#				
451	02	NJ	Evor Phillips Leasing	Old Bridge Township		R			
452	03	PA	Wade (ABM)	Chester		R	F	S	
453	03	PA	Lackawanna Refuse	Old Forge Borough		R			
454	06	OK	Compass Industries (Avery Drive)	Tulsa		R			

*: STATES' DESIGNATED TOP PRIORITY SITES
#: V = VOLUNTARY OR NEGOTIATED RESPONSE;
R = FEDERAL AND STATE RESPONSE;
F = FEDERAL ENFORCEMENT;
S = STATE ENFORCEMENT;
D = ACTIONS TO BE DETERMINED.

NPL RANK	EPA RG	ST	SITE NAME *	CITY/COUNTY	RESPONSE CATEGORY#				
			GROUP 10 (CON'T)						
455	02	NJ	Mannheim Avenue Dump	Galloway Township	V		F		
456	05	IN	Neal's Dump (Spencer)	Spencer			F	S	
457	02	NY	Fulton Terminals	Fulton		R			
458	03	PA	Westinghouse Elevator Co. Plant	Gettysburg		R	F		
459	01	NH	Auburn Road Landfill	Londonderry			F	S	
460	03	WV	Fike Chemical, Inc.	Nitro			F		
461	05	MN	General Mills/Henkel Corp.	Minneapolis				S	
462	05	OH	Laskin/Poplar Oil Co.	Jefferson Township	V	R	F		
463	05	OH	Old Mill	Rock Creek		R			
464	07	KS	Johns' Sludge Pond	Wichita	V		F		
465	05	WI	Stoughton City Landfill	Stoughton					D
466	09	CA	Del Norte Pesticide Storage	Crescent City		R			
467	02	NJ	De Rewal Chemical Co.	Kingwood Township			F		
468	03	PA	Middletown Air Field	Middletown					D
469	02	NJ	Swope Oil & Chemical Co.	Pennsauken	V	R	F		
470	04	GA	Monsanto Corp. (Augusta Plant)	Augusta	V				
471	01	NH	South Municipal Water Supply Well	Peterborough			F	S	
472	01	ME	Winthrop Landfill	Winthrop	V		F	S	
473	03	WV	Ordnance Works Disposal Areas	Morgantown			F		
474	06	AR	Cecil Lindsey	Newport		R			
475	05	OH	Zanesville Well Field	Zanesville	V			S	
476	02	NY	Suffern Village Well Field	Village of Suffern		R			
477	02	NY	Endicott Village Well Field	Village of Endicott		R			
478	05	MN	Kummer Sanitary Landfill	Bemidji		R			
479	05	OH	Sanitary Landfill Company (IWD)	Dayton					D
480	05	WI	Eau Claire Municipal Well Field	Eau Claire		R			
481	07	MO	Valley Park TCE	Valley Park					D
482	09	CA	San Fernando Valley (Area 4)	Los Angeles					D
483	04	GA	Powersville Site	Peach County		R			
484	05	MI	Grand Traverse Overall Supply Co.	Greilickville			F		
485	05	MI	Metamora Landfill	Metamora		R			
486	05	MI	Whitehall Municipal Wells	Whitehall		R			
487	05	MN	South Andover Site	Andover		R			
488	02	NJ	Diamond Alkali Co.	Newark	V	R	F	S	
489	03	VA	Avtex Fibers, Inc.	Front Royal					D
490	05	MI	Kentwood Landfill	Kentwood	V		F		
491	05	MI	Electrovoice	Buchanan					D
492	02	NY	Katonah Municipal Well	Town of Bedford		R			
493	02	PR	Fibers Public Supply Wells	Jobos					D
494	05	IN	Marion (Bragg) Dump	Marion		R			
495	05	OH	Pristine, Inc.	Reading		R	F		
496	05	WI	Mid-State Disposal, Inc. Landfill	Cleveland Township		R			

*: STATES' DESIGNATED TOP PRIORITY SITES
#: V = VOLUNTARY OR NEGOTIATED RESPONSE;
R = FEDERAL AND STATE RESPONSE;
F = FEDERAL ENFORCEMENT;
S = STATE ENFORCEMENT;
D = ACTIONS TO BE DETERMINED.

NPL RANK	EPA RG	ST	SITE NAME *	CITY/COUNTY	RESPONSE CATEGORY#				
			GROUP 10 (CON'T)						
497	04	TN	American Creosote (Jackson Plant)	Jackson		R			
498	08	CO	Broderick Wood Products	Denver	V		F		
499	05	OH	Buckeye Reclamation	St. Clairsville	V		F		
500	02	NY	Preferred Plating Corp.	Farmingdale					D
			GROUP 11						
501	06	TX	Bio-Ecology Systems, Inc.	Grand Prairie		R			
502	08	UT	Monticello Rad Contaminated Props	Monticello		R			
503	02	NJ	Woodland Route 532 Dump	Woodland Township	V	R		S	
504	05	IN	American Chemical Service, Inc.	Griffith			F		
505	01	MA	Salem Acres	Salem					D
506	01	VT	Old Springfield Landfill	Springfield	V		F		
507	02	NY	Solvent Savers	Lincklaen					D
508	03	VA	U.S. Titanium	Piney River			F	S	
509	05	IL	Galesburg/Koppers Co.	Galesburg				S	
510	02	NY	Hooker (Hyde Park)	Niagara Falls	V		F	S	
511	05	MI	SCA Independent Landfill	Muskegon Heights					
512	09	CA	MGM Brakes	Cloverdale				S	
513	06	LA	Bayou Sorrell	Bayou Sorrell			F		
514	05	MI	Duell & Gardner Landfill	Dalton Township					D
515	10	WA	Mica Landfill	Mica					D
516	02	NJ	Ellis Property	Evesham Township		R			
517	04	KY	Distler Farm	Jefferson County		R	F		
518	10	WA	Harbor Island (Lead)	Seattle					D
519	05	WI	Lemberger Transport & Recycling	Franklin Township		R			
520	05	OH	E.H. Schilling Landfill	Hamilton Township		R			
521	05	MI	Cliff/Dow Dump	Marquette			F		
522	02	NY	Clothier Disposal	Town of Granby		R			
523	03	PA	Ambler Asbestos Piles	Ambler	V	R	F	S	
524	10	WA	Queen City Farms	Maple Valley	V				
525	03	VA	L.A. Clarke & Son	Spotsylvania County		R			
526	05	WI	Scrap Processing Co., Inc.	Medford				S	
527	03	MD	Southern Maryland Wood Treating	Hollywood		R			
528	06	NM	Homestake Mining Co.	Milan	V		F		
529	09	CA	Beckman Instruments (Porterville)	Porterville					D
530	04	FL	Dubose Oil Products Co.	Cantonment				S	
531	05	MI	Mason County Landfill	Pere Marquette Twp		R	F		
532	05	MI	Cemetery Dump	Rose Center		R			
533	02	NJ	Hopkins Farm	Plumstead Township					D

*: STATES' DESIGNATED TOP PRIORITY SITES
#: V = VOLUNTARY OR NEGOTIATED RESPONSE;
R = FEDERAL AND STATE RESPONSE;
F = FEDERAL ENFORCEMENT;
S = STATE ENFORCEMENT;
D = ACTIONS TO BE DETERMINED.

NPL RANK	EPA RG	ST	SITE NAME *	CITY/COUNTY	RESPONSE CATEGORY#				
			GROUP 11 (CON'T)						
534	01	RI	Stamina Mills, Inc.	North Smithfield					D
535	05	WI	Lemberger Landfill, Inc.	Whitelaw				S	
536	05	IN	Reilly Tar (Indianapolis Plant)	Indianapolis			F		
537	01	ME	Pinette's Salvage Yard	Washburn		R			
538	06	TX	Harris (Farley Street)	Houston	V		F		
539	02	NJ	Wilson Farm	Plumstead Township					D
540	03	PA	Old City of York Landfill	Seven Valleys	V			S	
541	03	PA	Modern Sanitation Landfill	Lower Windsor Twp	V			S	
542	05	IL	Byron Salvage Yard	Byron		R			
543	05	MI	North Bronson Industrial Area	Bronson					D
544	03	PA	Stanley Kessler	King of Prussia			F		
545	02	NJ	Imperial Oil/Champion Chemicals	Morganville		R			
546	02	NJ	Myers Property	Franklin Township		R			
547	02	NJ	Pepe Field	Boonton		R			
548	10	WA	Northwest Transformer	Everson		R			
549	05	WI	Sheboygan Harbor & River	Sheboygan					D
550	05	MI	Ossineke Ground Water Contam	Ossineke					D
			GROUP 12						
551	03	WV	Follansbee Site	Follansbee	V		F		
552	02	NY	North Sea Municipal Landfill	North Sea		R			
553	09	CA	Koppers Co., Inc. (Oroville Plant)	Oroville				S	
554	09	CA	Louisiana-Pacific Corp.	Oroville					D
555	05	MI	South Macomb Disposal (Lf 9 & 9A)	Macomb Township					D
556	05	MI	U.S. Aviex	Howard Township	V		F		
557	03	PA	Walsh Landfill	Honeybrook Township		R	F		
558	02	NJ	Landfill & Development Co.	Mount Holly				S	
559	02	NJ	Upper Deerfield Township Slf	Upper Deerfield Twp					D
560	02	NY	Hertel Landfill	Plattekill					D
561	02	NY	Haviland Complex	Tow of Hyde Park		R			
562	05	MN	Adrian Municipal Well Field	Adrian		R			
563	06	NM	AT & SF (Clovis)	Clovis	V		F		
564	07	KS	Strother Field Industrial Park	Cowley County	V			S	
565	02	NJ	Fried Industries	East Brunswick Twp		R			
566	02	NY	American Thermostat Co.	South Cairo	V			S	
567	04	TN	Lewisburg Dump	Lewisburg					D
568	05	MI	McGraw Edison Corp.	Albion	V			S	
569	02	NY	Goldisc Recordings, Inc.	Holbrook	V				
570	04	KY	Airco	Calvert City	V				
571	03	PA	Metal Banks	Philadelphia	V		F		
572	02	NY	Sarney Farm	Amenia		R			
573	01	MA	Rose Disposal Pit	Lanesboro			F	S	
574	05	OH	Van Dale Junkyard	Marietta					D

*: STATES' DESIGNATED TOP PRIORITY SITES
#: V = VOLUNTARY OR NEGOTIATED RESPONSE;
R = FEDERAL AND STATE RESPONSE;
F = FEDERAL ENFORCEMENT;
S = STATE ENFORCEMENT;
D = ACTIONS TO BE DETERMINED.

NPL RANK	EPA RG	ST	SITE NAME *	CITY/COUNTY	RESPONSE CATEGORY#				
			GROUP 12 (CON'T)						
575	04	KY	B.F. Goodrich	Calvert City	V				
576	05	MI	Organic Chemicals, Inc.	Grandville				S	
577	02	NY	Volney Municipal Landfill	Town of Volney	V	R		S	
578	02	NY	FMC Corp. (Dublin Road Landfill)	Town of Shelby	V			S	
579	01	MA	Sullivan's Ledge	New Bedford		R	F		
580	04	KY	Smith's Farm	Brooks		R			
581	02	PR	Juncos Landfill	Juncos	V		F		
582	07	KS	Big River Sand Co.	Witchita		R			
583	05	IN	Bennett Stone Quarry	Bloomington	V		F		
584	04	FL	Munisport Landfill	North Miami					D
585	04	AL	Stauffer Chem (LeMoyne Plant)	Axis	V				
586	02	NJ	M&T Delisa Landfill	Asbury Park	V		F		
587	06	TX	Crystal City Airport	Crystal City		R			
588	04	SC	Geiger (C & M Oil)	Rantoules		R			
589	05	WI	Moss-American (Kerr-McGee Oil Co.)	Milwaukee		R	F		
590	05	WI	Waste Research & Reclamation Co.	Eau Claire				S	
591	10	OR	Gould, Inc.	Portland	V				
592	02	NY	Cortese Landfill	Vil of Narrowsburg	V			S	
593	05	MN	St. Louis River Site	St. Louis County		R			
594	05	MI	Auto Ion Chemicals, Inc.	Kalamazoo	V		F		
595	04	SC	Carolawn, Inc.	Fort Lawn	V	R	F		
596	07	IA	Midwest Manufacturing/North Farm	Kellogg					D
597	03	PA	Berks Sand Pit	Longswamp Township		R			
598	05	MI	Sparta Landfill	Sparta Township				S	
599	05	IL	ACME Solvent (Morristown Plant)	Morristown	V	R			
600	02	NJ	Pomona Oaks Residential Wells	Galloway Township		R			
			GROUP 13						
601	04	FL	Hipps Road Landfill	Duval County		R			
602	05	MN	Long Prairie Ground Water Contam	Long Prairie		R			
603	05	MN	Waite Park Wells	Waite Park		R			
604	09	CA	Intel Magnetics	Santa Clara					D
605	09	CA	Intel Corp. (Santa Clara III)	Santa Clara					D
606	04	FL	Pepper Steel & Alloys, Inc.	Medley		R	F		
607	01	ME	O'Connor Co.	Augusta	V	R			
608	05	WI	Oconomowoc Electroplating Co. Inc.	Ashippin		R			
609	05	MI	Rasmussen's Dump	Green Oak Township		R			
610	02	NY	Kenmark Textile Corp.	Farmingdale					D
611	03	PA	Westline Site	Westline		R			
612	04	KY	Maxey Flats Nuclear Disposal	Hillsboro		R			
613	08	MT	Mouat Industries	Columbus					D
614	02	NY	Claremont Polychemical	Old Bethpage	V			S	
615	05	OH	Powell Road Landfill	Dayton		R			
616	03	PA	Croydon TCE	Croydon					D

*: STATES' DESIGNATED TOP PRIORITY SITES
#: V = VOLUNTARY OR NEGOTIATED RESPONSE;
R = FEDERAL AND STATE RESPONSE;
F = FEDERAL ENFORCEMENT;
S = STATE ENFORCEMENT;
D = ACTIONS TO BE DETERMINED.

NPL RANK	EPA RG	ST	SITE NAME *	CITY/COUNTY	RESPONSE CATEGORY#				
			GROUP 13 (CON'T)						
617	07	IA	Vogel Paint & Wax Co.	Orange City				S	
618	05	MN	Kurt Manufacturing Co.	Fridley				S	
619	05	MI	Ionia City Landfill	Ionia	V		F		
620	06	TX	Koppers Co., Inc. (Texarkana Pit)	Texarkana	V		F		
621	08	CO	Lincoln Park	Canon City			F		
622	08	CO	Smuggler Mountain	Pitkin County	V		F		
623	05	IN	Wedzeb Enterprises, Inc.	Lebanon			F	S	
624	02	PR	GE Wiring Devices	Juana Diaz	V		F		
625	05	MI	Avenue "E" Ground Water Contamin	Traverse City				S	
626	05	OH	New Lyme Landfill	New Lyme		R			
627	02	NJ	Woodland Route 72 Dump	Woodland Township	V	R		S	
628	02	PR	RCA Del Caribe	Barceloneta					D
629	05	MN	Koch Refining Co./N-Ren Corp.	Pine Bend	V			S	
630	03	PA	Brodhead Creek	Stroudsburg		R	F		
631	05	WI	Fadrowski Drum Disposal	Franklin					D
632	10	OR	United Chrome Products, Inc.	Corvallis		R			
633	05	MI	Anderson Development Co.	Adrian		R			
634	05	MI	Shiawassee River	Howell					D
635	03	PA	Taylor Borough Dump	Taylor Borough		R			
636	03	DE	Halby Chemical Co.	New Castle					D
637	03	DE	Harvey & Knott Drum, Inc.	Kirkwood		R			
638	04	TN	Gallaway Pits	Gallaway		R	F		
639	05	OH	Big D Campground	Kingsville			F		
640	06	AR	Midland Products	Ola/Birta		R			
641	02	NY	Robintech, Inc./National Pipe Co.	Town of Vestal		R			
642	02	NY	BEC Trucking	Town of Vestal					D
643	03	DE	Wildcat Landfill	Dover		R			
644	05	MI	Burrows Sanitation	Hartford	V	R			
645	03	PA	Blosenski Landfill	West Cain Township			F		
646	03	VA	Rhinehart Tire Fire Dump	Frederick County	V	R	F		
647	03	DE	Delaware City PVC Plant	Delaware City	V		F		
648	03	MD	Limestone Road	Cumberland		R			
649	02	NY	Hooker (102nd Street)	Niagara Falls	V		F	S	
650	03	DE	New Castle Steel	New Castle County					D

*: STATES' DESIGNATED TOP PRIORITY SITES
#: V = VOLUNTARY OR NEGOTIATED RESPONSE;
R = FEDERAL AND STATE RESPONSE;
F = FEDERAL ENFORCEMENT;
S = STATE ENFORCEMENT;
D = ACTIONS TO BE DETERMINED.

			GROUP 14						
651	06	NM	United Nuclear Corp.	Church Rock			F		
652	06	AR	Industrial Waste Control	Fort Smith			F		
653	09	CA	Celtor Chemical Works	Hoopa		R			
654	01	MA	Haverhill Municipal Landfill	Haverhill					D
655	04	AL	Perdido Ground Water Contam	Perdido	V				
656	02	NY	Marathon Battery Corp.	Cold Springs		R			

NPL RANK	EPA RG	ST	SITE NAME *	CITY/COUNTY	RESPONSE CATEGORY#				
			GROUP 14 (CON'T)						
657	02	NY	Colesville Municipal Landfill	Town of Colesville					D
658	04	FL	Yellow Water Road Dump	Baldwin		R	F		
659	05	OH	Skinner Landfill	West Chester		R			
660	04	NC	Chemtronics, Inc.	Swannanoa	V	R			
661	05	IN	MIDCO II	Gary		R	F		
662	03	MD	Kane & Lombard Street Drums	Baltimore		R			
663	07	MO	Shenandoah Stables	Moscow Mills			F		
664	10	WA	Silver Mountain Mine	Loomis		R			
665	06	TX	Petro-Chemical (Turtle Bayou)	Liberty County		R			
666	05	OH	Republic Steel Corp. Quarry	Elyria					D
667	06	LA	Bayou Bonfouca	Slidell		R	F		
668	09	CA	Intel Corp. (Mountain View Plant)	Mountain View			F		
669	09	CA	Raytheon Corp.	Mountain View			F		
670	05	MN	Agate Lake Scrapyard	Fairview Township		R			
671	03	VA	Saltville Waste Disposal Ponds	Saltville		R			
672	01	MA	Shpack Landfill	Norton/Attleboro					D
673	03	PA	Kimberton Site	Kimberton Borough					D
674	01	MA	Norwood PCBs	Norwood		R			
675	03	MD	Middletown Road Dump	Annapolis		R	F		
676	10	WA	Pesticide Lab (Yakima)	Yakima					D
677	05	IN	Lemon Lane Landfill	Bloomington	V		F		
678	05	IN	Tri-State Plating	Columbus					D
679	10	ID	Arrcom (Drexler Enterprises)	Rathdrum		R			
680	01	NH	Coakley Landfill	North Hampton	V	R		S	
681	03	PA	Fischer & Porter Co.	Warminster	V		F		
682	09	CA	Jibboom Junkyard	Sacramento		R			
683	02	NJ	A. O. Polymer	Sparta Township		R			
684	05	WI	Wausau Ground Water Contamination	Wausau		R			
685	02	NJ	Dover Municipal Well 4	Dover Township		R			
686	02	NJ	Rockaway Township Wells	Rockaway					D
687	05	WI	Delavan Municipal Well #4	Delavan					S
688	07	MO	North-U Drive Well Contamination	Springfield		R			
689	09	CA	San Gabriel Valley (Area 3)	Alhambra		R			
690	09	CA	San Gabriel Valley (Area 4)	La Puente		R			
691	10	WA	American Lake Gardens	Tacoma	V	R	F		
692	10	WA	Greenacres Landfill	Spokane County		R			
693	10	WA	Northside Landfill	Spokane		R			
694	06	OK	Sand Springs Petrochemical Cmplx	Sand Springs		R	F		
695	06	TX	Pesses Chemical Co.	Fort Worth		R			
696	05	MN	East Bethel Demolition Landfill	East Bethel Township					D
697	06	TX	Triangle Chemical Co.	Bridge City		R			
698	02	NJ	PJP Landfill	Jersey City				S	
699	03	PA	Craig Farm Drum	Parker					D
700	03	PA	Voortman Farm	Upper Saucon Twp.		R			

*: STATES' DESIGNATED TOP PRIORITY SITES
#: V = VOLUNTARY OR NEGOTIATED RESPONSE;
R = FEDERAL AND STATE RESPONSE;
F = FEDERAL ENFORCEMENT;
S = STATE ENFORCEMENT;
D = ACTIONS TO BE DETERMINED.

NPL RANK	EPA RG	ST	SITE NAME *	CITY/COUNTY		RESPONSE CATEGORY#
			GROUP 15			
701	05	IL	Belvidere Municipal Landfill	Belvidere	R	
702	07	MO	Bee Cee Manufacturing Co.	Malden		D
703	03	PA	Lansdowne Radiation Site	Lansdowne	R	

*: STATES' DESIGNATED TOP PRIORITY SITES
#: V = VOLUNTARY OR NEGOTIATED RESPONSE;
R = FEDERAL AND STATE RESPONSE;
F = FEDERAL ENFORCEMENT;
S = STATE ENFORCEMENT;
D = ACTIONS TO BE DETERMINED.

United States
Environmental Protection Agency
Regional Offices

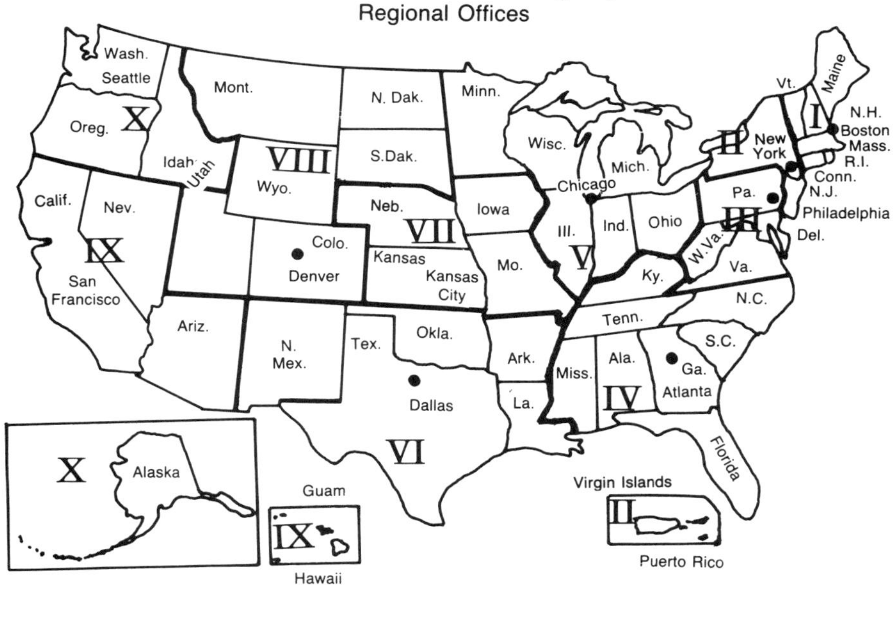

Fig. 6-2. EPA regions.

MOBILITY

Waste disposal systems can be permanent, mobile, or transportable. A permanent facility is one that is installed on-site and has a projected life of at least five years, with no anticipated salvage value.

Mobile systems are brought to the site and are designed to be removed from the site location at the conclusion of the clean-up. They will normally include all of the equipment and subsystems necessary for operation of the facility, such as electric power generation equipment, a fuel supply, and equipment to collect wastewater and to dispose of it. These systems will normally be designed for use on their trailers. If skid-mounted, they will be designed for use on their skids, which will normally be removed from their trailer chassis. A mobile system is normally applicable for clean-ups of no more than approximately 6-months duration.

Transportable equipment differs from mobile equipment in that they will require a significant installation effort. This equipment is provided in modular components and must be assembled before use. A process-water supply will be sought on-site and the wastewater discharge will be disposed of at the site location also, although water or wastewater discharge treatment facilities may be required. They will be constructed so that they can be dismantled, removed, and re-installed at another site later. This equipment if designed for a site clean-up effort lasting from one-half to five years.

The majority of technologies proposed for site clean-up are applicable to transportable systems. Portable-unit technology must carry with it all of its operational support, including utilities, laboratory facilities, personnel stations, etc., as well as the process equipment itself. This can result in ten to twenty trucks in a caravan, requiring many acres of land and perhaps two to three months for set-up. The water and wastewater discharge quantities may be too great to allow trucking on and off site. The ability for any thermal site clean-up system to be truly mobile is subject to these logistic constraints in addition to the permitting constraint. At the present time each site requires its own permits, including RCRA, TSCA (where PCB's are present), state and local permits. The permit process takes from 6 to 18 months, or more. Because the permits take longer than the time required for actual clean-up, the industry appears to be moving towards transportable, rather than truly mobile, units.

THERMAL DESTRUCTION TECHNOLOGIES

Thermal systems proposed for site clean-ups include the following technologies:

- USEPA or commercial rotary-kiln systems
- Fluid-bed reactor
- Circulating fluid-bed reactor
- Infrared incineration
- Rotary reactor
- High-temperature, fluid-wall reactor
- Electric pyrolyzer
- Tar-sands processor
- Plasma arc

Two of these systems, the fluid-bed reactor and the infrared incineration system, are discussed in Chapter 5. They have been adapted to transportable systems for the on-site clean-up of contaminated soils. The other systems will be discussed in detail in this chapter. Some of them are in the testing and development stage; however, a number of these have either been applied to site clean-up or have been available full-scale for other applications.

ROTARY-KILN SYSTEMS: USEPA MOBILE INCINERATOR

This system was built primarily to test the adequacy of present technology in destruction of organic contaminants in soils. It includes over a dozen trailers; the process equipment is contained on three separate trailers, as shown in Fig. 6-3.

Fig. 6-3. EPA mobile incineration system. Source: R. Miller, IT Corp.

The first trailer (see Fig. 6-4) carries the rotary kiln with a waste liquid/sludge feed nozzle, the solids feed unit, a combustion air blower, two hydraulic power units, and the main combustion-system control panel. The geometry of the rotary kiln and its ancillary equipment is maximized under the trailer size, weight, and axle-loading constraints imposed by over-the-road limitations and state highway regulations.

The kiln itself is a 6-inch refractory-lined, directly fired, parallel-flow unit designed to operate up to 1832° F with a nominal solids-retention time of one hour. The unit is fired with two 4-inch fuel-oil burners that

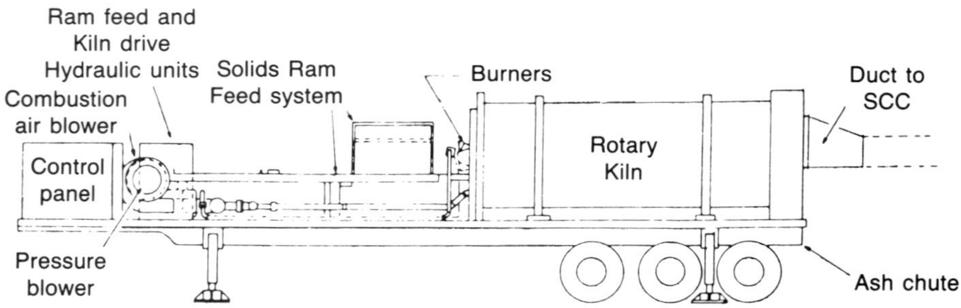

Fig. 6-4. Rotary-kiln furnace trailer. Source: Reference 2.

each provide high turbulence and short flames. Only one of these burners operates; the other is maintained as a spare. One of the two trailer-mounted hydraulic drive units rotates the kiln while the other unit powers the solids ram-feed system. The kiln seal was designed to resist the extreme wear and high temperatures generated in the kiln.

The hydraulically operated ram feed delivers solid wastes into the kiln. It includes an externally loaded hopper and is capable of feeding an adjustable volume and cycle rate of up to 2 cubic feet of solids in 30 seconds. Waste liquids and sludges are fed to the rotary kiln through a feed system that is not mounted on the trailer. The waste liquid is atomized with compressed air supplied through the kiln feed nozzle. To provide a heat sink for temperature control, atomized water is fed to the kiln through an adjacent nozzle. The combustion-air blower provides atomizing and combustion air to both burners as well as excess air to the combustion chamber. The main control panel for both the primary combustion chamber and the secondary combustion chamber (mounted on the second trailer) is mounted on the first trailer. Ash or sterilized soil from the kiln drops into a removable container and is kept dry.

The secondary combustion chamber, combustion-air blower, and the quench elbow are mounted on the second trailer, as shown in Fig. 6-5. The chamber has a 52-inch inside diameter and is 36 feet long. It is lined

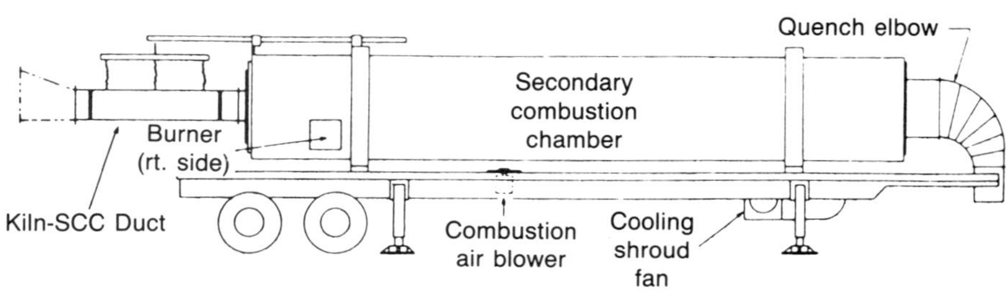

Fig. 6-5. Secondary combustion furnace trailer. Source: Reference 2.

with 6 inches of castable refractory and is designed to provide over 2 seconds retention time at 2200° F for completion of the combustion process. Two tangentially mounted oil-fired burners, similar to those in the rotary kiln, are located at the inlet of the secondary combustion chamber. They are designed to maintain the chamber temperature at 2200° F. An Inconel 601-alloy register with swirl vanes, located at the entrance of the burners, promotes gas mixing.

Atomization, combustion, and excess air are provided by a combustion air blower. To control the shell temperature, the secondary combustion chamber is enclosed in a shroud through which air is blown by a second trailer-mounted blower.

A wetted-throat venturi quench elbow is located on the exit of the secondary combustion chamber. Flue gases are cooled from 2200° F to approximately 190° F with recycle and makeup water from a series of eleven water spray nozzles. Excess water from the water spray drains to the quench elbow sump and is recirculated to the quench elbow at approximately 80 gpm.

A skid-mounted quench-surge sump and recycle pump is located on the ground between trailers 2 and 3. The quench system is the initial stage of the air pollution control (APC) system. Cooling and saturation of the gases preconditions them for the latter stages of the APC system. Through pH adjustment with an alkaline solution, the excess water in the quench removes part of the acid gases in the flue gas stream. If the quench water pH drops below 7, alkaline solution is added to the quench sump from auxiliary alkali supply tanks. Some particulate is also removed by the quench.

Mounted on the third trailer, shown in Fig. 6-6, are the particulate scrubber, mass-transfer scrubber, fan-drive engine, flue-gas stack, instrument-air compressor and control panel. The particulate scrubber is a commercially available, cleanable, high-efficiency air filter (CHEAF) constructed of Inconel 625, for improved wet corrosion resistance. It operates with a 30-inch WC pressure differential across a wetted fiberglass filter mat to remove submicron particulate from the flue-gas stream.

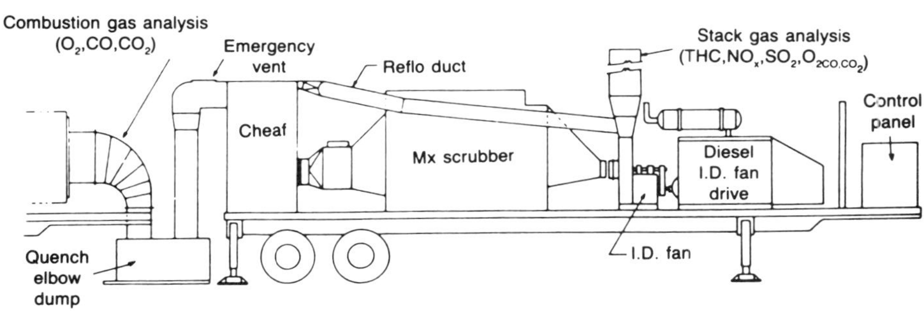

Fig. 6-6. Exhaust-gas cleaning equipment trailer. Source: Reference 2.

The mass-transfer scrubber is a horizontal, crossflow, irrigated packed-bed absorber tower designed and reinforced for this specific mobile application. The packed bed has a 25-square-foot cross-section and is over 8 feet long; the last 9 inches of packing act as a de-mister. The scrubbing medium is kept alkaline by automatic pH adjustment with an off-trailer alkaline-solution feed system. Sumps are located in the bottom of both the mass-transfer scrubber and the CHEAF. Recycle pumps are also provided for both scrubber units.

An induced-draft fan is also placed on the third trailer. It is driven by a 155-horsepower diesel engine which has its own fuel-oil tank and operates at a nominal speed of 3350 rpm, maintaining a negative pressure of 43 inches WC. The fan is a single-stage, heavy-duty industrial unit with a 304 stainless-steel housing, an Inconel 625 shaft, and 36-inch diameter wheel.

An exhaust stack is mounted on top of the fan and is hinged so that it will stay locked in a horizontal position during transport. A sound attenuator is incorporated within the stack and is designed to reduce the sound pressure broadcast by the stack to 85 decibels, measured 5 feet from the stack outlet. The overall height of the operating stack is 30 feet above ground level.

All controls required to operate the equipment on the third trailer are located in a panel at the side of the trailer. An air compressor located on the third trailer provides instrument air for all three trailers.

This mobile incineration system uses electrical relay logic and conventional industrial-process instrumentation and hardware. Instrumentation is designed to monitor process conditions; provide data for assuring compliance with regulatory requirements; and assure appropriate process response and control, operational flexibility, and safety interlocking and shutdown features. The safety interlocks and shutdown features compose a major portion of the control system.

Safety shutdown responses are relayed to various equipment when certain process limits are reached or are not met. In general, the process parameters that alert and initiate response to alarm conditions are:

- High and low kiln temperature;
- High and low secondary-combustion chamber temperature;
- Low secondary combustion chamber outlet oxygen level;
- Low flow in the quench, particulate scrubber, or mass transfer scrubber pumps;
- Very low level in the quench, particulate scrubber, or mass transfer scrubber sumps;
- High gas temperature or pressure in the quench section;
- High pressure at the induced-draft fan inlet;
- High vibration of the induced-draft fan; and
- Insufficient burner air or fuel supply.

The burner system fuel oil burner system includes an internal interlock that shuts down the fuel flow if any of the following occur:

- A flame is detected during pre-ignition,
- The pilot fails to ignite,
- The burners fail to ignite,
- There is a loss of flame after ignition.

A shutdown of the burners automatically stops waste feed.

During an alarm condition, waste and fuel-oil feeds are immediately stopped. When required, the induced-draft fan is shut down and the emergency vent located between the quench sump and the CHEAF opens. All recycle and makeup flows are maintained, if possible, to prevent overheating in the mass-transfer scrubber. The safety interlock system is designed to provide protection for operating personnel and for the incineration equipment.

Since the mobile incineration system was designed to safely destroy or de-toxify a wide range of hazardous wastes, an important aspect of its design was providing a monitoring system which analyzes the flue and stack gases for constituent emissions (carbon monoxide, oxides of sulfur and nitrogen, oxygen, etc.). The stack monitoring system principally performs two critical functions. It provides the operator with current data on the performance of the incineration and gas cleaning processes, and it generates and records accurate data on gas emissions. This ensures operator safety and compliance with operating-permit requirements. Functional requirements are met by the selection of a dual gas-chromatograph (GC) system that has a high degree of reliability and versatility, and conforms to the mobile nature of the incineration system.

The stack gas monitoring system consists of the three subsystems: gas sampling and conditioning; gas analysis calibration; and system control and reporting of results.

A specialty probe was designed which was expected to provide reliable operation under severe conditions.

ROTARY-KILN SYSTEMS: COMMERCIAL SYSTEM

A number of firms have developed mobile and transportable rotary-kiln systems for site clean-up. One example of this technology was used for the clean-up of the Sydney Mine site near Tampa, Florida, in 1985.

This abandoned mine was used for the dumping of waste oils and sludges from the surrounding industrial areas for a period of approximately 20 years. A separate lagoon on the site was used for septic waste.

In response to fears that these wastes may reach groundwater supplies, a decision was made to clean up the area in-place, without moving the wastes.

The waste profile of the lagoon, which is typical of many sites around the country, is shown in Fig. 6-7. Lighter oils float on top of an aqueous phase, which is a combination of rainwater and the water component of the original waste. Heavier sludges settle to the bottom of the lagoon. Soil beneath the lagoon and the soil making up the sides of the lagoon were contaminated and had to be treated also.

The quantity of soil was approximately equal to the amount of other wastes found at this site. In total, 10,000 cubic yards of material had to be processed through the incinerator. This does not include the aqueous phase of the lagoon, which was passed through a set of carbon filters for contaminant removal.

The equipment layout on site is shown in Fig. 6-8. The arrangement of these trailers is shown in Fig. 6-9. There are five major elements, or modules, of the system, as follows:

- Rotary-kiln incinerator
- Afterburner and liquids incinerator
- Waste-heat boiler
- Emissions control system and prime mover
- Process control and laboratory facility

The kiln, or solids module, consists of a trailer-mounted rotary kiln; solids preparation and charging equipment; a supplemental fuel burner; a combustion-air fan; and an air-discharge system. The charging system includes a conveyor belt and a kiln-charging hopper with an integral ram feeder. During operation, contaminated solids are placed on the conveyor belt and are subsequently fed to the charging hopper. An intermittently operated ram feed mechanism located in the feed chute pushes the material into the kiln. The rotary kiln is lined with castable refractory for its initial six feet of length. The balance of the kiln is lined with refractory brick. Ash generated from incineration is discharged to the end breeching

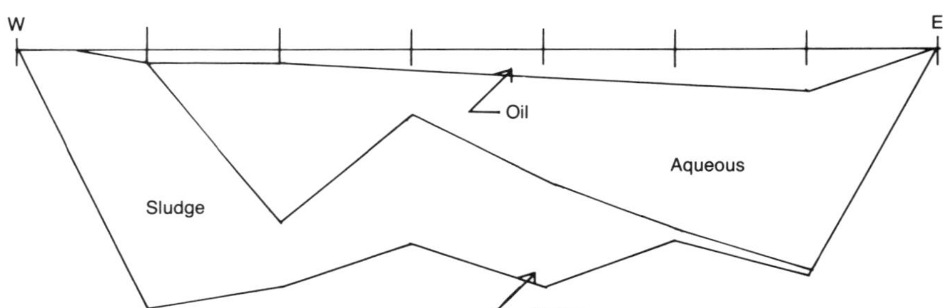

Fig. 6-7. Sydney mines oil pond profile.

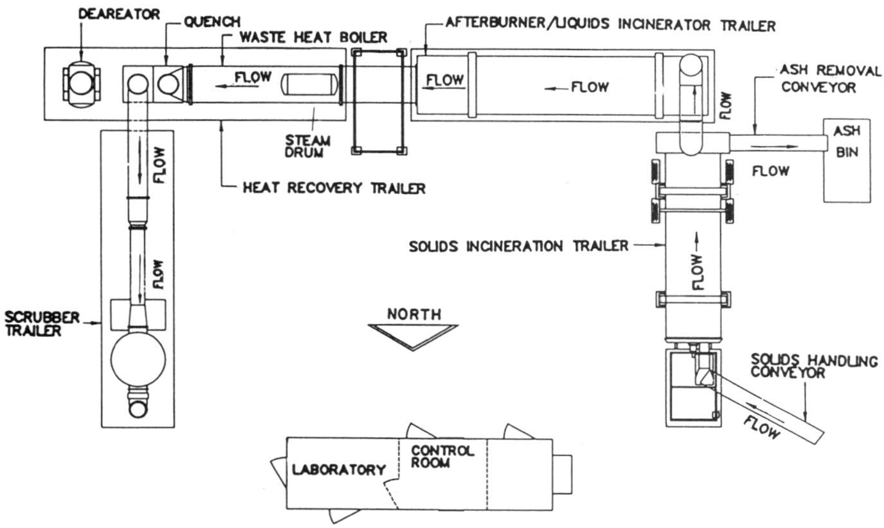

Fig. 6-8. Mobile incineration system equipment layout.

Fig. 6-9. Sydney mine clean-up. Source: J. Martin, ENSCO-Pyrotech.

where it falls into a dry-ash discharge sump. A water-cooled screw conveyor subsequently carries the wetted ash to a storage bin, from which samples are taken for analysis. If residual organics are found in the ash, the entire load is recycled through the incinerator. If not, the ash is taken to a landfill.

It has been shown that in the temperature range of 1300° F to 1400° F, organic contaminants volatilize from the soil. Some of these organics are destroyed in the kiln; the remaining contaminants are carried within the gas stream to downstream equipment.

To reduce the carryover of residual soil in the gas stream, cyclones are installed between the kiln and the secondary combustion chamber. The cyclones remove over 70 percent of the airborne residual in the exhaust gas.

The secondary combustion chamber, or afterburner, is a refractory-brick-lined horizontal cylinder, sized to provide a residence time of 2 seconds at a temperature of 2200° F. The volatiles and combustion products exiting the kiln pass through the afterburner, where air is injected and where a supplemental fuel burner brings the gas to a sufficiently high temperature to destroy any organics present.

A trailer-mounted fire-tube boiler is provided to generate the steam required by the exhaust-gas cleaning system. A boiler feedwater-treatment package is included within this mobile facility to increase system reliability. The de-aerator and water polisher included in this package reduces boiler fouling and subsequent boiler maintenance. Excess steam can be discharged through a silenced steam vent. During normal operation, more steam is generated than can be used by the system.

Gases exit the waste heat boiler at a temperature of approximately 450° F, and are quenched to a temperature of 170° F to 180° F in a water bath immediately downstream of the boiler. The quench includes an Inconel elbow with water sprays and a wet FRP sump.

Steam from the waste-heat boiler is used to drive a scrubber/ejector system located on another trailer. Steam and water are injected through an ejector nozzle system, which creates a negative pressure, or draft. This draft is sufficient to draw gases from the kiln and through the cyclones, afterburner, waste heat boiler, quench elbow, and quench tank. In addition to producing the draft required as a prime mover for the system, the ejector system creates small, atomized water particles which absorb particulate matter from the gas stream. The saturated gas stream is directed to a mist eliminator, where entrained water drops out of the stream. The spent scrubber water is directed to a lamella-type sludge processor. The overflow is recycled to the ejector; the sludge that collects at the bottom of the processor is conveyed to a container. The containerized sludge is brought to the rotary-kiln feed belt, dumped on the belt, and fed to the incinerator.

The contaminants at the Sydney mine site did not have significant chloride content, and acid gas was not generated. Where chlorinated organics are present and acid gas is generated from the burning of these materials, a packed tower could be added to the system, downstream of the scrubber. A caustic solution would be recirculated through the packed tower to neutralize the acid gases.

The control room and the laboratory occupy the same trailer. The incinerator combustion process and ancillary processes are monitored and controlled by an integrated system of analog and digital readouts, single-loop controllers, and a computer system from within this trailer. The computer system is employed in a multi-function role, providing data acquisition and display, as well as operational process control. Single-loop controllers are used to regulate combustion-air flow, fuel flow and boiler-drum level. Single parameter readouts, mounted on a graphic panel, are used to monitor process temperatures, wastewater flow, quench water flow, and scrubber/ejector water flow. Calculated parameters, such as combustion efficiency and gas flows, are also generated, displayed, and logged by the computer. Analyzers and recorders monitor and display flue gas constituents (CO, oxygen, and NO_x).

The laboratory is primarily used to provide chemical and heating-value analyses of the material to be incinerated. Samples of the effluent waste streams are also taken and analyzed for contaminants with the gas chromatograph and atomic-absorption spectrophotometer located within the laboratory. This allows on-site confirmation of the absence or presence of contaminants in the residual ash or other system effluents.

These five modules are designed to be operated seated on their trailers. Other mobile systems require that skids be removed from the trailer, and that the trailers not be used for equipment mounting.

The kiln trailer should be placed on a concrete pad, but it is not necessary to place the other trailers on pads. The kiln and afterburner are not moved with refractory brick in place. These components are the largest, with regard to highway regulations and constraints, that can be moved on a truck trailer. With refractory brick, these trailers would be too heavy. An additional consideration is that were brick installed, it would be prone to damage when moved from site to site. Refractory brick is installed on-site and is removed before the trailers are moved off-site.

The throughput of this system is nominally 2 tons per hour of soil at 15 percent moisture content (by weight).

CIRCULATING FLUID BED

The circulating fluid-bed incineration system has been proposed for site clean-up as a transportable unit. Components of the system would be transported to a site, then assembled and erected.

The circulating fluid-bed concept is distinct from conventional fluidized-beds. Conventional reactors have a fixed bed depth and operate within a narrow range of gas velocities (the minimum and maximum fluidization velocities), from 1.5 to 4.5 ft/sec. At velocities above the maximum, the reactor freeboard and off-gas become permeated with carryover of unburned particles from the bed. At velocities below the minimum, the bed may slump and lose fluidization.

In the circulating-bed concept, illustrated schematically in Fig. 6-10, combustible waste is introduced into the bed along with recirculated bed material from the hot cyclone. A high air velocity (from 15 to 20 ft/sec) elutriates both the bed and the combustible waste, which rise through the reaction zone to the top of the combustion chamber (freeboard) and pass through a hot cyclonic collector. Hot gas passes through the cyclone; the majority of solids drop to the bottom of the cyclone and are re-injected into the bed of the furnace. The hot flue gases pass to a convective gas cooler, then to a baghouse cooler for removal of residual particulate.

Solid or sludge waste is fed between the cyclone and the bed of the reactor through a feeding bin. A metering screw conveys waste from the bin to the feed leg. A pump is used to meter liquid or slurry waste from a stirred tank to the reactor. No atomizers or nozzles are required for introduction of fluid wastes into the sand bed. The waste feed rate is automatically adjusted to maintain a pre-set oxygen concentration in the flue gas.

Lime can be added to the waste feed through a lime metering system to neutralize acid generating constituents of the waste. As shown in Fig. 6-10, lime and make-up sand are both added to the reactor through the waste-feed metering screw.

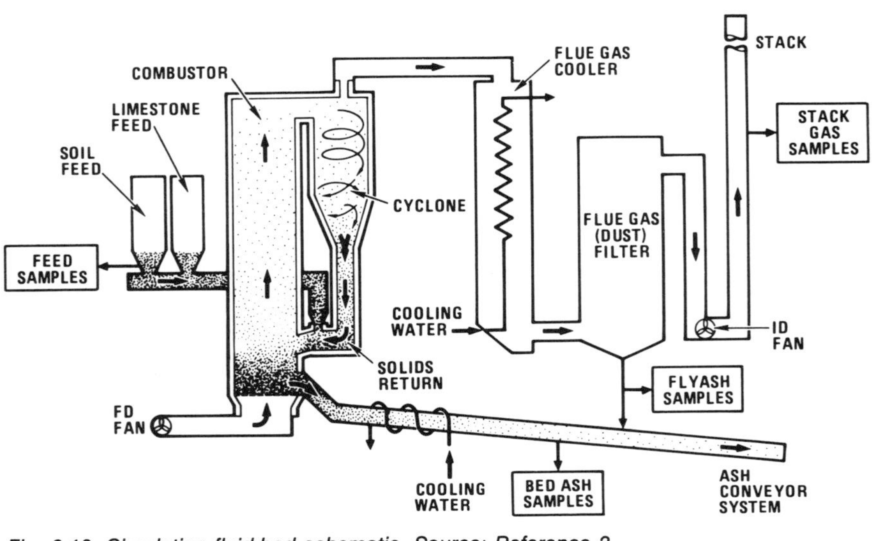

Fig. 6-10. Circulating fluid-bed schematic. Source: Reference 3.

Refractory in the reactor is 12 inches thick, comprising a dual wall of insulating castable refractory against the shell and dense castable refractory, for erosion resistance, in contact with the waste and hot gases.

The design operating temperature is normally 1600° F, although the system can withstand temperatures up to 2000° F on a continuous basis.

A combustion-air fan provides air to the bed for fluidization and oxidation. The furnace draft is maintained by an induced-draft (ID) fan downstream of the cyclone.

Retention time of material within the system is controlled by the discharge from the ash cooler. The cyclone bottom ash discharges to the reactor, but this ash flow can also be removed from the system through a water-cooled ash conveyor. By increasing the speed of this conveyor, material is removed more rapidly from the furnace system, and the residence time within the system (bed and reactor) is less than if the speed of the ash conveyor were slower. At a lower conveyor discharge rate, material would be discharged more slowly from the system, and the solids retention time would be increased.

Flue gas exiting the cyclone passes through a conventional exhaust-gas treatment system which removes particulate and other undesirable constituents from the gas stream.

A major feature of the circulating fluid-bed system is its ability to control the residence time of wastes from seconds to minutes. By increasing the residence time over that of conventional incineration systems, the temperature required for destruction can be substantially reduced. In this system, destruction of organics generally occurs at temperatures below 1600° F rather than at the 2000° F+ temperatures required in conventional afterburners. This lower temperature translates to lower supplemental fuel requirements, less (or insignificant) quantities of NO_x produced, less refractory maintenance, more favorable bed eutectics, etc.

ROTARY REACTOR

The rotary reactor is adaptable to chassis mounting and can be moved from one site to another. It is a hollow, three-compartment, horizontal cylinder that rotates at 10 to 30 revolutions per minute. It acts as a horizontal fluid-bed reactor with a hot inert medium such as sand.

As shown in Fig. 6-11, solid and semi-solid wastes, in addition to the resident sand, are mechanically lifted on internal radial fins and cascade through the combustion gases in the combustion zone of the reactor. This cascading action provides excellent mass transfer and efficient heat transfer. The recycling feature of this system provides relatively high residence time, which allows operation at lower temperatures than more conventional incineration equipment.

The normal operating temperature for this reactor is approximately 1600° F. Lime can be added to neutralize acid gases generated by the

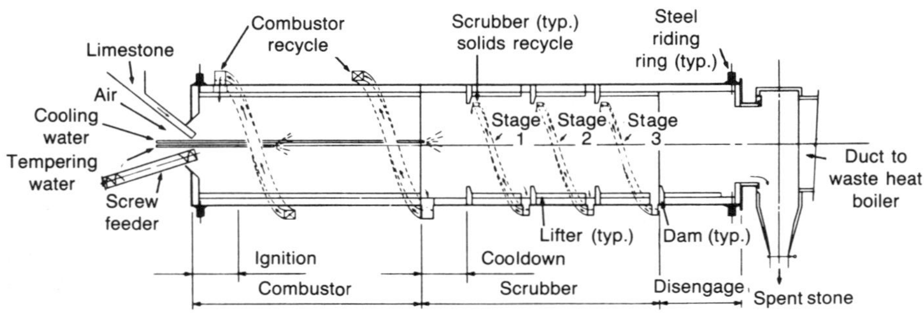

Fig. 6-11. Rotary reactor process schematic. Source: Reference 4.

burning of halogenated or sulfonated organics. This eliminates the need for special acid-resistant construction materials within the reactor, as well as the need for acid scrubbing equipment in the exhaust gas train.

During the operation of this reactor, the solids, which can include sand, waste feed, lime, ash, and lime-reaction products, follow a relatively complex route. Sand, lime, and waste are fed into the front of the combustor, as shown in Fig. 6-11. The organic portion of the waste is burned, and the tumbling action carries the solids to the far end of the combustion zone. Here part of the solids load is carried back to the front of the combustion zone through the combustor recycle duct, a variation of Archimedes' screw.

HIGH-TEMPERATURE, FLUID-WALL REACTOR

The high-temperature, fluid-wall reactor (HTFWR) has been developed for the treatment of soils contaminated with organic materials. This system is adapted as a transportable unit that can be assembled and installed in a few weeks.

As shown in Fig. 6-12, contaminated soil is dropped through a metered screw feeder into the reactor. The organic contaminants volatilize from the soil's surface and dissociate into their elements. The heated soil falls through a post-reactor zone where it begins to cool. It then drops into a solid-waste bin, where it is collected for ultimate disposal. An induced-draft fan downstream of the reactor generates the draft required to draw the gases produced.

A cyclone followed by a baghouse removes particulate from the exhaust-gas stream. If halogenated or sulfonated contaminants are present in the waste soil, the exhaust gas can be diverted to a caustic scrubber for acid neutralization. Activated carbon filters help assure the capture of any organics that may be present in the gas stream because of operating upsets.

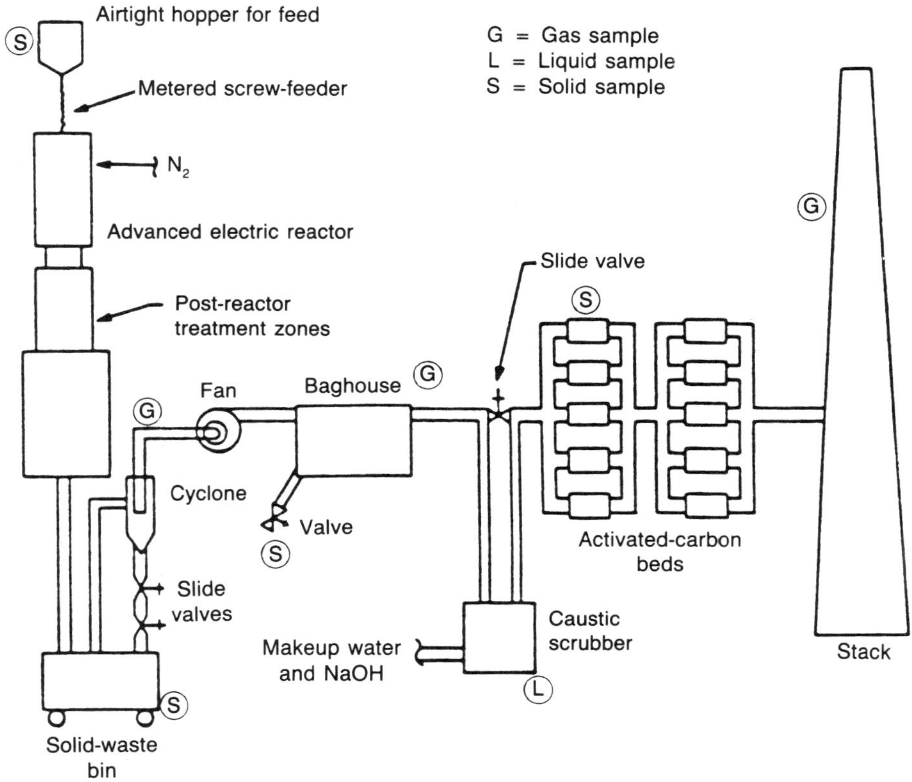

Fig. 6-12. High-temperature, fluid-wall reactor process-flow diagram. Source: The Huber Corp.

The heart of the reactor is a vertical graphite cylinder, the porous core indicated in Fig. 6-13. It is surrounded by a set of electrodes, which are encased within a cooled and insulated jacket. The electrodes provide the thermal energy required to heat the core to radiant temperatures. Temperatures in the range of 4000° F to 5000° F are developed.

A nitrogen purge is introduced within the porous graphite core to act as a blanket, or fluid wall. It tends to prevent the soil from coming in contact with the graphite surface. The graphite is sensitive to plugging and fouling; the nitrogen purge keeps the core clean.

The heat flux developed within the graphite core rapidly heats the surface of the falling soil. Contaminants present are typically adsorbed onto the soil's surface. These organic contaminants will volatilize at the high temperatures present. The organic contaminants released from the soil will then dissociate into their elemental constituents (carbon, hydrogen, oxygen, and, when present in the feed, nitrogen, chlorine, and other elements).

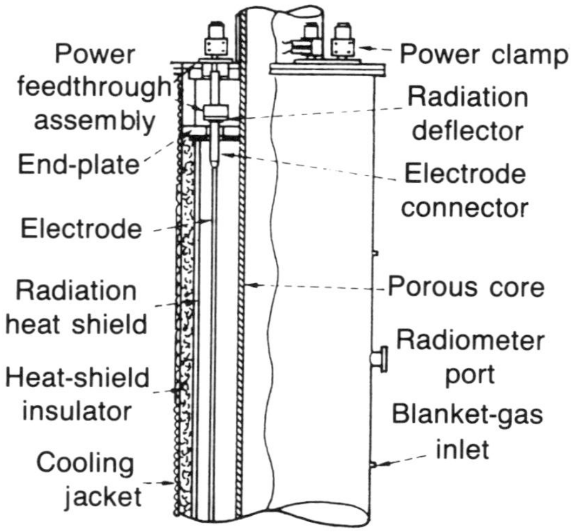

Fig. 6-13. High-temperature, fluid-wall reactor cross-section. Source: The Huber Corp.

The soil itself does not have sufficient residence time within the reactor to increase in temperature by more than a few hundred degrees. The energy required for this process, therefore, need only be sufficient to volatilize the surface contaminant components of the soil particle, not heat the soil itself to a temperature of thousands of degrees.

These systems have been built as pilot units, in 3-inch and 8-inch nominal sizes (the nominal size refers to the diameter of the graphite core). They will be built full-scale in 16-inch nominal size, which can handle as much as 5 tons of soil per hour.

The reactor requires significant soils preparation. The process is sensitive to surface area of the soil particles; for effective contaminant release, the soil must be ground to pass a 32 mesh. Soil moisture content is also a factor in effective unit operation. The soil moisture must be reduced to less than 8 percent by weight to aid in the grinding of the soil and to promote effective volatilization of soil contaminants.

ELECTRIC PYROLYZER

This system has been developed to process soils contaminated with organics. It has been built in a pilot size, as a fully mobile facility, shown in Fig. 6-14.

Waste requires minimal processing for introduction into this facility. Process feed must pass through a 2-inch grizzly (2-inch open mesh). Soil can have a moisture content of up to 25 percent by weight before drying is necessary. Solid waste is dropped, by gravity, through the reactor, or pyrolyzer. Liquid feed is injected into the reactor.

Fig. 6-14. Electric pyrolyzer. Source: W. Reed, Westinghouse Electric Corp.

The pyrolyzer system, shown in schematic in Fig. 6-15, promotes the release of volatiles from the surface of the soil or material. As waste is dropped into the unit, it passes through a high temperature zone, where the majority of the organics volatilize. The soil, or other solid waste, drops to the bottom of the unit, which is maintained at a high enough temperature to keep the soil and other inorganics in a molten state.

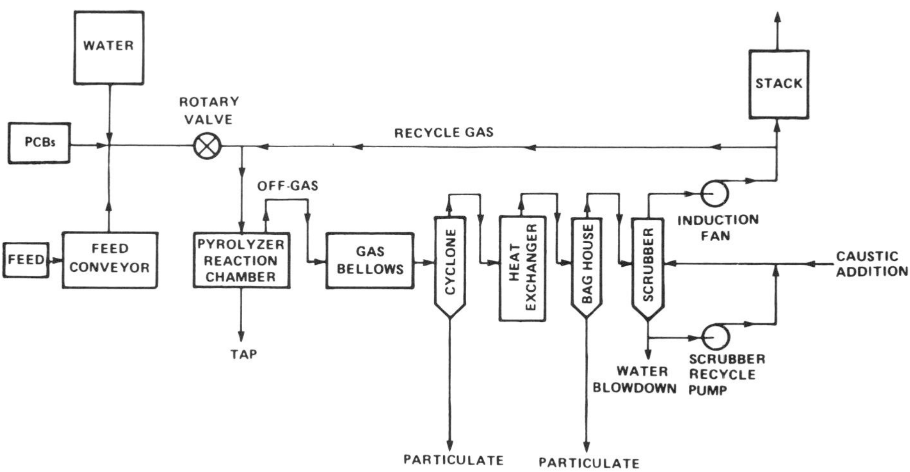

Fig. 6-15. Electric pyrolyzer process. Source: Westinghouse Electric Corp.

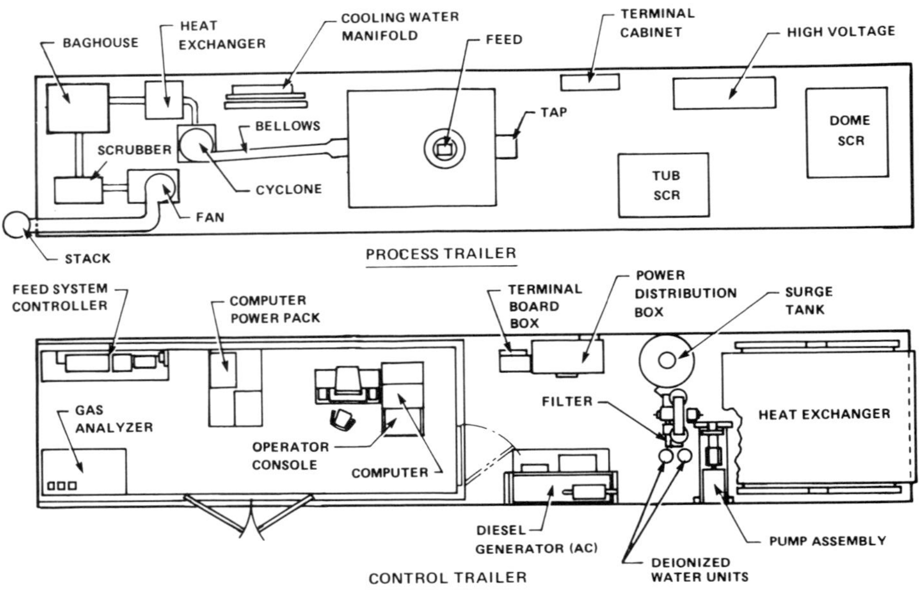

Fig. 6-16. Electric pyrolyzer general arrangement. Source: Westinghouse Electric Corp.

Supplemental electrodes with the melt, shown in Fig. 6-16, assure that its temperature will be sustained at a relatively high level, uniform throughout the melt. Any metals present will be found either in their elemental form or as a salt, and will be removed from the melt on a continuous basis from an appropriately placed tap. Other taps are located at other levels of the reactor wall to provide a means for discharge of slag and other materials generated by the process. The tapped materials fall into a water bath where they cool immediately. The cooled residual has the appearance of dark glass.

Any organics that have not volatilized as the soil dropped through the reactor will be destroyed within the melt. With the presence of a melt, the size of the soil particles and soil moisture content is not critical. Surface area of the soil particles is a factor in the evaporation of organics from the particle surface, but if volatilization does not occur as a surface phenomenon, the melt will provide the medium of destruction.

The electrodes generate a temperature on the order of 4000° F. The melt is maintained at a much lower temperature. The chemistry of the melt can be controlled by additives such as lime, salts, or other compounds. Adjusting the melt constituents can neutralize acid gas components of the waste and control the properties of the slag.

The off-gas from the reactor passes through a cyclone where the majority of particulate that may be suspended in the gas stream is removed.

A baghouse removes the balance of particulate matter. A wet scrubber placed downstream of the baghouse removes any halogenated or sulfonated gases that were not neutralized within the reactor.

The pilot unit, with a capacity of 5 to 10 tons of soil per day, is mounted on a single trailer; utility equipment is on a second trailer. The equipment trailer requires a pad, but the utility trailer can be placed on level ground.

TAR-SANDS PROCESSOR

This system is being developed as a transportable facility. A pilot unit with a processing capacity of 10 to 15 tons per day of tar sands has been built. This unit can also be applied to the incineration of contaminated soils.

Figure 6-17 is a simplified schematic of the reactor. Waste is deposited in the reactor, a horizontal cylinder rotating at a speed which ranges from 5 to 10 revolutions per minute. The rector contains four separate zones, as follows:

- Preheat
- Reaction
- Combustion
- Cooling

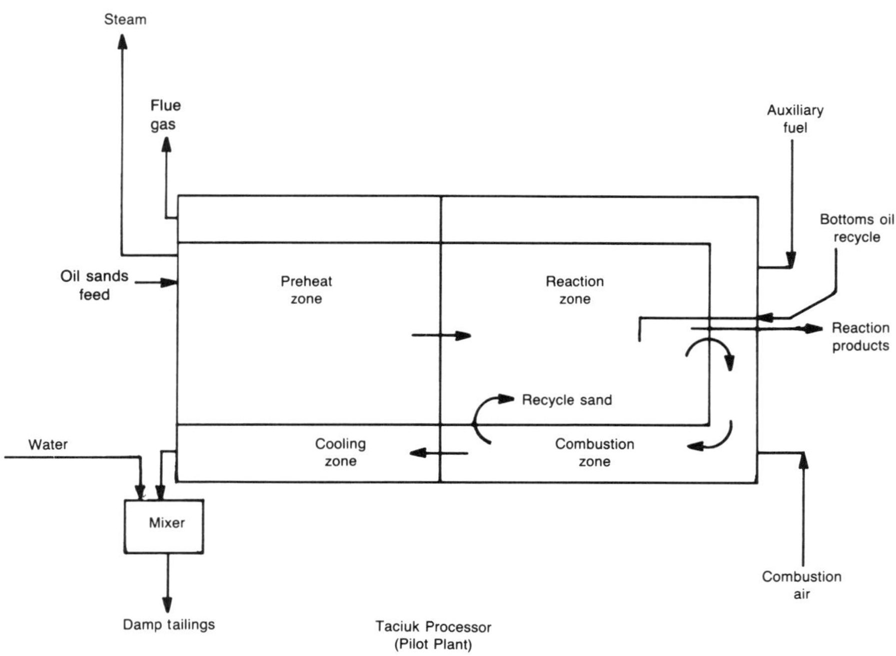

Fig. 6-17. Tar-sands thermal processor. Source: Taciuk/Umatak Industrial Processes Ltd.

Waste is first indirectly preheated by heat from the exhaust gas exiting the process, or from waste residual exiting the reactor. In the reaction zone, organics are released from the waste materials (soils) and generate a vapor stream. Some of the waste is allowed to re-enter the combustion zone. In the combustion zone, air is added to the stream for firing a portion of the waste to generate the temperatures required to sustain the process.

Soils residual exiting the combustion zone is essentially free of organics. It enters a reactor cooling zone, where much of its heat content is used to preheat the entering waste feed. There are provisions for internal recycling of these materials from one zone of the reactor to another.

The off-gas of the process contains a significant amount of organic material. This gas stream is passed through a condenser where the majority of organics condense (condensables). The liquids are separated into organic and aqueous phases. The gas stream also contains non-condensables, and it may be possible to fire this gas in an afterburner to destroy its organic components. The gas passes through an air emissions control system before discharge to the atmosphere.

The reactor is designed for an operating temperature of 1000° F to 1400° F. The solids residence time is variable. It is controlled by the amount of residue that is recycled to the combustion chamber.

PLASMA ARC

Plasma systems use the extremely high temperatures developed within a plasma stream to destroy hazardous organic wastes. A pilot unit has been developed; it is completely mobile on a self-contained trailer. This system has been designed for leachate destruction and has been demonstrated to have a capacity of from 1½ to 2½ gallons per minute. It is not applicable for solids, but may eventually be adapted to destroy sludge wastes containing fine solid particles.

The principle of plasma-arc technology is the breaking of the chemical bonds between the elements of the organic constituents. This occurs in an atomization zone, where a series of co-linear electrodes generate an electric arc, or plasma, which is stabilized by field-coil magnets. As a low-pressure air stream passes through the arc, the electrical energy is converted to thermal energy through the ionization of the oxygen and nitrogen molecules. The temperature of the plasma arc will exceed 50,000° F. When the excited atoms and molecules relax to lower energy states, intense ultraviolet energy is emitted. This energy from the decaying plasma is transferred to the injected waste stream.

As shown in Fig. 6-18, waste is injected directly into the reactor. As the waste stream passes through the reactor, the plasma dissociates it into its elemental constituents, which may recombine to simple molecular form by the time they exit the reactor. Residence time within the reactor

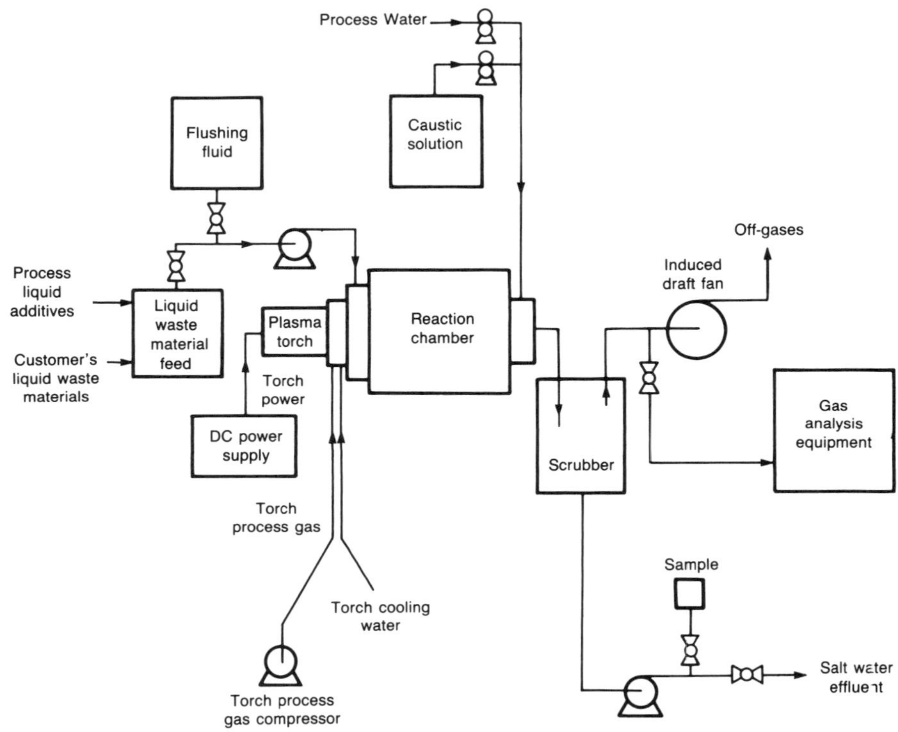

Fig. 6-18. Plasma-arc system process-flow schematic. Source: Westinghouse Electric Corp.

is on the order of one second. The gas exiting the reactor is condensed in a sump where caustic is added to neutralize halogenated or sulfonated components.

The gas may contain hydrogen, which can be flared, as shown in the schematic diagram. Carbon filters remove any residual carbon in the exhaust-gas stream before discharge to the atmosphere.

REFERENCES

1. C. R. Brunner, *Incineration Systems: Selection & Design* (Reston, VA: Incinerator Consultants Incorporated, 1988).

2. J. Yezzi, Jr., et al., "The EPA-ORD Mobile Incineration System," *Proceedings of the Solid Waste Processing Conference,* American Society of Mechanical Engineers, June 1982.

3. *The Hazardous Waste Consultant,* May/June 1986.

4. W. Philipbar, "New Developments in Solid Hazardous Waste Management," *Proceedings of the Energy Sources and Technology Conference and Exhibition,* American Society of Mechanical Engineers, February 1985.

7

Waste Destruction in Industrial Processes

Many high-temperature industrial systems may be applicable to the destruction of hazardous wastes. Some, such as cement kilns and industrial boilers, have been the subject of hazardous-waste-destruction testing programs and are being used for the incineration of hazardous-waste streams at the present time. Other processes have the potential for waste incineration but, generally, have not yet served this function on a large scale.

FEDERAL AND STATE REGULATIONS

When waste is used in an industrial process as a source of fuel, waste disposal may be included in the category of recycle, reuse, or reclamation, as described in Chapter 2, and is not subject to regulations governing burning hazardous waste. States have the authority, however, to regulate the disposal of wastes and, regardless of federal action, many states have their own set of air-emissions and hazardous-waste regulations governing the firing of wastes in combustion units.

Table 7-1 lists permit requirements state-by-state for facilities that burn hazardous wastes in combustion units.

GENERAL CONSIDERATIONS

Organic liquid wastes may be good candidates for destruction in an industrial furnace, but contaminants may restrict their applicability. Solids,

Table 7-1. State Regulations for Facilities that Burn Hazardous Wastes in Combustion Units.

STATE	Air Regulations: Permit to Construct	Air Regulations: Permit to Operate	Hazardous-Waste Regulations: Permit Exemption For Reuse Recycle or Reclaim	Hazardous-Waste Regulations: Permit Criteria For Recovery Facilities	COMMENTS
Alabama	Y	Y	Y	N	A combustion unit which is burning conventional fuels must obtain a permit if hazardous waste is added to the fuel.
Alaska	Y	Y	Y	N	No one may construct, modify, reconstruct, operate or cause the operation of an incinerator (with a rated capacity of 1,000 lb/hr or more) without obtaining a permit to operate.
Arizona	Y	Y	Y	N	Regulations are the same as federal regulations.
Arkansas	Y	Y	O	O	Facilities which burn hazardous waste generated on-site do not require a hazardous-waste permit but may need an air permit. Facilities accepting waste from off-site may be required to get a storage permit.
California	O	O	N	Y	The state Air Resources Board has abrogated its permitting authority to 43 air pollution control districts which may or may not require permitting. Hazardous-waste management regulations classify energy recovery facilities as part of the incineration process and require that they obtain a permit.
Colorado	Y	Y	Y	N	The state Air Pollution Control Division requires permits for all stationary sources of air contaminants unless exempted. The permit criteria set technical modeling, monitoring requirements, and performance tests as may be deemed necessary for a facility.
Connecticut	Y	Y	N	Y	The state considers any burning of hazardous waste a dis-

STATE	Air Regulations: Permit to Construct	Air Regulations: Permit to Operate	Hazardous-Waste Regulations: Permit Exemption For Reuse Recycle or Reclaim	Hazardous-Waste Regulations: Permit Criteria For Recovery Facilities	COMMENTS
					posal method and does not exempt facilities from hazardous-waste regulations due to recycle, reclaim or reuse.
Delaware	Y	Y	Y	N	The Air Resources Section considers combustion units that burn hazardous waste for energy recovery as waste-heat boilers and regulates them on a case-by-case basis.
Florida	Y	Y	Y	N	Permits are required for all sources of pollution in the state air regulations.
Georgia	Y	Y	Y	N	All sources of air contaminants are required to apply for permits unless specifically exempted by the Director of the Environmental Protection Division.
Hawaii	Y	Y	Y	N	The state has no hazardous waste program. Presently, EPA Region IX enforces federal hazardous-waste regulations.
Idaho	Y	N	Y	N	All non-exempt facilities must supply information on types and amounts of waste they will produce and the control equipment they will use.
Illinois	N	Y	N	Y	A facility has to meet certain criteria to obtain a permit to burn hazardous waste as fuel, such as demonstrating the economic feasibility of any such burning.
Indiana	Y	Y	N	Y	The state defines hazardous waste facilities more broadly than RCRA to include those that recover hazardous waste. Thus, the beneficial exemption under RCRA does not apply. RCRA permitting requirements (Part A and B) are applicable though no such permits have been issued yet.

STATE	Air Regulations: Permit to Construct	Air Regulations: Permit to Operate	Hazardous-Waste Regulations: Permit Exemption For Reuse Recycle or Reclaim	Hazardous-Waste Regulations: Permit Criteria For Recovery Facilities	COMMENTS
Iowa	Y	N	Y	N	The Iowa air pollution regulations specify permit requirements for new or existing stationary air contaminant sources.
Kansas	Y	Y	Y	N	The state is developing a consolidated permit approach that effectively integrates construction and operating permits plus multiple environmental regulations (e.g., air, state, hazardous-waste management) through a lead administrative agency.
Kentucky	Y	Y	Y	O	Facilities which apply for an exemption from hazardous waste permitting are reviewed on a case-by-case basis.
Louisiana	Y	Y	Y	N	Air permits are required for all air contaminant sources. Air permit criteria require the inclusion of the composition and amount of air contaminants to be emitted.
Maine	N	N	O	O	Air permits are not required for combustion units burning hazardous waste. All facilities must adhere to some degree of control through the hazardous waste division.
Maryland	O	O	N	Y	The burning of acutely hazardous waste must be done in a hazardous waste incinerator, not in a boiler, and incinerator operators must obtain a hazardous-waste incinerator permit. Energy recovery facilities must obtain limited facility permits.
Massachusetts	Y	Y	O	O	Every source of air contamination must be registered with the air quality control section. Generator and user (if different) must apply for an exemption on the basis of intent to reuse the material.

STATE	Air Regulations: Permit to Construct	Air Regulations: Permit to Operate	Hazardous-Waste Regulations: Permit Exemption For Reuse Recycle or Reclaim	Hazardous-Waste Regulations: Permit Criteria For Recovery Facilities	COMMENTS
Michigan	Y	Y	Y	N	The state has the legal ability to review (and potentially revoke) permits to operate when fuel switching occurs, e.g., blending any waste in fuel mixtures.
Minnesota	Y	Y	O	O	Combustion units burning hazardous waste as fuel are regulated under air regulations. Generators which recycle hazardous wastes on site are exempted from hazardous waste permits, all others must obtain a hazardous-waste facility permit.
Mississippi	Y	Y	O	Y	Permit procedures for a facility to burn hazardous waste as fuel require an analysis of the heat content and halogen content of the waste. Each facility is regulated on a case-by-case basis. All air contaminant sources must follow air permitting procedures.
Missouri	Y	Y	O	O	Certification as a resource recovery facility is required by the Department of Natural Resources. In some situations, hazardous-waste permit applications and review procedures must be followed. All sources of air contamination must have a permit unless exempted.
Montana	Y	Y	Y	N	All fuel burning equipment must have an air quality permit. Each facility is reviewed on a case-by-case basis. Exemptions from permit requirements are based on fuel types.
Nebraska	Y	Y	O	N	Although hazardous waste which is beneficially reused, recycled or reclaimed is exempt from hazardous-waste

STATE	Air Regulations: Permit to Construct	Air Regulations: Permit to Operate	Hazardous-Waste Regulations: Permit Exemption For Reuse Recycle or Reclaim	Hazardous-Waste Regulations: Permit Criteria For Recovery Facilities	COMMENTS
					regulations, generators must go through notification procedures.
Nevada	N	Y	Y	N	Air permit regulations require the registration of air pollution sources and a description of the processes employed, fuels used, nature of emissions and any other information necessary to prevent, abate or control air pollution. Hazardous wastes burned as fuel must have a DRE of 99.99% to be exempt from regulation.
New Hampshire	Y	Y	O	Y	The state's definition of incinerators includes heat recovery facilities and these units must obtain permits.
New Jersey	Y	Y	N	O	If a combustion unit burning hazardous waste meets one of the seven exemptions, only an air permit is required.
New Mexico	Y	N	Y	N	There are no specific permitting requirements for facilities that burn hazardous wastes as fuel.
New York	Y	Y	N	Y	Energy recovery facilities are regulated as incinerators.
North Carolina	Y	Y	Y	O	The Solid and Hazardous Waste Section of the Department of Human Resources requires that all facilities that burn hazardous wastes sign a form which specifies the state's interpretation of legitimate recycle or beneficial reuse. All sources of air contaminants must obtain a permit.
North Dakota	Y	Y	N	Y	Permits are required for treatment facilities, including energy recovery facilities.
Ohio	Y	Y	Y	N	There are permitting criteria (air permits) for all sources of air contaminants.

STATE	Air Regulations: Permit to Construct	Air Regulations: Permit to Operate	Hazardous-Waste Regulations: Permit Exemption For Reuse Recycle or Reclaim	Hazardous-Waste Regulations: Permit Criteria For Recovery Facilities	COMMENTS
Oklahoma	Y	Y	Y	O	There are regulations for recycle operations and permit requirements for process facilities which do not qualify for exemption under the recycle operations requirements.
Oregon	Y	Y	Y	O	Facilities which generate and use hazardous waste generated on-site are only responsible for the notification phase of permitting. Facilities which use hazardous waste recycled off-site must adhere to all phases of permit criteria.
Pennsylvania	Y	Y	O	O	At a minimum, resource recovery facilities must adhere to notification manifesting, and reporting requirements of hazardous-waste permit criteria. All non exempt air contaminant sources must obtain air permits.
Rhode Island	Y	N	O	O	The Director of the Department of Environmental Management may, on a case-specific basis, determine those materials which are beneficially used or reused, or which of those legitimately recycled or reclaimed are not considered waste. Generators who use, reuse or recycle their own waste on-site are exempt, while all others must adhere to hazardous-waste permit criteria.
South Carolina	Y	Y	N	Y	The Hazardous Waste Management Division classifies combustion units that burn hazardous wastes for fuel as hazardous waste TSD facilities.
South Dakota	Y	Y	Y	N	For those facilities not exempted, the air permit procedure requires analysis of waste streams, types and amounts of expected emissions, types and

STATE	Air Regulations: Permit to Construct	Air Regulations: Permit to Operate	Hazardous-Waste Regulations: Permit Exemption For Reuse Recycle or Reclaim	Hazardous-Waste Regulations: Permit Criteria For Recovery Facilities	COMMENTS
					amounts of fuels to be used, and the nature of the process(es) which affect emissions.
Tennessee	Y	Y	Y	N	Facilities exempted from hazardous waste regulations may come under state air regulatory guidelines.
Texas	Y	Y	O	O	Facilities which generate and process (energy/recovery included) hazardous waste onsite are exempt from hazardous-waste permit requirements, while all others must go through permit procedures.
Utah	Y	N	Y	N	There are no specific hazardous-waste requirements for resource recovery facilities.
Vermont	Y	N	N	Y	Combustion units that burn hazardous wastes must obtain a hazardous waste permit and comply with thermal treatment operations.
Virginia	Y	Y	Y	N	The state air pollution control board regulates existing combustion units that burn hazardous wastes under fuel-burning equipment regulations.
Washington	Y	Y	O	O	All sources of air contaminants are required to register and report to an activated local authority or the state air pollution control authority. Rules on reuse and recycle exempt some resources recovery facilities, while others must adhere to hazardous-waste permit criteria.
West Virginia	Y	N	Y	N	There are no permit requirements for the operation of boilers.
Wisconsin	Y	Y	O	O	All fuel burning equipment requires an air permit to construct and operate. Generators burning hazardous waste on

STATE	Air Regulations: Permit to Construct	Air Regulations: Permit to Operate	Hazardous-Waste Regulations: Permit Exemption For Reuse Recycle or Reclaim	Hazardous-Waste Regulations: Permit Criteria For Recovery Facilities	COMMENTS
					site may be exempt. All others must obtain a hazardous-waste license.
Wyoming	Y	Y	Y	N	The state has not promulgated regulations which specifically deal with hazardous waste management under RCRA.

Note: Y = Yes, N = No, O = Other, see comments.

vaporize, but differences in specific gravities can cause the mixtures to stratify in holding tanks and fuel lines. Chemical reactions can cause precipitation or polymerization, which may affect feed rates. Solids suspended in liquid waste can also erode or clog burner tips and control devices.

Metals create problems in both solid and liquid waste fuels. If a burner operates at temperatures greater than the fusion point of the metal, or its oxides or carbonates, slagging and fouling can occur. Metals can also form eutectic mixtures on refractory surfaces and promote corrosion at temperatures much lower than would otherwise be expected.

The use of wastes in a production process requires a careful evaluation of the possible effects of the disposal of that waste on the process and the effect that waste may have on the product.

INDUSTRIAL-BOILER SYSTEMS

Regulations under RCRA are being promulgated governing the burning of hazardous wastes in industrial boilers. These regulations will probably include DRE requirements, as do regulations governing permits for incinerators used for hazardous-waste destruction.

There are over 40,000 industrial boilers in the United States with capacity in excess of 10,000,000 Btu/hr, and over 5000 of these generate steam equivalent in excess of 100,000,000 Btu/hr (Merrill 1981). These larger boilers are candidates for use in the destruction of hazardous wastes, although it may also be practical to fire industrial waste streams in the smaller units. The larger boilers are essentially waterwall types, as shown in Fig. 7-1. Smaller boilers include firetube units, shown in Fig. 7-2.

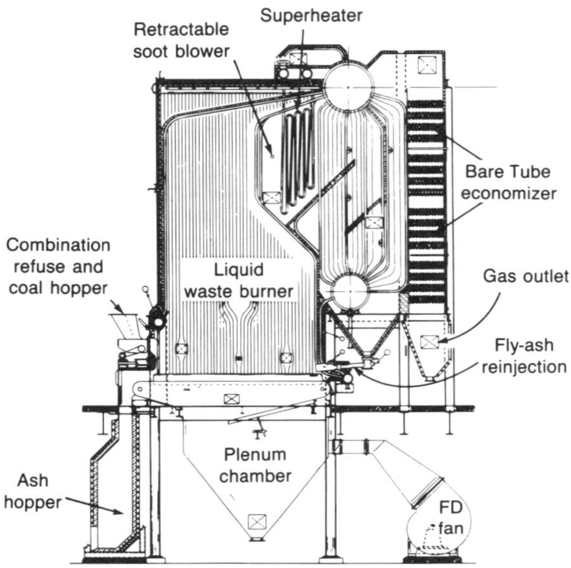

Fig. 7-1. Industrial boiler system.

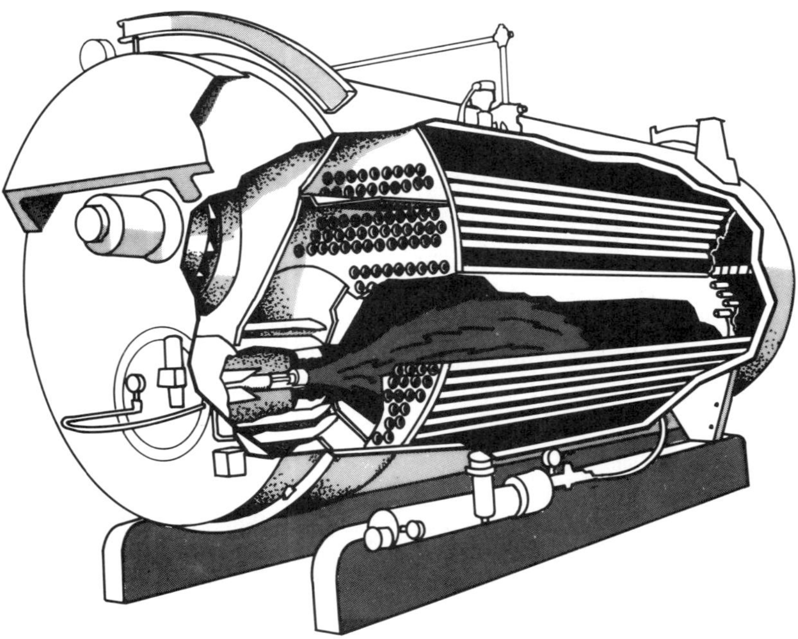

Fig. 7-2. Firetube boiler.

The primary purpose of an industrial boiler is to generate steam, and the introduction of hazardous waste must be compatible with steam generation. Fuels used for boiler firing are relatively consistent, with uniform heating value and firing characteristics. Wastes will not have uniform properties, and this may be the greatest difference between the use of fossil fuels versus wastes used as fuels.

Hydrocarbons, such as waste oil and waste solvent, are likely candidates for combustion in industrial boilers. Waste oils, however, are apt to contain contaminants which pose environmental as well as combustion problems. Approximately two-thirds of the waste oil in the United States comes from the crank cases of automobiles and trucks, and contains an estimated 1 percent lead, although, with the increasing use of lead-free gasoline, this lead content is decreasing. Treatment prior to use is generally required because waste oils frequently contain sludge, grit, solvents, water, heavy metals, and processing additives. This pretreatment, however, rarely encompasses more than the removal of water and solid particles by gravity or filtration.

Waste solvents are usually contaminated with suspended or dissolved solids, metals, organo-metallics, or other solvents. Normally, the contaminants comprise less than 10 percent of the solvent. Industries that produce waste solvent include the drug industry, solvent refineries, vegetable-oil extraction, polymerization processes, and industrial cleaning operations. Fuels derived from these hydrocarbon solvents are markedly different from petroleum oil in chemical and physical properties. They differ in flow characteristics, required air-to-fuel ratios, pumping requirements, and pressure requirements. Since they often have a lower viscosity than fuel oils, liquid hydrocarbons require lower pressure for movement and atomization; this can affect any preset feed rate. The heat content also varies and, often, in order to achieve the temperatures and steam pressures required for plant processes, more of the waste must be used.

The single largest category of solvents used in industry are halogenated solvents. Chlorinated or fluorinated solvents represent over 25 percent of all solvents manufactured in this country. The heat content of these solvents tends to decrease as the weight of the halogen constituent increases. Their benefit as a fuel supplement decreases as their halogen content increases.

The use of halogenated waste as a boiler fuel also presents potential corrosion problems. High or low temperature corrosion may occur, as described in Chapter 17.

The types of boilers in current use vary widely in design and capacity. A specific boiler must be evaluated for a specific waste to determine whether waste destruction criteria will be met.

Table 7-2. Typical Industrial Boiler Parameters.

Boiler	Volume cu. ft.	Flow SCFM	Temperature, F Maximum	Temperature, F Exit	Heat Input Million Btu/hr
Firetube, #2 Fuel Oil	74	2850	2615	2215	15
Firetube, Natural Gas	74	2490	2810	2425	15
Package Watertube, #6 FO	2400	28000	2600	1875	150
Package Watertube, #2 FO	2400	28000	2630	1955	150
Package Watertube, NG	2400	24900	2810	2200	150
Watertube, Coal	7500	31070	2450	1925	150
Watertube, Coal	11200	38470	2770	2120	200

Source: Reference 1.

Table 7-2 lists critical parameters for seven different boiler operations. Two of these are small, packaged, marine-type boilers and three are packaged waterwall systems, a "D" design, with two drums. In addition, parameters for two stoker-fired coal boilers are included.

These boiler parameters were evaluated to determine the time-temperature profile through the furnace, by the methods discussed in Chapter 15 for determining bulk residence time. The profile of each of these furnaces was plotted against the destruction requirements of three typical hazardous-waste compounds, pyridine, hexachlorobenzene, and nitrobenzene, in Figs. 7-3 through 7-9.

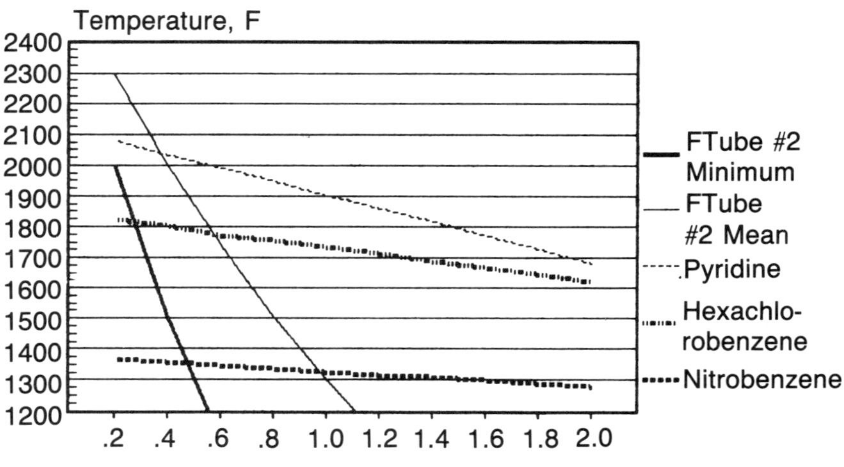

Fig. 7-3. Firetube boiler firing #2 fuel oil, time-temperature requirements for 99.99% destruction, 15 MBH.

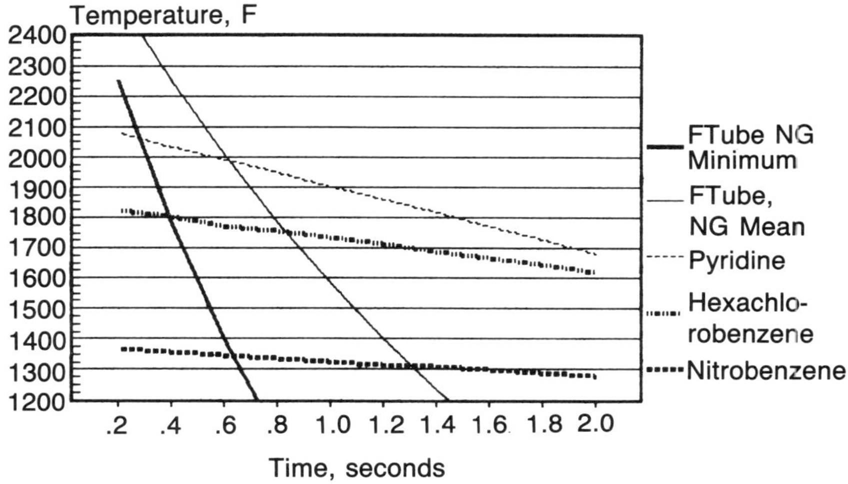

Fig. 7-4. Firetube boiler firing natural gas, time-temperature requirements for 99.99% destruction, 15 MBH.

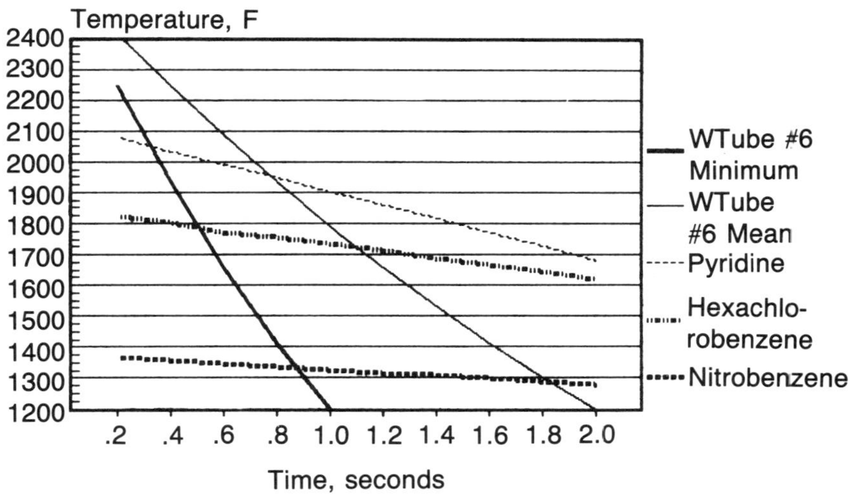

Fig. 7-5. Package watertube boiler firing #6 fuel oil, 150 MBH, time-temperature requirements for 99.99% destruction.

These plots indicate that in most cases, there is a regime of the furnace wherein 99.99 percent destruction of all three constituents will take place. Note that where the curve representing destruction of a hazardous constituent is beneath the curve representing boiler operation, destruction will occur.

From Fig. 7-3, which characterizes a small, marine-type boiler firing #2 fuel oil, it can be seen that destruction of hexachlorobenzene and

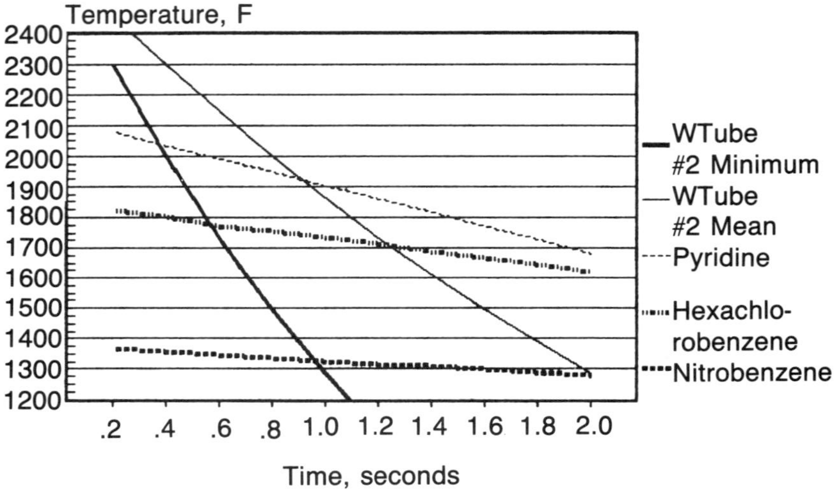

Fig. 7-6. Package watertube boiler firing #2 fuel oil, 150 MBH, time-temperature requirements for 99.99% destruction.

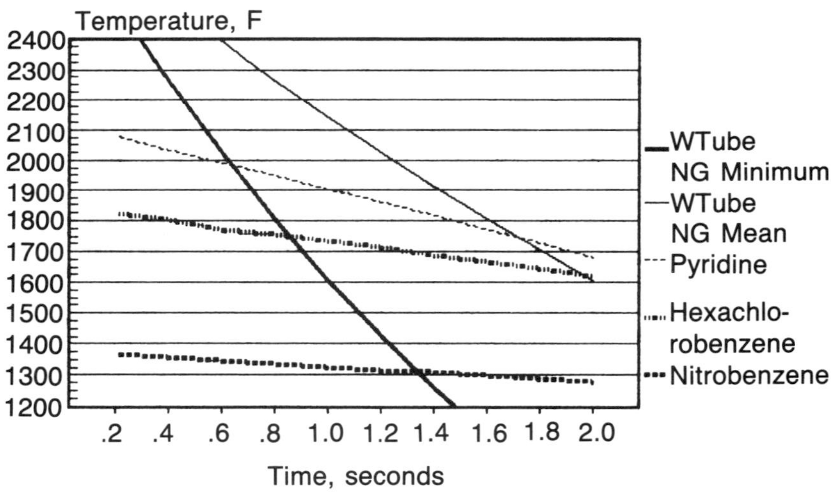

Fig. 7-7. Package watertube boiler firing natural gas, 150 MBH, time-temperature requirements for 99.99% destruction.

nitrobenzene is likely. The destruction characteristics of pyridine, however, are beyond the fast path (minimum detention) line. This indicates that 99.99 percent destruction of this waste might not occur in this furnace.

From Fig. 7-4, representing a small, marine-type, firetube furnace firing natural gas, the destruction of each of the hazardous compounds appears likely. Although small, the residence time of the constituent compounds within the furnace is long enough (approximately 0.2 seconds at

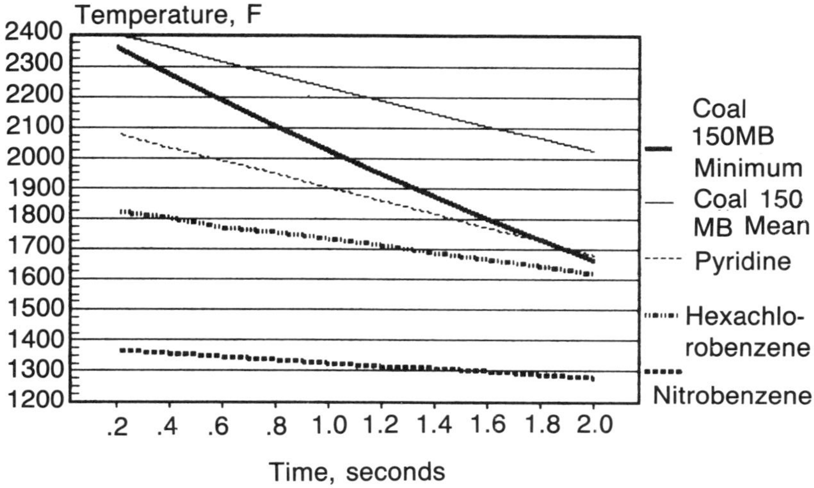

Fig. 7-8. Field-erected watertube boiler firing coal, 150 MBH, time-temperature requirements for 99.99% destruction.

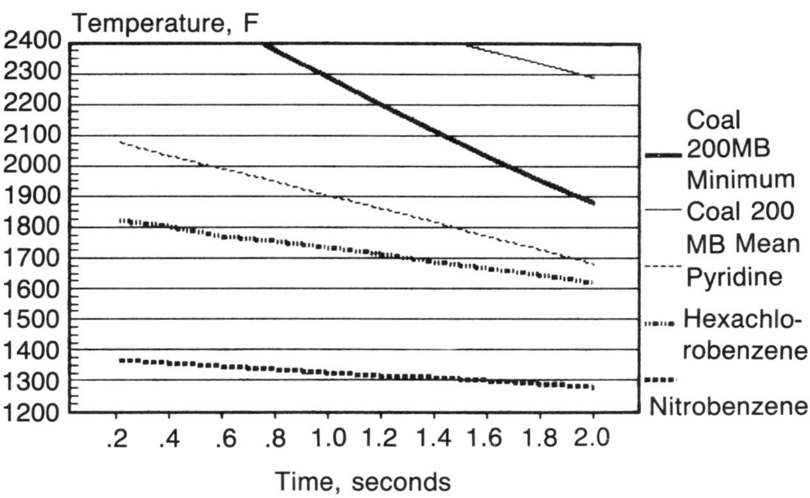

Fig. 7-9. Field-erected watertube boiler firing coal, 200 MBH, time-temperature requirements for 99.99% destruction.

2050° F) so that, even considering the fastest path through the chamber, all of the listed wastes will decompose in the regime of the boiler above 2040° F.

Figure 7-5, representing a package watertube boiler firing number 6 fuel oil; Fig. 7-6, representing a package watertube boiler firing number 2 fuel oil; and Fig. 7-7, representing a package watertube boiler firing natural gas, indicate these boilers will have sufficient residence times and temperatures to destroy the three wastes.

The coal-fired boilers, whose characteristics are illustrated in Fig. 7-8 and Fig. 7-9, are excellent candidates for destruction of these wastes.

CEMENT-KILN SYSTEM

Cement kilns are ubiquitous. The high temperatures, long residence times, and high degree of turbulence within these kilns have been found to lead to excellent destruction of the most toxic and unmanageable wastes.

There are approximately 350 cement kilns in the United States, located, generally, in areas of industrial activity or near quarries. There are a number of reasons for their growing use in the thermal destruction of hazardous industrial wastes, including the physical location of these plants. Some of the features of these systems are:

- Chlorinated waste streams are neutralized by the alkali present in the cement clinker.
- Conversely, the chlorine component in a chlorinated waste fulfills the need for chlorine in the clinker to reduce its alkali content.
- Investment in capital equipment is minimal because the cement kilns are already in place.
- Organic wastes with a significant heating value, in excess of 12,000 Btu/lb, such as waste oils, will allow a fuel savings for a cement kiln. Fuel costs can run as high as 65 percent of the operating cost of a cement kiln.
- Ash from destruction of the waste is absorbed in the clinker. Heavy metals tend to be trapped within the clinker as trace contaminants. This contamination has been found to be insignificant in cement quality.
- There is a huge thermal inertia within the kiln. The mass of clinker processed through the system tends to eliminate the possibility of rapid swings in temperature or other changes in the process. The process requires thermal stability to insure product quality.
- The cement process requires that temperatures in the kiln be maintained at 2500° F to 2800° F. Higher temperatures may be developed in some kilns. These temperatures will destroy almost all organic wastes.
- Air emissions control equipment is already in place for the collection of particulate from the kiln.
- The cement kiln operates under negative pressure, or draft. This is a requirement for kilns used as hazardous waste incinerators, which must be maintained at a negative pressure to prevent fugitive emissions.

A number of concerns must be addressed in the use of a cement kiln for waste disposal. Process design features and institutional factors can affect the use of a cement kiln. Typical of these considerations are the following:

- The location of the waste feed must be carefully considered. If waste is introduced at the low temperature end of the kiln, volatiles may be released too soon, without sufficient residence time to completely burn-out. If waste is placed in the kiln too close to the product discharge, there may be insufficient contact time to ensure homogeneous mixing of waste residual; clinker and product quality may deteriorate.
- Excess chloride degrades cement quality. An excessive level is difficult to predict and must be determined by tests on the actual waste stream. A chloride content in the clinker (dry basis) of less than 0.7 percent dry basis has been found to be acceptable.
- Conventional cement kilns are run with very little operator attention required. Hazardous waste requires continuous operator attention to feed and product quality control, which represents an increase in the cost of operation.
- The storage and feeding of hazardous wastes at a cement kiln requires that extensive personnel safety procedures be implemented, which represents another increase in the cost of operation of the facility.

In balance, the use of existing cement kilns has been found to be effective in the disposal of hazardous waste. There is a potential cost savings associated with this process, which may promote greater industrial and public acceptance in the future.

Kiln Description

Kilns used for cement production are large, up to 24 feet in diameter and 750 feet in length. A typical kiln system is shown in Fig. 7-10. They typically process thousands of tons of clinker per day or millions of tons per year.

The kilns are built in sections, and welded together to form a single unit. In wet-process systems, heavy chains are normally provided within the kiln, in the front third of the unit (the end of the kiln closest to the introduction of clinker). The chains are attached to the kiln's inside surface and, as the kiln rotates, the chains fall and wipe the inside kiln surface. These chains promote good clinker mixing and uniformity, and promote the drying process. They also serve to increase heat transfer between the hot gases and the cooler clinker.

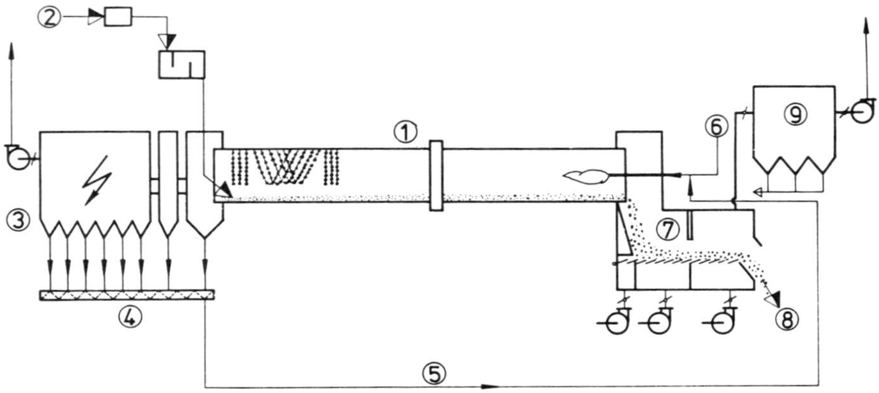

Fig. 7-10. Typical cement kiln.

The kiln is supported on a series of trunnions every 50 to 100 feet. Normally, the kiln is turned through a sprocket or chain drive near the kiln center.

Scoop systems are often provided at points along the kiln length for the introduction of additives or other materials. Air ports may also be installed along the kiln length for the admission of combustion air at points beyond the hot zone of the kiln.

Usually coal is used as the kiln fuel. It is ground or pulverized by on-site equipment adjacent to the kiln and is pneumatically injected into the back end of the kiln, or it can be added to the kiln feed. The back end of the kiln is the end at which product is discharged.

The capacity of a cement kiln can be approximated by the following expression:

$$T = (L \times D^2) / 100$$

where T is cement in tons per day of product (current), L and D are kiln length and outside diameter, in feet, respectively. As an example, for a 15-foot diameter kiln, 200 feet long, a cement production rate of 450 tons per day can be expected.

Cement Manufacturing

Cement is a complex mixture of four principal constituents, dicalcium silicate, tricalcium silicate, tricalcium aluminate and tetracalcium aluminate, as listed in Table 7-3. A small amount of gypsum (calcium sulfate, $CaSO_4 \cdot 2H_2O$) is added to the final product to prevent a too-rapid set-up when mixed with water.

Table 7-3. Cement Constituents.

NAME OF COMPOUND	CHEMICAL FORMULA	ABBREVIATION
Tricalcium Silicate	$3\ CaO \bullet SiO_2$	C_3S
Dicalcium Silicate	$2\ CaO \bullet SiO_2$	C_2S
Tricalcium Aluminate	$3\ CaO \bullet Al_2O_3$	C_3A
Tetracalcium Aluminoferrite	$4\ CaO \bullet Al_2O_3 \bullet Fe_2O_3$	C_4AF

Portland cement (named for its resemblance to a stone quarried near Portland, England, where modern cement was invented early in the last century) is the predominant cement produced in this country. Its manufacture requires three steps: preparation of the meal, or raw mixture; production of the clinker; and final preparation of the cement.

The raw materials are chalk or limestone, the source of calcium carbonate, and clay or shale, the source of aluminum silicates.

Where the raw materials available are chalk and clay, the wet process is used for cement manufacture. Over half the cement kilns in operation in the United States utilize the wet process.

Water is added to the chalk before it is blended with clay, forming a slurry containing 30 to 40 percent water. The blended slurry is sized by being passed through a grinder before being pumped to the rotary kiln.

The dry process is used where the raw materials are limestone and shale. This process was developed in an effort to reduce operating costs by reducing the fuel requirement.

Raw materials are blended and ground prior to being fed to the raised end of the kiln. The retention time of feed in the kiln is 1 to 4 hours, and the temperature range at the lower end of the kiln is 2500° F to 3000° F. From the kiln emerge nodules roughly ¾ inch in size, and are cooled with air in a clinker cooler.

Gas from the kiln flows countercurrent to clinker. The exhaust passes through an electrostatic precipitator (ESP) or fabric filter; the dust is normally recycled in the kiln.

Temperatures in the kiln are well above those required for the destruction of three hazardous wastes discussed in this chapter (pyridine, hexachlorobenzene, and nitrobenzene).

THE BRICK AND TILE INDUSTRY

The brick and tile industry includes those facilities engaged in the manufacture of common brick, refractory brick, and structural clay tiles. Modern brick and tile plants use tunnel kilns in their manufacturing process, as shown in Fig. 7-11. Unfired product (ware) is loaded into a car and is charged at the left end of the kiln. As the car progresses along the length of the kiln, it passes through zones of increasing temperature.

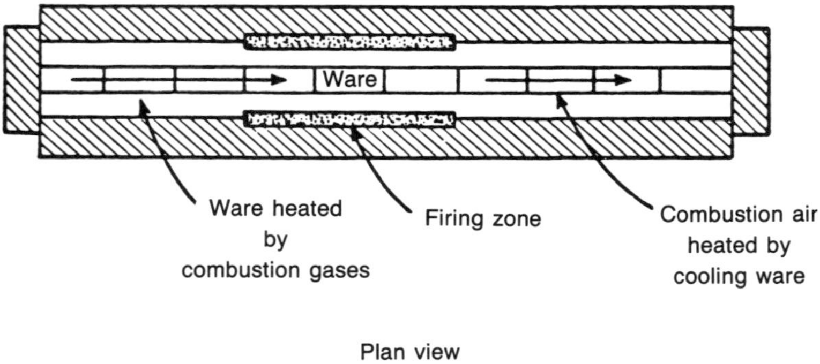

Fig. 7-11. Tunnel kiln. Source: Reference 6.

In most cases, the kiln has two sets of burners. The main (or upper) fuel-fired burners are located along the middle of the kiln and provide the major portion of heat necessary for firing. Secondary (or lower) burners are located between the preheating zone and the firing zone. Air is passed through the kiln against the movement of the ware. As the ware leaves the combustion zone, cold air forced in from the right end of the kiln passes over the charge, cooling it. Some of this air is withdrawn for use as preheated primary combustion air. The remaining air continues to the left, enters the combustion zone, mixes with the combustion gases, and then passes over the incoming charge, where it loses much of its heat. Air locks located at both ends of the kiln prevent flow conditions from being disturbed by the entrance or exit of cars.

Because of its wide application, relatively long residence time, and moderately high-temperature range, the tunnel kiln appears to offer an alternative for the disposal of hazardous waste. New tunnel kilns are being designed to burn more than one fuel type. This dual-fuel design would be advantageous if hazardous wastes were to be introduced. The waste burners should be located along the upper row, near the burner firing the main fuel.

Characteristics of a typical tunnel kiln are listed in Table 7-4. Figure 7-12 is a graphic representation of the ability of this typical kiln to provide 99.99 percent destruction of three typical hazardous wastes. It shows that hexachlorobenzene and nitrobenzene will be destroyed; however, pyridine destruction will not occur under fast-path conditions.

CARBON BLACK

Carbon black is the major ingredient of tires. It is generated through the oil-furnace process, illustrated in Fig. 7-13.

The heart of the oil-furnace process is the reactor, where incomplete combustion of a liquid hydrocarbon feedstock occurs. The materials put

Table 7-4. Manufacturing Industry Process Parameters.

Industry	Volume cu. ft.	Flow SCFM	Temperature, F Maximum	Exit
Brick & Tile	11833	18000	2150	500
Carbon Black	340	4000	2550	1400
Copper Smelting	20000	100000	2600	2400
Glass	5000	16800	2700	1140
Iron & Steel				
Blast Furnace	20800	196000	3400	2000
Open-Hearth Furnace	7000	30000	3250	2700
Lead Smelter	4940	10500	2200	1300
Lightweight Aggregate	35343	73170	2100	700
Lime Process	19792	26460	3340	2115
Zinc				
Fluid-Bed Roaster	3500	2500	1830	1740
Multiple-Hearth Roaster	9420	3700	1800	400
Suspension Roaster	4000	3500	1840	1700

Source: Reference 9.

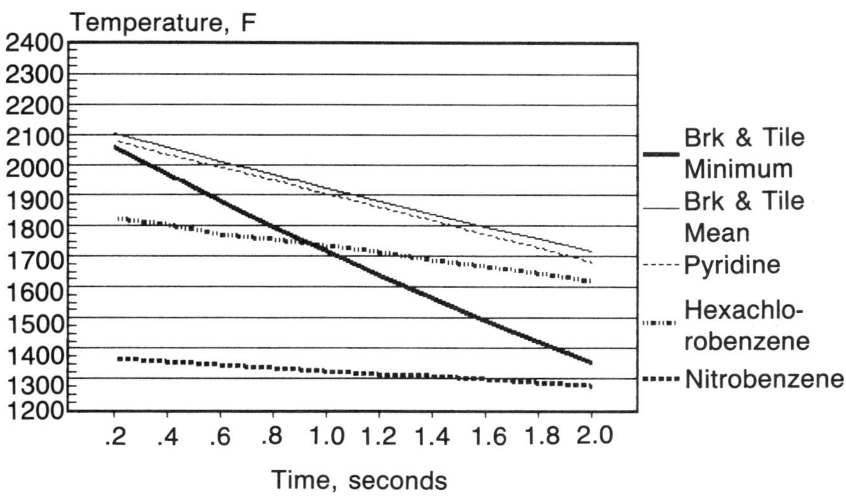

Fig. 7-12. Tunnel kiln, time-temperature requirements for 99.99% destruction.

into the reactor include a preheated oil feedstock, preheated air, and natural gas for heat and product quality control. Refinery by-product residual tars and decant oil from gasoline production are often used as feeds.

With less-than-stoichiometric oxygen provided to the system, the process is endothermic, and a considerable amount of energy is required to vaporize and heat the feedstock to reaction temperatures.

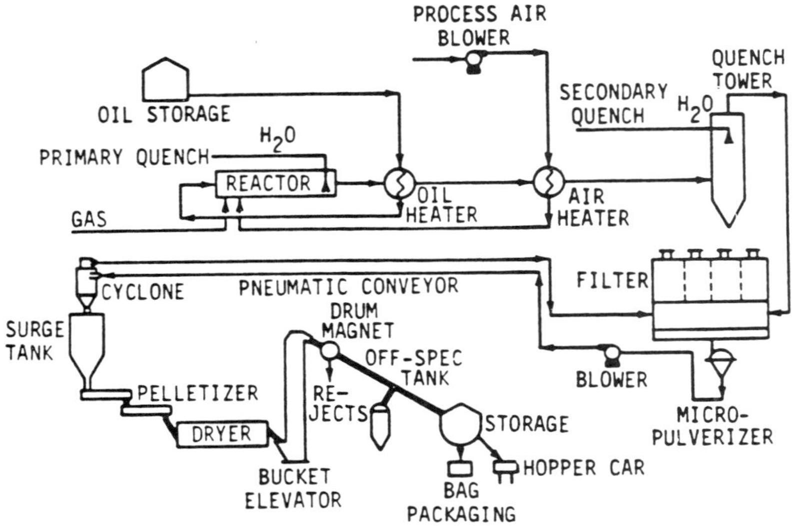

Fig. 7-13. Oil furnace process. Source: Reference 6.

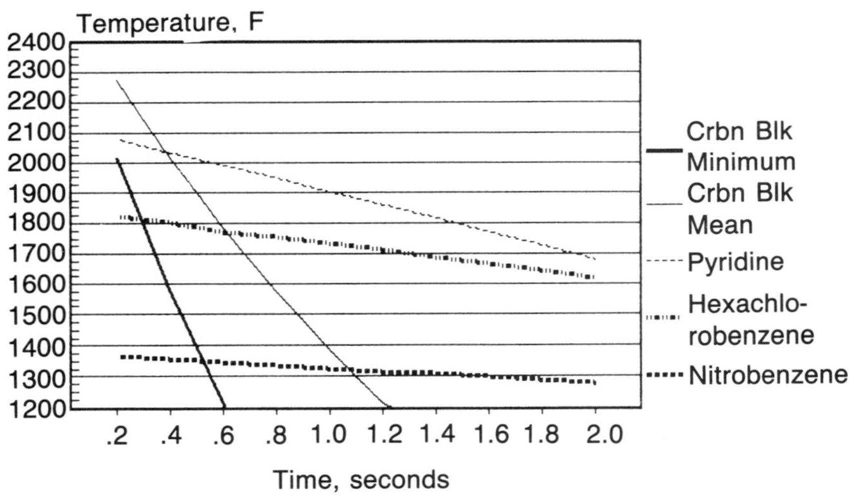

Fig. 7-14. Oil furnace process, time-temperature requirements for 99.99% destruction.

Primary and secondary quenching reduces the temperature of the off-gas for entrance into the bag filter. Carbon black is separated in the filter, collected, and mechanically processed into the product, usually in pellet form.

Hazardous waste streams may be candidates for destruction in an oil-furnace reactor. Based on the characteristics of a typical reactor (Table 7-4), Fig. 7-14 illustrates destruction capabilities for pyridine, hexachlorobenzene, and nitrobenzene. There is limited ability of this process to dispose of these three hazardous wastes.

COPPER SMELTING

Copper smelting is the process of removing iron and other impurities from a roasted or dried ore concentrate and producing a molten mixture that can be efficiently processed by a copper converter into a refined copper. This process is most often accomplished with a reverberatory furnace, shown in Fig. 7-15.

The term *reverberatory* refers to the configuration of the flame, which enters the chamber from one end, reverberates off the roof, and strikes the charge from above. As the charge melts, it separates into two layers: *matte* and slag. The matte is rich in copper and iron; the slag consists of silicates and compounds of calcium, magnesium, and aluminum. Slag is drained periodically from a skimming bay at one side of the furnace. Matte is withdrawn periodically through tap holes in the lower furnace wall at the other end of the furnace.

Natural gas, pulverized coal, and oil are used as fuel. Liquid hazardous wastes can be fired with fuel and solids or other non-liquid wastes could be charged into the furnace. Based on the characteristics of a typical copper reverberatory furnace, from Table 7-4, the time-temperature requirements graphed in Fig. 7-16 indicate that each of the three candidate hazardous wastes will be effectively destroyed.

GLASS INDUSTRY

Of the four segments of the glass industry (glass containers, flat glass, pressed and blown glass, and fiberglass) the glass-container manufacturing process is most applicable to hazardous waste disposal. Regenerative furnace systems, shown in Fig. 7-17, are used in the manufacture of container glass. The furnace consists of a double chamber filled with brick

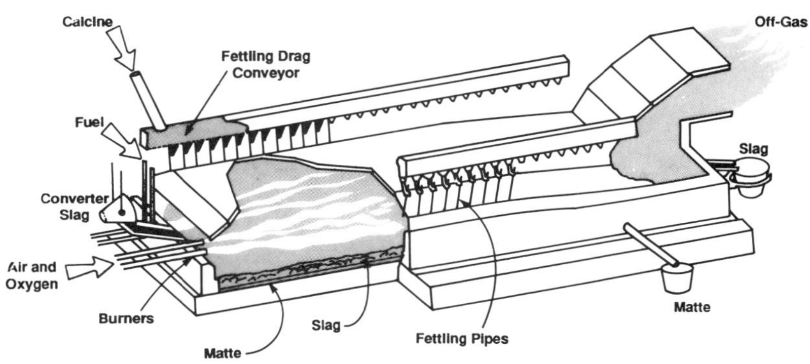

Fig. 7-15. Copper reverberatory furnace.

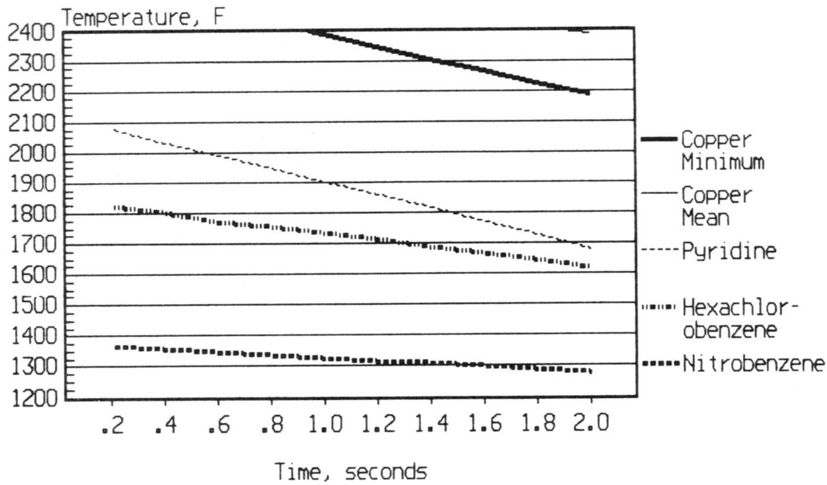

Fig. 7-16. Copper reverberatory furnace, time-temperature requirements for 99.99% destruction.

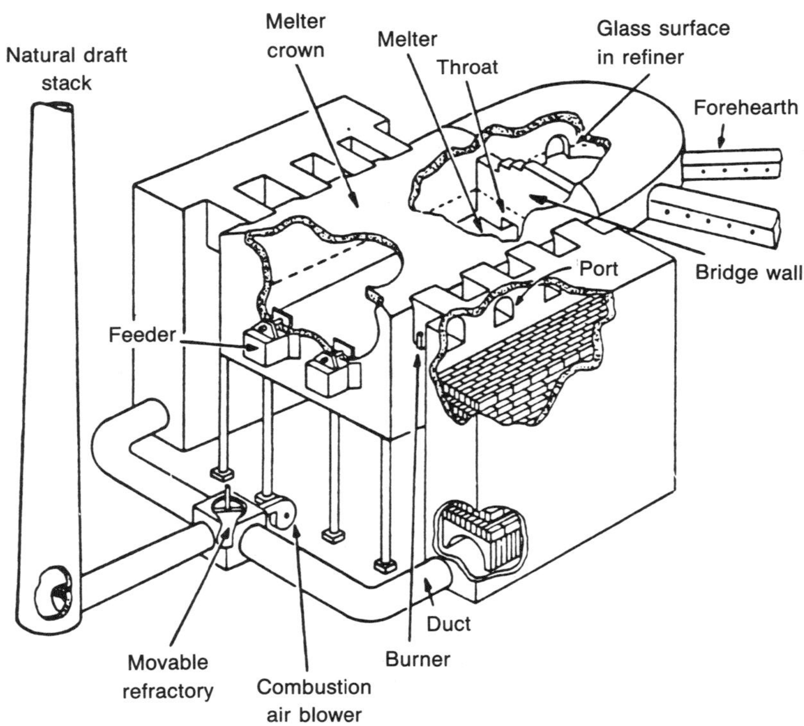

Fig. 7-17. Glass-melting furnace. Source: Reference 7.

checkerwork. While the products of combustion from the melter pass through and heat one chamber, combustion air is preheated in the other chamber. The functions of each chamber are interchanged by reversing the flow of air and combustion products every 15 to 20 minutes.

The furnace is fired by gas or oil. Feed is introduced into the furnace by screw or reciprocating feeders. Solid material floats on top of the molten glass until it melts. As the carbonates introduced in the feed decompose, they release carbon dioxide. Particles suspended in gas bubbles and volatilized components of the feedstock are swept from the furnace by the hot gases passing across the molten surface. These materials are either deposited in the back checkerwork and refractory lined passages or are discharged to the atmosphere. In the refining section, the glass is allowed to cool until it is viscous enough to form. It is then drawn from the furnace and sent to forming machines.

Liquid hazardous wastes are more compatible with oil than gas fuel because they can be blended with oil and introduced into the melting furnace through the atomizing burners provided for firing fuel oil. Glass quality, however, is sensitive to the introduction of metals to the melt and therefore, the compatibility of hazardous waste is a function of the presence of metals in the waste.

Using the characteristics of a typical glass furnace, listed in Table 7-4, the time-temperature requirements, Fig. 7-18, indicate that the likelihood of waste destruction is high for each of the three listed hazardous wastes. Maintaining temperatures in excess of 2000° F in the furnace ensures the destruction of all three hazardous wastes.

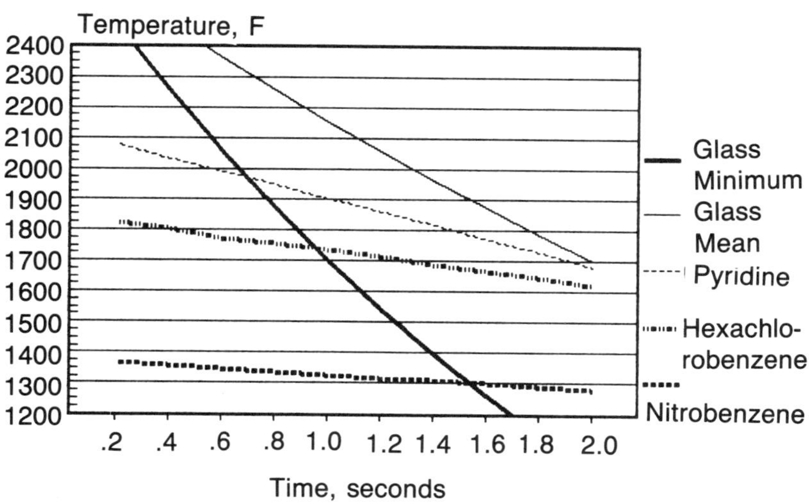

Fig. 7-18. Glass-melting furnace, time-temperature requirements for 99.99% destruction.

IRON AND STEEL INDUSTRY: THE BLAST FURNACE

Blast furnaces are used for the reduction of iron ore to basic iron. They are tall, cylindrical, countercurrent flow reactors, as shown in Fig. 7-19. Iron ore, pellets, sinter, limestone, and coke are charged into a blast furnace from the top. Hot air is introduced at the bottom of the furnace through tuyeres. The coke burns at the tuyere level, and the hot gases rise through the descending solids, transferring heat to the descending raw materials. Gases exit the top of the reactor and are directed through an air emissions control system.

The reactions that take place as the raw materials descend through the furnace are predominantly endothermic. Iron ore is progressively reduced to iron oxide as the solids descend to the bosh area, where the final reduction to iron takes place. A reducing atmosphere prevails throughout the descent of the materials in the furnace.

In most blast-furnace installations, a supplemental fuel is injected through the tuyeres to reduce coke consumption. The fuels used for this

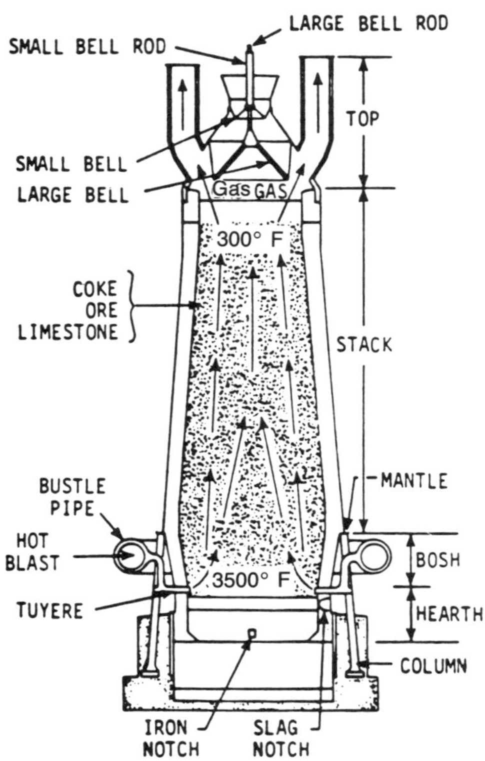

Fig. 7-19. Blast furnace.

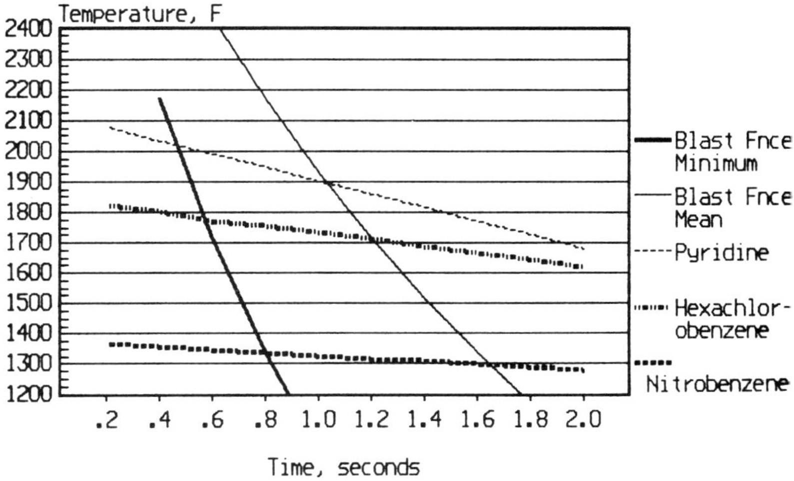

Fig. 7-20. Blast furnace, time-temperature requirements for 99.99% destruction.

purpose include oil, tar, pitch, natural gas, coke-oven gas, and pulverized coal. This is an area where hazardous wastes can be utilized.

In a typical blast furnace, the time-temperature relationship for destruction of hazardous waste (Fig. 7-20) indicates that waste destruction may be marginal. Destruction along the fast path will be assured only in the regime of the furnace where the temperature is above 2050° F.

IRON AND STEEL INDUSTRY: THE OPEN-HEARTH FURNACE

The open-hearth furnace is used to refine scrap and hot metal into steel. Up to 100 percent steel scrap or a combination of scrap and molten iron can be charged in this process. Limestone and fluospar are used as fluxes to absorb impurities. Fuels (coke-oven gas, natural gas, oil, tar, or a combination of gas and liquid fuel) are alternately fired from both ends of the furnace. As the combustion gases pass through brick checkerwork, they heat the brickwork.

As indicated in Fig. 7-21, the incoming combustion air is preheated by passing through the checkerwork. Every half-hour the firing is reversed to recover the heat through the brickwork.

Scrap and flux are charged to the furnace through the doors of specially-designed charging boxes. Molten metal is poured into the furnace through the hydraulically operated doors by means of a hot-metal chute. Steel and slag are drawn from the tap hole at the rear of the furnace, opposite the charging doors.

Liquid hazardous wastes are candidates for replacement of some of the fossil fuel normally fired. Time-temperature requirements for hazard-

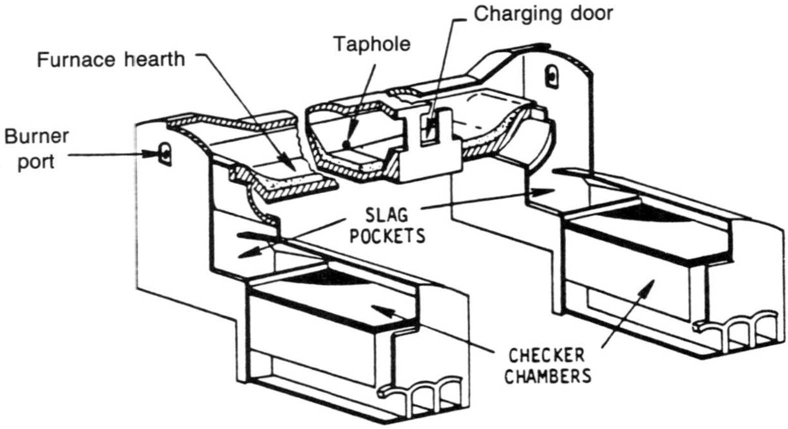

Fig. 7-21. Open-hearth furnace.

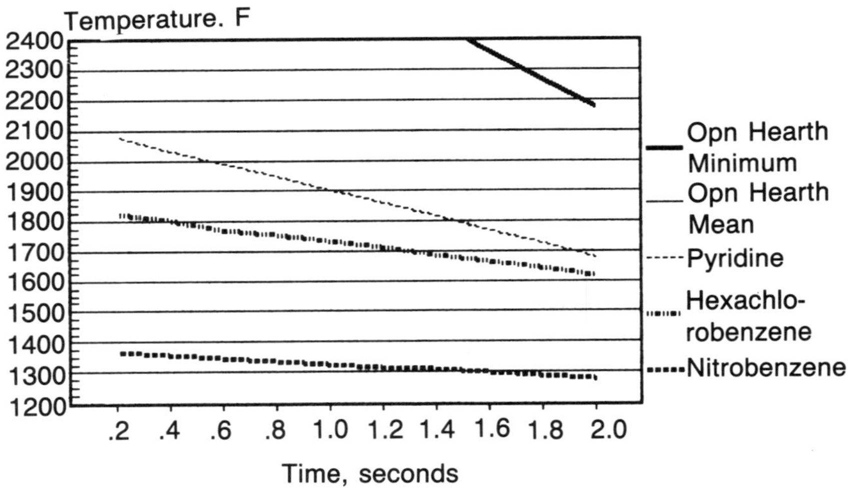

Fig. 7-22. Open-hearth furnace, time-temperature requirements for 99.99% destruction.

ous-waste destruction, Fig. 7-22, based on typical furnace characteristics, indicate that the open-hearth furnace will effectively destroy the three selected hazardous wastes.

LEAD PRODUCTION

The production of lead begins with sintering the ore. *Sintering* is the blending of the ore, its pelletization, and roasting, including conversion of lead sulfide to a lead oxide. A blast furnace (Fig. 7-23) is used to reduce sintered feed to a crude lead bullion.

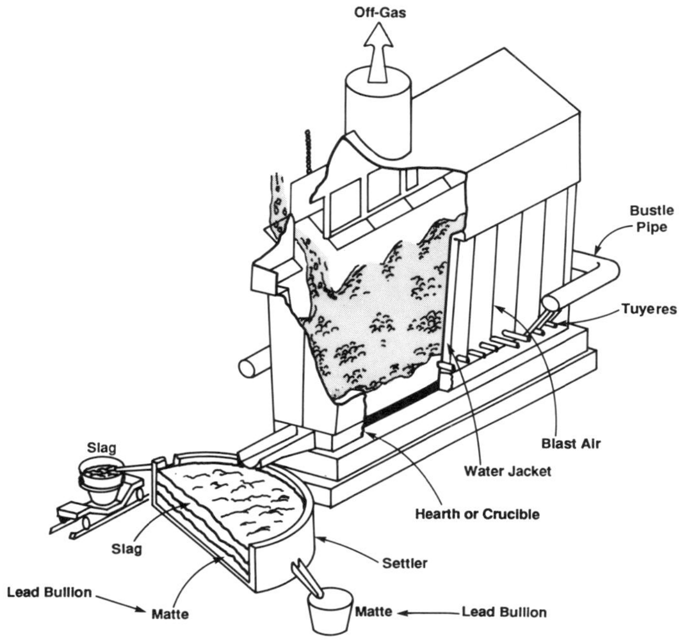

Fig. 7-23. Lead smelter.

The charge, which consists of sinter, metallurgical coke, and fluxing compounds, is fed to the furnace through a water-jacketed shaft. Combustion air is injected through tuyeres located near the bottom, and evenly spaced on either side of the furnace.

Metallurgical coke is the primary fuel for this process. Because this fuel is one of the reactants in the process, hazardous waste cannot be substituted for any part of the fuel, but it can be introduced in addition to the coke.

From the characteristics of a typical lead smelter, the derived destruction characteristics, Fig. 7-24, indicate that destruction is likely for hexachlorobenzene and nitrobenzene. Destruction of pyridine is not as certain.

LIGHT-WEIGHT AGGREGATES

Light-weight aggregates encompass a variety of raw materials that provide the bulk of concrete products. Raw material such as clay, shale, or slate that are expandable and contain small quantities of gas producing compounds can be used to produce light-weight aggregates. The raw materials are fed to a rotary kiln, as shown in Fig. 7-25. In the kiln, the feed is initially preheated by hot combustion gases.

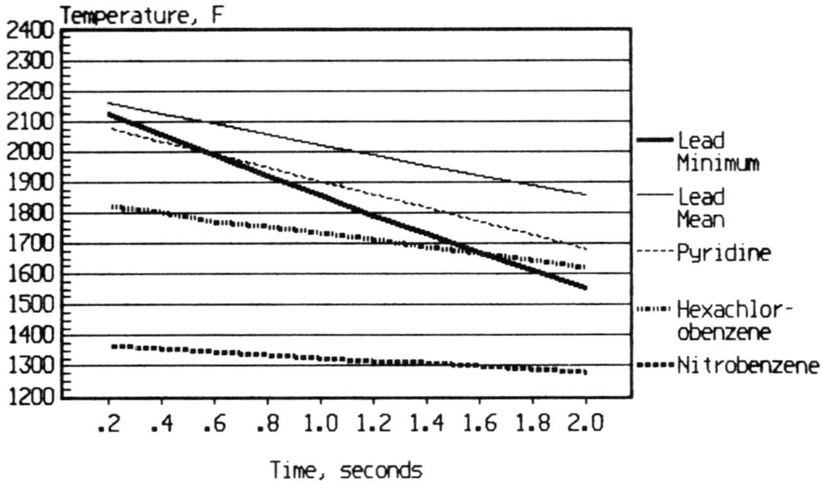

Fig. 7-24. Lead smelter, time-temperature requirements for 99.99% destruction.

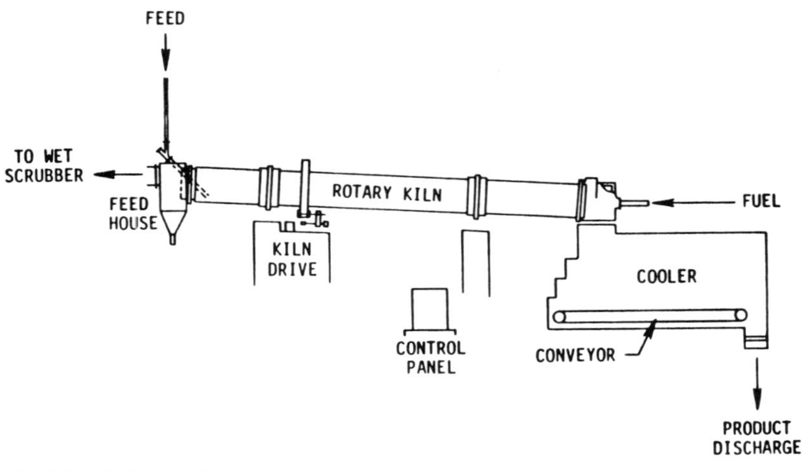

Fig. 7-25. Light-weight aggregate process kiln.

Passing to a second furnace zone, the raw materials melt to a semiplastic and begin to generate the expanding agent. In this zone, specific compounds (hematite, calcite, dolomite, pyrites, and lemonite) begin to decompose and form gases that eventually trigger the expanding action. The expanding gases formed during decomposition include sulfur dioxide, sulfur trioxide, carbon dioxide and oxygen. As temperatures reach their maximum level in the firing, or expanding zone, the semiplastic raw materials become sufficiently viscous to contain the expanding gases. This expanding action produces a mass of small, unconnected air cells, which the material retains after it cools and solidifies.

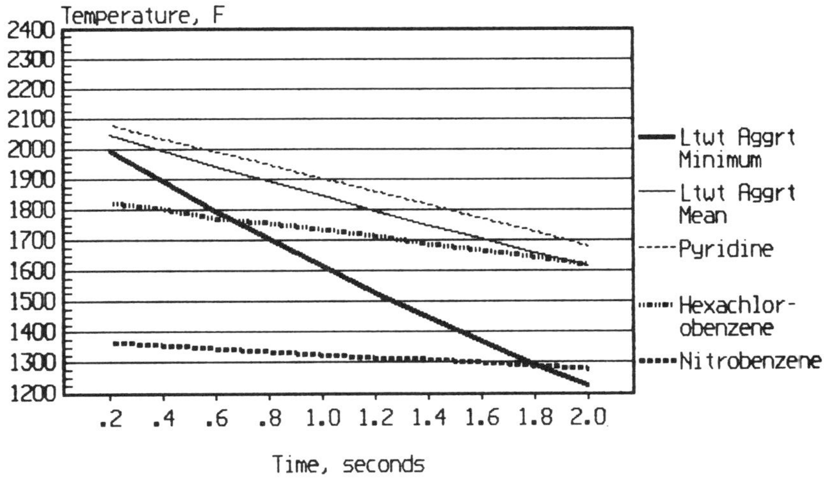

Fig. 7-26. Light-weight aggregate process kiln, time-temperature requirements for 99.99% destruction.

Liquid hazardous wastes are candidates for replacement of a portion of fuel required for this process. For the process parameters listed in Table 7-4, the temperature-requirements-for-destruction curves in Fig. 7-26 indicate that pyridine would not be destroyed. Nitrobenzene would be thoroughly destroyed; the destruction of hexachlorobenzene would be less complete.

LIME PRODUCTION

The production of lime includes quarrying raw limestone, then crushing, sizing, and calcining it. Either a long kiln or a short kiln with a stone preheater is used for calcining.

Long and short kilns operate similarly. The kilns are made of heavy steel plate, lined with refractory brick. Although most kilns are fired with pulverized coal, they can be adapted to firing fuel oil or gas fuel. As shown in Fig. 7-27, limestone is fed to the elevated end of the kiln and is discharged as quicklime at the lower end.

Hazardous wastes can be used to supplement the fuel put into the system. Typical characteristics of a lime kiln plot far above the destruction characteristics of pyridine, hexachlorobenzene, and nitrobenzene. This kiln is, therefore, an excellent tool for the destruction of these three wastes.

ZINC PRODUCTION: FLUID-BED ROASTER

The primary purpose of the roasting process is the removal of sulfur from the concentrated zinc ore. The fluid-bed roaster, shown in Fig. 7-28,

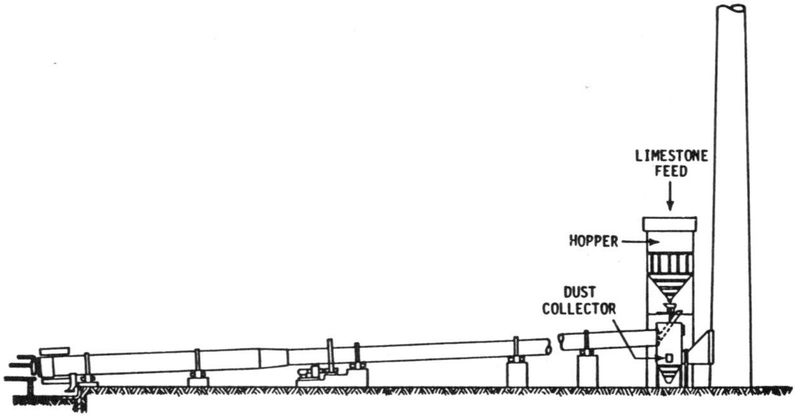

Fig. 7-27. Lime kiln. Source: Reference 6.

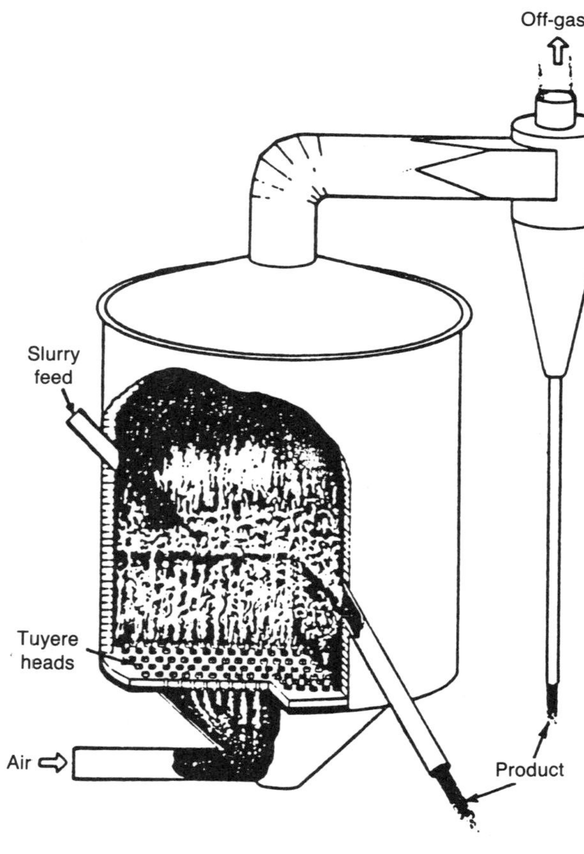

Fig. 7-28. Zinc Fluid-bed roaster. Source: Reference 8.

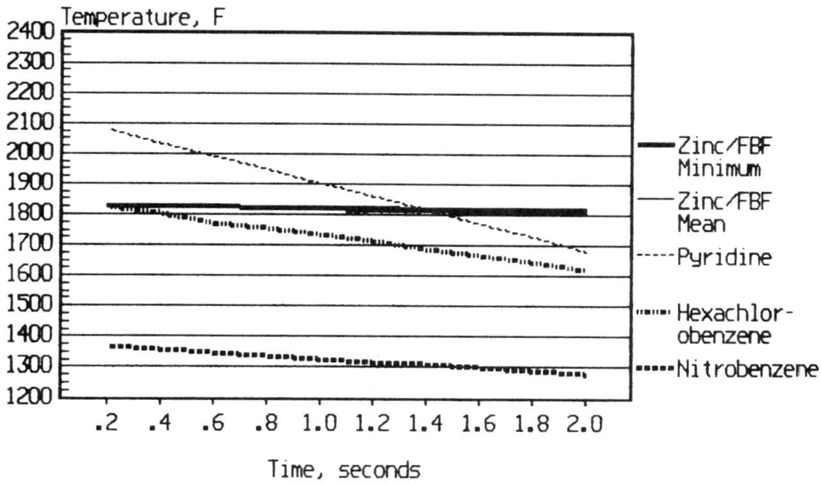

Fig. 7-29. Zinc fluid-bed roaster, time-temperature requirements for 99.99% destruction.

can be charged with dry or slurry feed. Natural gas, oil, or coal is used to bring the feed to reaction temperatures; once reaction temperatures are reached, the exothermic oxidation of sulfur maintains the bed temperature.

Hazardous waste can be injected into the furnace once it has reached operating temperature. The characteristics of a typical fluid-bed roaster indicate that the maximum temperature within the unit and the temperature at the exit of the unit are relatively close. This is reflected in the time-temperature curves, Fig. 7-29. The mean-path and fast-path lines are coincident. These curves indicate pyridine would not be a candidate for destruction in this furnace whereas both hexachlorobenzene and nitrobenzene would be.

ZINC PRODUCTION: MULTIPLE-HEARTH ROASTER

This furnace is essentially the same type of unit as the incinerator used for the disposal of sludge waste, as illustrated in Fig. 7-30. Feed enters the top of the furnace and is discharged as a calcine, zinc oxide, for refining in another furnace, or by electrolysis. In most cases, supplemental fuel is required to maintain the reaction. Liquid hazardous waste could be used to supplement this fuel.

Based on the characteristics of a typical multiple-hearth roaster and the time-temperature requirements for 99.99 percent destruction, Fig. 7-31 indicates that not all of the three selected hazardous wastes will be effectively destroyed. Nitrobenzene would be a good candidate for destruction, but the other two hazardous wastes would not be.

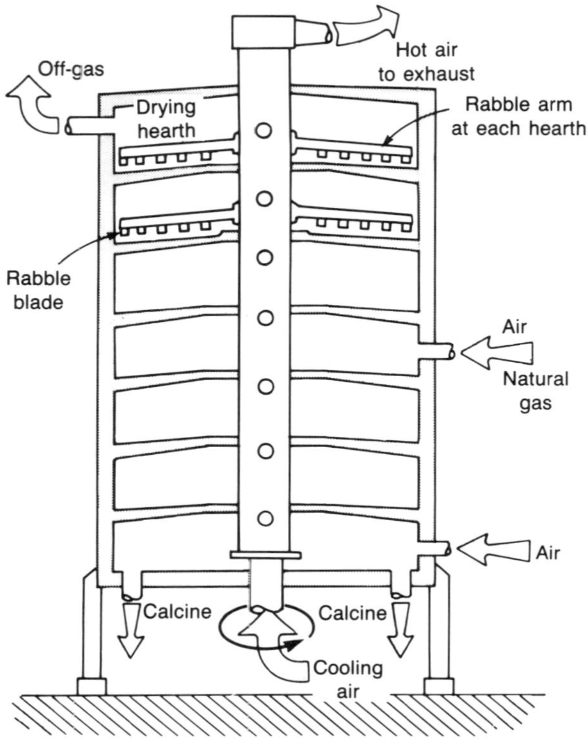

Fig. 7-30. Zinc multiple-hearth roaster. Source: Reference 8.

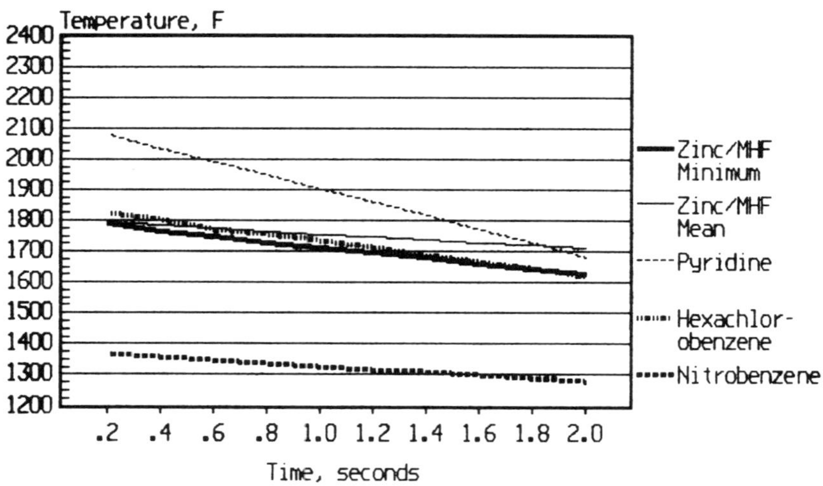

Fig. 7-31. Zinc multiple-hearth roaster, time-temperature requirements for 99.99% destruction.

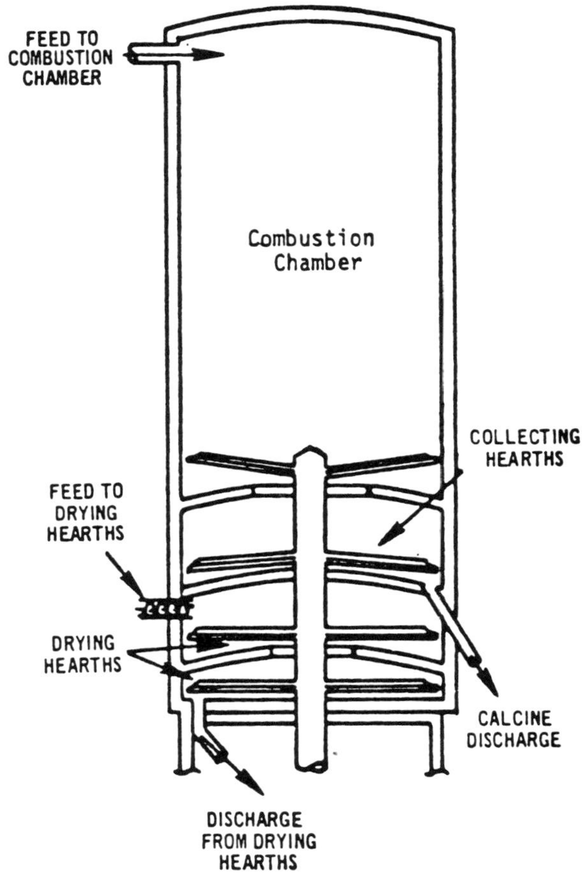

Fig. 7-32. Zinc suspension-roasting furnace. Source: Reference 8.

ZINC PRODUCTION: SUSPENSION ROASTING

Suspension (or *flash*) *roasting* is a process for rapid sulfur removal and conversion of the zinc charge to a calcine (zinc oxide) by allowing concentrates to fall through a heated oxidizing atmosphere or blowing them into a combustion chamber, as shown in Fig. 7-32. As with the fluid-bed and multiple-hearth roasters, there is often sufficient sulfur in the feedstock to maintain autogenous combustion in the furnace. Liquid hazardous wastes can be injected into the combustion chamber of the suspension-roasting furnace to supplement combustion.

As with the fluid-bed roaster, the temperature at the exit of the unit is close to the maximum temperature in the furnace. This results in the

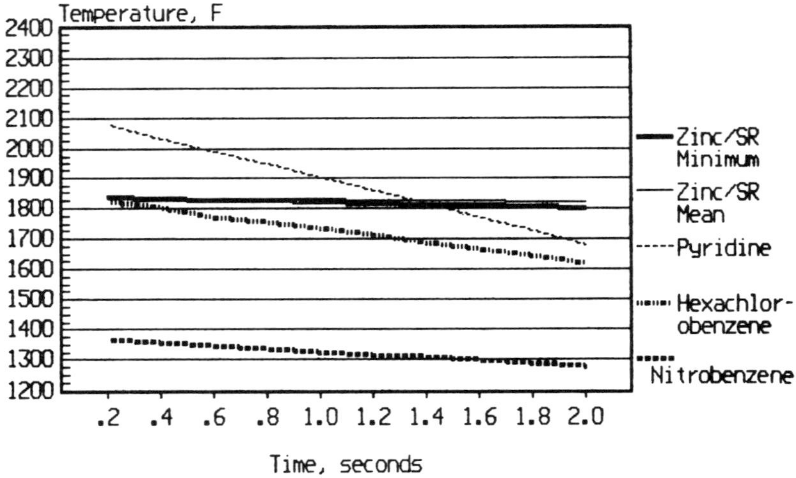

Fig. 7-33. Zinc suspension-roasting furnace, time-temperature requirements for 99.99% destruction.

mean-path and fast-path lines, shown on the time-temperature requirements graph, Fig. 7-33, being very close to each other. These curves also show that pyridine and hexachlorobenzene would not be good candidate wastes, but nitrobenzene would.

REFERENCES

1. R. Merrill, *Technical Overview of the Concept of Disposing of Hazardous Wastes in Industrial Boilers,* United States Environmental Protection Agency Industrial Environmental Research Laboratory, Office of Research and Development, March 1981.

2. D. Hazelwood, F. Smith, and E. Gartner, *Assessment of Waste Fuel Use in Chimney Kilns,* USEPA-600/S2-82-013, October 1982.

3. M. Black and J. Swanson, "Destruction of PCB's in Cement Kilns," *Pollution Engineering,* June 1983.

4. G. Higgins and A. Helmstetter, *Evaluation of Hazardous Waste Incineration in a Dry Process Cement Kiln,* Systech Corporation, 1981.

5. L. MacDonald, et al., "Burning Waste Chlorinated Hydrocarbons in a Cement Kiln," *Fisheries and Environment Canada Report,* EPS 4-WP-77-2, March 1977.

6. K. Othmer, *Encyclopedia of Chemical Technology,* Third Edition, Volume 5 (New York: John Wiley & Sons, 1979).

7. J. Stockham, "The Composition of Glass Furnace Emissions," Presented at the 63rd Annual Meeting of the Air Pollution Control Association, 1970.

8. United States Environmental Protection Agency, "Background Information for New Source Performance Standards, Primary Zinc and Lead Smelters," *Proposed Standards,* Volume I, EPA-600/2-80-170, July 1980.

9. United States Environmental Protection Agency Industrial Environmental Research Laboratory, *Feasibility of Destroying Hazardous Waste in High-Temperature Industrial Processes,* Draft Report, Contract Number 68-03-3036, March 1982.

8

Incineration at Sea

Northern Europe has been disposing of organic wastes with incinerator ships for nearly twenty years. In the United States, the use of ship-mounted incinerators has not extended past the testing phase.

OCEAN INCINERATION

Ocean incineration is the burning of wastes on ocean-going vessels, in waters remote from land.

In a typical vessel, liquid waste incinerators are mounted at the rear of a ship, as shown in Fig. 8-1, and wastes are stored within the ship. The ship is an ocean-going vessel that is capable of both transporting wastes and burning these wastes at a site far removed from human populations and from vulnerable sources of fresh water.

BULK VERSUS CONTAINERIZED STORAGE

Two types of ship systems are currently being considered in the United States. One provides bulk tanks on board; wastes are pumped to these tanks from shore facilities. A ship with integral tanks is illustrated in Fig. 8-1. The second concept is the storage and placement of containerized waste storage tanks on the ship, or on a specially constructed barge. Liquids would not be unloaded, neither at the port facility nor from the dock to the ship. A waste container, standardized as an intermodal tank (IM) container with a capacity of 5,000 gallons, is shown in Fig. 8-2.

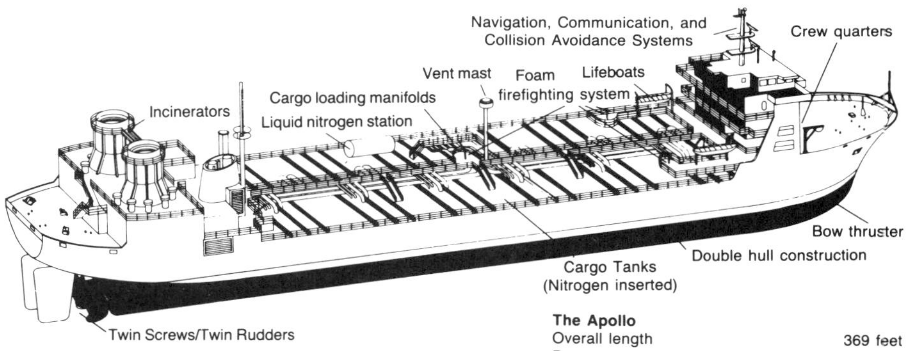

The Apollo	
Overall length	369 feet
Beam	60 feet
Gross tonnage	4,850 tons
Liquid cargo capacity	1.33 million gallons
Number of cargo tanks	12
Sustained sea speed	10.7 knots
Range with full load	6,940 nautical miles

Fig. 8-1. Typical incinerator ship. Source: At-Sea Incineration, Inc.

Fig. 8-2. Intermodal tank container.

Table 8-1 lists characteristics of existing and proposed incinerator vessels. Five utilize bulk storage of wastes below decks; the other two, not yet under construction, will be designed to carry waste in IM containers on or above deck.

IM containers can be transported by rail, truck, barge, and ship. They are sealed by the generator and are not opened in transit. No blending or handling of wastes will occur at the port facilities. Aboard ship, wastes will be pumped from the IM containers to a feed tank directly connected to the incinerator. After the tanks are cleaned, they will be returned to

Table 8-1. Characteristics of Existing and Proposed Incinerator Ships.

Vessel	Owner/Operator	Status	Cargo	Capacity (gallons)	Feed Rate (gallons/hr)	Burn (days)
Vulcanus I	Waste Management	Operating	8 Tanks	800,000	3,300	10.1
Vulcanus II	Waste Management	Operating	8 Tanks	800,000	4,950	6.7
Apollo I	Tacoma Boat/At-Sea Incineration	Completed	12 Tanks	1,300,000	5,500	9.9
Apollo II	Tacoma Boat/At-Sea Incineration	Under Construction	12 Tanks	1,300,000	5,500	9.9
SeaBurn*	Stolt-Nielsen	Planned	Containers	720,000	3,240	9.3
EOS	Environmental Oceanic Svcs.	Planned	Containers	80,000	unknown	unknown
Vesta	Lehnkering AG	Operating	9 Tanks	290,000	2,640	4.6

*SeaBurn is proposing to utilize an ocean going barge-tug combination.

the generator for re-use. The IM container is encased in a rigid protective frame, as in Fig. 8-2, that substantially reduces the risk of tank breaching in a transportation accident. If there is an accident, particularly one involving the incinerator vessel, the relatively small size of each container limits the potential spill. Another feature of IM containers is that wastes from different generators are not mixed or blended before they are incinerated. This allows isolation of one generator's waste from another and reduces each generator's potential liabilities.

The use of IM containers in lieu of bulk storage will create certain hardships. Each container will have to be sampled before incineration, which places a large burden on the shipboard testing facility. When containers are switched, there is the potential for incompatible wastes coming in contact. An elaborate means of isolation, including purge-cycling between loads, must be employed to preclude this possibility. While port activities are reduced with IM containers, shipboard activity is greatly increased. Container discharge must be switched every few hours, which creates the potential for spills and increases danger to operating personnel. With frequent tank switching, incinerator feed will probably not be continuously consistent and incinerator performance will be more difficult to control.

APPLICABILITY OF OCEAN INCINERATION

Ocean incineration is applicable to liquid organic-waste destruction when these wastes have sufficient heat value to burn without supplemental fuel and when the wastes have negligible metal content. With intrinsic

Table 8-2. Types of Waste Incinerated at Sea and Country of Origin, 1982.

Country	Type of Waste	Quantity
Australia	Vinyl Chloride and PCB wastes	5,302
Austria	Organohalogen wastes	539
Belgium	Organohalogen wastes	11,509
Finland	Organohalogen wastes	3,025
France	Organohalogen wastes	7,240
Germany	Organohalogen wastes	58,026
Italy	Organohalogen wastes	3,774
Japan	Oily sludges	1,637
The Netherlands	Organohalogen wastes	10,336
Norway	Organohalogen wastes	8,800
Spain	Organohalogen wastes	231
Sweden	Organohalogen wastes	7,062
Switzerland	Organohalogen wastes	4,082
United Kingdom	Organohalogen and organo-phosphorus wastes	6,813
	TOTAL	128,376

NOTE:
1. All quantities are in tons.
2. These are the quantities for which permits were granted. The actual amount of wastes incinerated may have been less than these figures.

Source: Reference 2.

heat value, the wastes will not require the ship to haul supplemental fuel. In the absence of a scrubber system, described below, metals discharge from the incinerator will be uncontrolled, and that is unacceptable in terms of potential environmental and ecological damage.

The majority of wastes that were incinerated at sea during 1982 are listed in Table 8-2, which includes the country of origin of each stream. Most of these wastes were halogenated. The off-gas from the incinerator is not neutralized within the incinerator facility, as is required of land-based installations. The hydrogen chloride generated by burning chlorinated wastes, which accounts for the vast majority of halogenated compounds, must be scrubbed out of the exhaust to prevent the formation of hydrochloric acid, which will affect animal and plant life, and structures. In the ocean, the natural alkalinity of seawater neutralizes these acid emissions; therefore, the incineration process does not have to provide for acid gas neutralization.

The cost of acid neutralization is a major factor in seeking to dispose of these wastes at sea, where no acid scrubbing is required. For this reason organo-halogen wastes predominate those incinerated at sea. They would require acid gas scrubbing if they were incinerated at land-based facilities.

PAST AND CURRENT PRACTICES

During the past 20 years, six incinerator ships have been used for the disposal of organic wastes. These vessels, their dates of service, number of incinerators, cargo and displacement capacity are listed in Table 8-3. All of these ships had or have been operating in the North Sea. Only one of these ships, Vulcanus I, has ever operated off the United States coast. This operation concentrated on demonstration burns of Department of Defense materials, such as agent orange.

Quantities of waste incinerated at sea from 1969 through 1984 are shown in Fig. 8-3, exclusive of trial burns off the United States coast.

REGULATORY ACTIVITY

On February 20, 1985, the United States EPA proposed rules for the regulation of liquid hazardous-waste incineration at sea. The proposed regulations provide specific criteria for evaluating ocean incineration permits for research and emergency disposal and for operating ocean incineration vessels. They prohibit ocean incineration of radioactive materials, synthetic or natural materials that float or remain in suspension in the ocean, and materials produced for biological or chemical warfare, among other substances. The rules' criteria are based upon standards mandated by the 1972 Marine Protection, Research, and Sanctuaries Act; the 1972 London Dumping Convention; RCRA; and the 1976 Toxic Substances Control Act.

Performance standards for ocean incineration under the proposed rules are:

Table 8-3. Incineration Vessels Employed Overseas, 1969 through 1986.

	Matthias I	Matthias II	Matthias III	Vulcanus I	Vulcanus II	Vesta
Dates of Service	**1968-76**	**1970-83**	**1975-77**	**1972-present**	**1982-present**	**1979-present**
Site of Operation	Exclusively in the North Sea			North Sea United States Pacific Australia	North Sea	North Sea
Incinerators	1	1	1	2	3	1
Total cargo (tons)	605	1,320	16,500	3,850	3,520	1,540
Gross tons	438	999	12,636	3,100	3,100	999

Source: Reference 3.

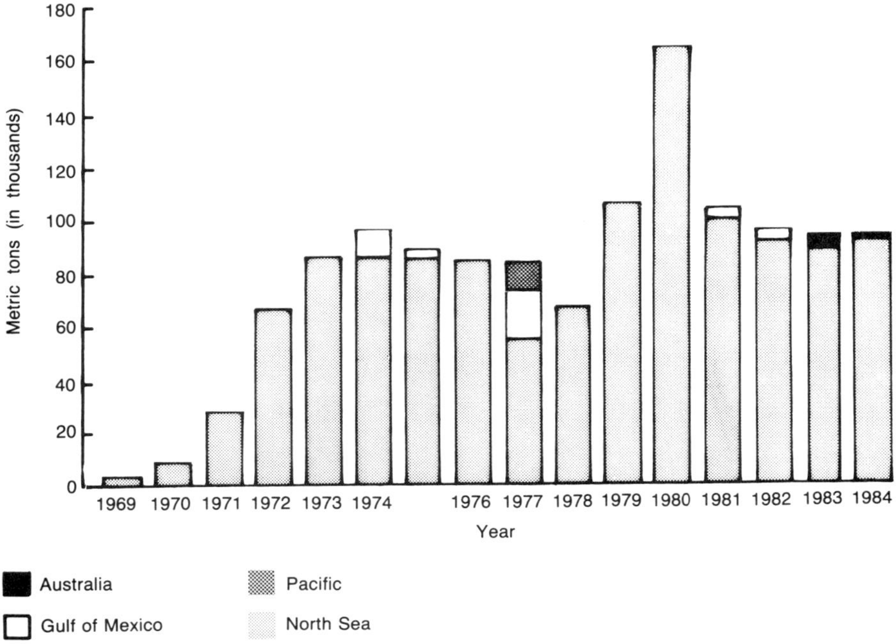

*This waste was generated in Australia and incinerated while the ship was en route to Singapore.

Fig. 8-3. Quantities of waste annually incinerated at sea, 1969-1984. Source: Reference 1.

- A combustion efficiency of 99.95 percent must be obtained.
- A destruction and removal efficiency of 99.99 percent must be maintained, except for polychlorinated biphenyls (PCBs), dioxins, or dibenzofurans, which must be incinerated with a destruction and removal efficiency of 99.9999 percent.
- Stack emissions must meet or exceed existing marine water quality standards.
- Incinerator residue (ash) must be carried back to land for appropriate land disposal (or re-use, if that becomes an option).

Table 8-4 lists proposed ocean incinerator performance standards compared to performance standards for land-based incineration systems.

As part of a permit application, waste analyses are required prior to a trial burn and prior to each incineration ship voyage. Emissions during incineration must be monitored for surviving organic compounds, incomplete products of combustion, and concentrations of oxygen, carbon monoxide, and carbon dioxide. The regulations will require a USEPA employee to sail on each incineration ship voyage and will empower this representative to stop the flow of wastes to the incinerator if the operator of the

Table 8-4. Performance Standards Applicable to Land-based and Ocean Incineration.

Performance Standard	Land-based Incineration	Ocean Incineration
Combustion Efficiency (CE)	99.9% for PCB's (TSCA) No CE specified for any other wastes (RCRA)	$99.95 \pm .05\%$, all waste
Destruction and Removal Efficiency (DRE)	99.99% DRE except 99.9999% for PCB's, dioxins, furans	99.99% DRE exept 99.9999% for PCB's, dioxins, furans
Hydrogen chloride (HCl)	If >4 lb/hr control to larger of either ≤ 4 lb/hr or 1% of total HCl	After initial mixing change in sea-water alkalinity must be $\leq 10\%$
Particulate emissions	≤ 0.08 gr/DSCF when corrected to 50% excess air	Metal Emissions less than amount exceeding marine water quality

vessel violates permit requirements or if there are unreasonable risks to human health or to the environment.

The USEPA will issue research permits for up to 6 months under the proposed regulations for research on new incineration techniques and their environmental effects. Operating permits will last up to 10 years, with a 5 year renewal period, and require a trial burn, a 45-day public comment period for data from the trial burn, and certification by the operator that the incinerator meets performance standards. Under the proposed regulations, the USEPA could issue an emergency permit in a situation requiring a marked degree of dispatch to protect the public health and when ocean incineration presented the only feasible method of waste disposal.

There are a number of statutory restraints on the incineration of wastes at sea in addition to those promulgated by the EPA. Table 8-5 lists governmental agencies with regulatory interest in ocean incineration and their areas of activity.

SAFETY SYSTEMS

Concern that the discharge of hazardous wastes into vulnerable waterways can cause catastrophic damage to marine life and can have a deleterious effect on shore communities has been addressed by unique features of incinerator ship design and operation. Included in these features are the following:

Table 8-5. Federal Regulatory Framework for At-Sea Incineration.

Statue/Regulation	Activity	Agency
Resource Conservation & Recovery Act	USEPA	Waste Storage, Waste Content Land Transportation, Residuals Disposal
Marine Protection, Research & tation Act	USEPA, Coast Guard	Ocean Incineration PCB Incineration
Toxic Substances Control Act Coastal Zone Management Act	USEPA States	Activities affecting land or water use in the coastal zone
Hazardous Materials Transpor- Sanctuaries Act	Department of Transportation	Hazardous waste transportation by truck or rail
Port & Tanker Safety Act	Coast Guard Coast Guard	Transportation by water Design, construction, certification, operation of incineration vessels
Port & Waterways Safety Act	Coast Guard	Vessel movement through ports, waste storage, transfer at waterfront
Endangered Species Act	USEPA	Compatibility of designated sites with protection of wildlife
Clean Water Act	USEPA Coast Guard	Cleanup of spills in territorial waters
Comprehensive Environmental Response, Compensation and Liability Act (Super Fund)	USEPA Coast Guard	National Contingency Plan, Cleanup of spills

- Rotary cup burners are proposed for use on most incinerator ships to be operated out of U.S. ports. They can atomize waste within a wide range of viscosity and are not affected by viscosity changes during operation. Also, changes in waste feed pressure are not critical. System clogging is minimal, and there is not a significant problem with abrasion and wear, as is found with nozzle-type burners. Because the feed lines are at low pressure when compared to conventional nozzle-type burners, the threat of leaks and potential explosions is significantly reduced.
- The incinerator combustion system has a number of automatic shut-off devices. If the temperature falls below the minimum permitted level, waste feed is automatically shut off. Shut-off devices are also activated if the carbon monoxide level exceeds 100 ppm or if excess oxygen falls below 5 percent. If there is a flame-out, waste feed to that burner is automatically shut off.

- On incinerator ships where waste is pumped to integral shipboard tanks, compatibility tests are run. Before waste is transferred into the shipboard storage tank, a sample of that waste is mixed with a sample of the contents of the tank in appropriate proportions to determine if any chemical or physical reaction takes place. In addition, on the *Vulcanus II* for example, a computer model is used, which allows entering the critical parameters of the waste streams received and then examines the resulting mixture if it were to be added to the several tanks available. The computer model will automatically announce if any of the resulting critical parameters fall out of specification. This allows a determination as to which storage tank this waste stream can best be added.
- Incineration vessels are unique in their construction. They must have a Type 2 classification, which signifies chemical-tanker construction, intended for product transport with appreciably significant measures taken to prevent the escape of such cargo. The International Maritime Organization (IMO) Code requires incineration vessels to meet the Type 2 requirement even though many of the waste materials that are accepted for ocean incineration may otherwise only require a Type 3 classification, which is less restrictive.
- Incineration vessels with integral waste-storage tanks must have double-hull construction, with bulkheads defining independent tank sections. A typical ship contains eight integral tanks, with 25 to 30 secure void spaces to provide flotation when required. According to the U.S. Coast Guard, in a reasonably-foreseeable worst-case situation, such as a hard grounding on a rocky bottom or a broad side collision, a maximum of two cargo tanks could be ruptured and, if tanks emptied totally, a maximum of one-quarter of the cargo would be lost.

Typical of the proposed incinerator ships is *Vulcanus II* (owned and operated by Chemical Waste Management), at approximately 3,300 gross tons. The typical Type 2 chemical tanker regularly engaged in operation in or near United States waters is approximately 30,000 gross tons, or roughly nine times the size, with nine times the capacity, of *Vulcanus II* (shown in Fig. 8-4). The Coast Guard estimates that there are approximately 10,000 movements involving chemical products in and out of Gulf ports each year. A large percentage of this volume is by barges not built to the high standards of chemical tankers. This traffic is estimated to involve 390,000,000 metric tons of cargo annually.

A comparison among the sizes of a proposed incineration ship, a typical oil tanker, and a common ferry boat is illustrated in Fig. 8-5.

Most of the chemical tankers trading through U.S. ports are foreign-flagged. As such, they are not inspected by the U.S. Coast Guard to the

Fig. 8-4. Vulcanus II. *Source: G. Nassos, Chemical Waste Management.*

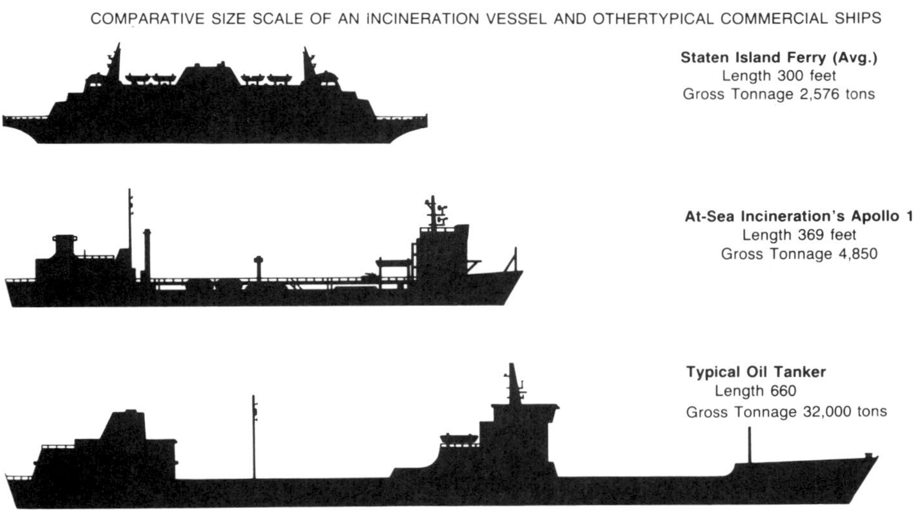

Fig. 8-5. Comparative size of an incineration vessel and other typical commercial ships. Source: At-Sea Incineration, Inc.

extent that an American-flagged ship is inspected. All of the incinerator ships proposed to date, however, are American-flagged, subject to the requirements and standards established by the U.S. Coast Guard and other U.S. government agencies.

REFERENCES

1. *Ocean Incineration: Its Role in Managing Hazardous Waste,* Congress of the United States, Office of Technology Assessment (Washington, D.C.: U.S. Government Printing Office, August 1986).

2. "Consideration of Report on Dumping," *Draft Report of Permits Issued in 1982,* International Marine Organization, Document LDC/SG.8/INF.3, December 1984.

3. *15 Years of Waste Incineration at Sea* (Rotterdam: Ocean Combustion Service, February 1985).

9

European Technology

Incineration technology developed earlier in Europe than in the United States. Current practice in Europe is different than American practice both administratively and technically. European technology and methods will be discussed in this chapter.

HISTORICAL FACTORS

In the United States 80 to 90 percent of hazardous wastes generated by industry are disposed of by land disposal or deep-well injection. This contrasts sharply with Europe. In Denmark, for instance, no untreated industrial wastes are dumped on the land, and in West Germany approximately 60 percent of the industrial wastes generated are detoxified instead of dumped. Table 9-1 lists waste generation in both western Europe and the United States.

Incineration technology applied to hazardous waste disposal derives from European experience in refuse disposal. Europe has taken the lead in the development and use of incineration technology in the treatment and disposal of industrial and municipal wastes for a number of reasons. Some of these reasons are political; others reflect the availability or lack of resources on the continent and in Great Britain. For instance:

- Land is relatively scarce. Incineration requires much less area than does land disposal or land application.

Table 9-1. Waste Generation in Western and Northern Europe.

Country	Hazardous Waste Generation Rate tons/year	Population
Austria	300,000	7,600,000
Denmark	100,000	5,100,000
Finland	71,000	4,800,000
France	380,000	55,000,000
Netherlands	1,000,000	15,000,000
Norway	120,000	4,100,000
Sweden	480,000	8,500,000
United States	200,000,000	225,000,000
West Germany	30,000,000	62,000,000

NOTE: The above figures for waste generation rate are based on projections for 1984. Population figures are 1984 estimates.

Source: Reference 2.

- Groundwater is the prime source of potable water on the European continent. Over 98 percent of drinking water in Denmark, for instance, comes from groundwater.
- In most of Europe there is a tendency towards centralized control of economic activity and decision-making. The siting of incineration facilities, for instance, does not undergo the same type of public scrutiny that occurs here in the United States, where these decisions are made on a local level.
- Utilities in Europe are owned and operated by the government, as are municipal-waste disposal facilities and many industrial-waste disposal facilities. This common ownership leads to greater cooperation between these sectors. In the United States, however, utilities are generally privately owned and operated. Often there is little incentive for the privately-owned power company to cooperate with a waste disposal facility in, for instance, the purchase of electric power generated by these facilities.
- The cost of fossil fuel is and has been greater in Europe than in the United States. The development of waste-to-energy plants has been encouraged by this cost factor. This higher cost of fuel has encouraged the design of larger facilities, where advantages of scale predominate.
- European governments help fund hazardous-waste facilities. In the United States, government participation in the financing of such facilities is politically untenable.
- Industry in the United States tends to be suspicious of government-run enterprise, particularly when such activity affects business con-

duct. Where government ownership or operation of hazardous waste facilities has been proposed in the United States, it has been rejected. In Europe, the government is considered a desirable partner and is encouraged to provide such services.

- European governments have the authority to require that all wastes be sent to a particular facility. This concept, referred to as *flow control*, is not readily available in the United States. Where municipalities have tried to legislate and enforce these laws they have been subject to litigation. Municipalities have usually lost. With flow control, waste quantity can be forecast with a fair degree of accuracy. This allows European planners to size and finance their facilities with reliability that is not possible in the United States.

While current practice in Europe tends to favor a more deliberate approach to waste disposal, a more effective waste-treatment policy is certainly not a conclusive reason for abandoning the American economic system or the priorities of American life. The various pressures within the American system, such as an emphasis on private, rather than public solutions to problems; decision-making on a local level; the availability of legal redress to every citizen; and the competitive spirit and philosophy, contribute to the dominance of the United States in world industrial production.

CURRENT PRACTICE

Most hazardous-waste incinerator facilities in the United States are single-user facilities, located on the site of the generator. There are a number of privately-owned contract incinerators licensed for hazardous waste, but they process a fraction of the hazardous wastes incinerated. In Europe, there is a tendency towards central disposal facilities, which are wholly- or partially-owned by the national or local government.

A typical hazardous-waste facility in Europe will receive wastes from a variety of industries. This typical facility comprises approximately 20 acres, additional land if a landfill or treatment lagoon is located on site. It affords secure pretreatment storage facilities for a variety of wastes, including sufficient storage space for approximately one year's waste receipts, and facilities and equipment for the thermal, chemical, and biological treatment of wastes. Energy is recovered by the thermal processing equipment (incinerator) in the form of steam and electric power generation. A secure waste landfill is available, either on- or off-site, for disposal of post-treatment and post-reclamation residues, or ash. Analytical laboratory facilities, safety, and maintenance facilities, as well as plant management, security, contingency, and data management facilities are on the premises.

The following European facilities will be described subsequently:

- Beibesheim, West Germany (HIM)
- Ebenhausen, West Germany (GSB)
- Herten, West Germany (AGR)
- Hombourg Center, France (TREDI)
- Leverkusen, West Germany (Bayer)
- Limay Center, France
- Mitry-Compans Center, France (GEREP)
- Norrtorp, Sweden (SAKAB)
- Nyborg, Denmark (Kommunekemi)
- Riihimaki, Finland
- Rotterdam, Netherlands (AVR)
- Saint Vulbas Center, France (TREDI)
- Sandouville, France (SERPE/SIDIBEX)
- Schwabach, West Germany (SMM/ZVSMM)
- Slemmestad, Norway (Norcem)
- Vienna, Austria (EBS)

BEIBESHEIM, WEST GERMANY

The Beibesheim facility, in Hesse, started operation in 1981. It is owned and operated by a non-profit industry/government partnership, Hessische Industriemüll Gmbh. (HIM). Two incinerator lines are located at this site. HIM operates chemical/physical plants and other disposal facilities at other locations.

The incineration facility consists of two rotary-kiln incinerators, each with an afterburner, waste-heat boiler, and a flue-gas treatment system. One of the rotary kilns is shown in Fig. 9-1. Residue from the incinerator

Fig. 9-1. Beibesheim rotary kiln. Source: J. McClure, Jr.

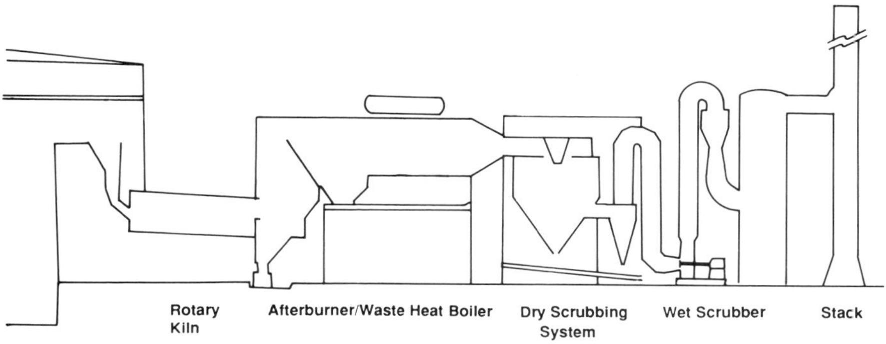

Fig. 9-2. Beibesheim incineration facility.

(ash) and the flue-gas treatment system (fly-ash and spent lime) are disposed of at an underground disposal facility, in an inactive salt mine. The total capacity of the facility is 60,000 tons per year.

The incineration lines can operate independently. As shown in Fig. 9-2, solid waste is deposited in a storage pit. A grapple charges the kiln hopper. Drums can be charged to the kiln from a dedicated drum elevator. Waste oil is fed to the kiln or to the afterburner through a separate burner; other liquid wastes, including aqueous wastes, are injected into the kiln or the afterburner through dedicated lances. A piston pump injects waste sludges into the kiln.

The rotary kilns are operated at temperatures in the range of 2000° F to 2400° F. At these temperatures, ash is retained in the kiln as slag; the slag overflow is continuous, discharging to a wet sump.

The temperature maintained in the afterburner is 1700° F. A secondary air supply is provided in the secondary combustion chamber to ensure complete combustion of the injected liquid wastes and the organics remaining in the gas stream, and to help control the afterburner temperature. A tertiary air supply reduces the temperature of the gases exiting the afterburner to between 1500° F and 1550° F to reduce the potential for deposition of accretions in the waste-heat boiler.

The waste-heat boiler is the convection type with a maximum gas velocity of 14 feet per second. This low velocity reduces the erosive effect of the dirty gases in the boiler tubes and encourages the fall-out of fly-ash from the gas flow. The temperature of the gas leaving the waste-heat boiler ranges from 430° F to 530° F.

Each of the two air emissions-control systems includes a dry scrubber and a wet scrubber, shown in Fig. 9-3. A lime slurry is injected into the dry scrubber through a quickly spinning atomizer disk. This neutralizes the acidic component of the flue gas and quenches the flue gas to

Fig. 9-3. Beibesheim acid-gas scrubbing system. Source: J. McClure, Jr.

a temperature of 320° F to 360° F. A cyclonic reactor, located immediately downstream of the dry scrubber, collects the larger particulate. The scrubber discharge and the cyclone discharge, as shown in Fig. 9-2, are tied together for collection and disposal. Hydrogen chloride, hydrogen fluoride, and sulfur oxides in the gas stream neutralize to their respective calcium salts within the dry scrubber; these salts predominate the dry scrubber and cyclone discharges.

Figure 9-4 is a schematic diagram of flow through the scrubbing system. The dry scrubber and cyclone hoppers are equipped with heaters to prevent condensation, which can be corrosive, or will lead to the unwanted caking of salts or lime. From the cyclone, the gas stream passes through a venturi scrubber, where most of the particulate matter, including the lime injected in the dry scrubber, is wetted. The subsequent wet scrubber, a tray tower, removes the majority of particulate from the gas stream, and encourages the separation of the entrained particulate and salt-laden mist from the gas.

All of the water injected into the gas stream is controlled to maintain a pre-selected pH. Both lime (CaO) and caustic soda (NaOH) are used as neutralizing agents in an effort to increase the removal efficiency of all anticipated acid components from the gas stream. The cleaned gas exiting the tower is discharged to the atmosphere at approximately 160° F, through an induced-draft fan.

The emissions standards imposed on the Beibesheim incinerators by the Hesse authorities are comprehensive, including limitations for gaseous constituents as well as metals. These standards are listed in Table 9-2.

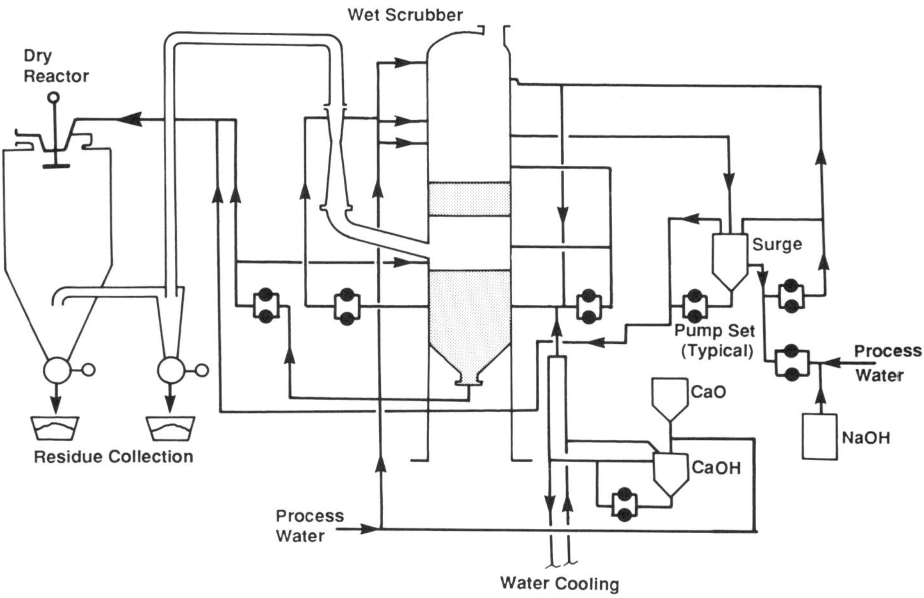

Fig. 9-4. Beibesheim gas-scrubbing system.

EBENHAUSEN, WEST GERMANY

This facility is the largest in West Germany. It is owned and operated by GSB, a non-profit waste management company set up in 1976 in Bavaria. GSB owners include the State of Bavaria, over 75 industrial generators, and three local-government associations. GSB operates 10 regional stations that feed Ebenhausen, collecting wastes from over 10,000 generators.

Ebenhausen consists of two rotary-kiln incinerators, a chemical/physical treatment plant, and a wastewater treatment plant. GSB also operates a secure landfill approximately 20 miles from Ebenhausen for the disposal of incinerator ash and other residues from the plant.

The two incinerators burn solid, liquid, and semi-solid hazardous industrial wastes. They are two similar rotary-kilns discharging to a common afterburner, waste-heat boiler and air emissions-control system. Waste liquids can be fired in either the kilns or the afterburner.

Each kiln is 12 feet in diameter and 40 feet long. The incinerator line is designed for a capacity of approximately 72,000 tons per year waste. The kiln temperature is maintained at approximately 2200° F and the afterburner temperature is 1800° F. The residence time in the kiln at 2200° F under design loading is 3 seconds. This combination of operating parameters allows the facility to incinerate PCBs. The scrubbing system can meet acid gas-emissions standards if the organic chlorine in the feed is limited to 10 percent by weight.

Table 9-2. Beibesheim, West Germany, Emissions Standards.

Contaminant	Limitation, mg/NCM
Class I substances	0.2 total
Cd and Cd compounds	
Hg and Hg compounds	
Ti and Ti compounds	
Class II substances	1.0 total
As and As compounds	
Co and Co compounds	
Cr + 6 and Cr + 6 compounds	
Ni and Ni compounds	
Se and Se compounds	
Te and Te compounds	
Class III substances	5.0 total
Cu and Cu compounds	
Cyanide compounds	
Fluorine compounds	
Mn and Mn compounds	
Pb and Pb compounds	
Sb and Sb compounds	
V and V compounds	
Unclassified substances	0.1 total
Benzo(a)pyrene	
Dibenzo(a,b)anthracene	
Be and Be compounds as Be	
Gaseous emissions	
CO	100
Total organic carbon	50
Cl compounds	100
F compounds	5

Note: Emissions related to 50% excess air.

The fields around Ebenhausen were recently sampled for heavy metals. Relatively high levels of mercury, cadmium, and lead were found. A new scrubber will be installed shortly to improve removal of these and other heavy metals.

HERTEN, WEST GERMANY

The AGR hazardous-waste disposal facility is located in Herten, in Westphalia. It is owned and operated by a governmental authority.

This incinerator system is a single rotary kiln with an afterburner, waste-heat recovery boiler, and a flue-gas cleaning system. It was built in 1982, with a waste capacity of approximately 30,000 tons per year. The incinerator burns solids, liquids, and sludges, including PCBs and hospital (pathogenic) wastes. Steam is sold for district heating in the winter; when there is no steam market, electric power is generated and sold.

Scrubber water is collected in evaporation ponds. The plant has been denied required permits for discharge into the adjacent Emscher river.

HOMBOURG CENTER, FRANCE

The private firm TREDI has been operating this hazardous-waste treatment facility since 1974. Beside an incinerator, the facility includes a chemical/physical treatment plant, sludge stabilization equipment and an ion-exchange resin-regeneration facility.

The incinerator has a capacity of approximately 3,000 tons per year. It is a liquid-injection furnace that was designed for the incineration of organic liquids, solvents, and pumpable organic sludges.

LEVERKUSEN, WEST GERMANY

The Bayer Chemical Works in Leverkusen is one facility within the Bayer organization, which includes dozens of plants manufacturing thousands of diversified products. This facility generates over 550,000 tons of solids waste, sludges, slurries, and liquid wastes per year. Approximately 110,000 tons are incinerated; the balance goes to landfills.

There are two incinerator plants at this facility, both rotary-kiln incinerator installations. Plant Number 1, the oldest of the two, has no provisions for acid-gas scrubbing. It handles approximately 50,000 tons per year of non-chlorinated, non-sulfonated waste streams. Incinerator Plant Number 2 does include acid-gas scrubbing and accepts approximately 30,000 tons per year of wastes that cannot be handled by Plant Number 1. Through 1985, over 20,000 tons per year of chlorinated wastes (such as PCB contaminated liquids) had been shipped from the plant to one of two incinerator ships operating in the North Sea. Public opposition has placed the future of at-sea incineration off the German coast in question.

Plant Number 2 is shown schematically in Fig. 9-5. Solid waste is trucked to the enclosed waste-disposal building and is dumped into a large storage pit. Waste is dropped, by grapple, into a charging hopper with two control gates. The horizontal, sliding gate near the top of the charging hopper opens when the inclined gate at the bottom of the hopper closes.

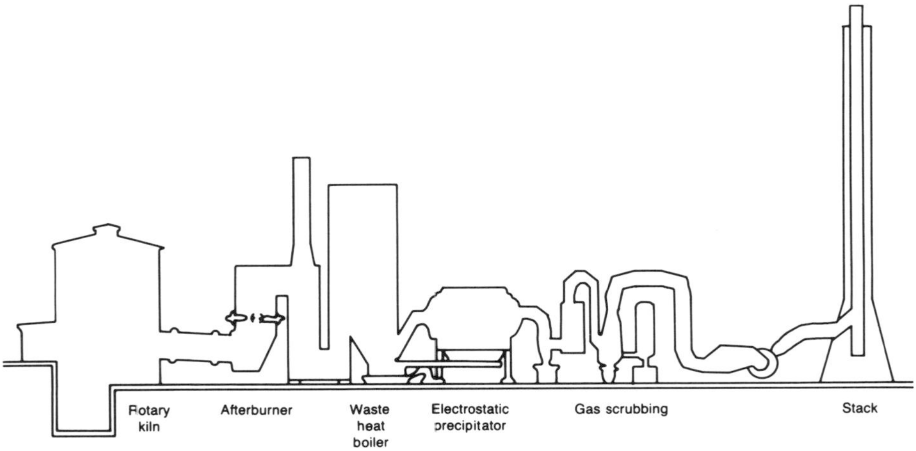

Fig. 9-5. Bayer incinerator line number 2.

The inclined gate is located at the drop chute feeding the rotary kiln. By alternately opening and closing these gates, a seal is maintained, preventing air flow into the kiln from the charging system. This seal also tends to prevent *flashback*, or fires, in the charging system.

Liquid wastes are stored in a tank farm outside the incinerator building. They are either pumped to the tank farm, injected to the tanks with pressurized nitrogen, or are transported to the tank farm by truck. One burner is located on the face of the rotary kiln, and three within the secondary combustion chamber, for firing of liquid wastes. Both steam and compressed air are available for atomization, depending on the nature of the waste material.

Small containers, filled with sludge, solid, or other wastes, can be dropped into the kiln. An elevator, roller conveyor, air-lock system, and tilting table are provided for containerized materials.

The rotary kiln is 11 feet, 6 inches in diameter and approximately 45 feet in length. It is lined with 10 inches of refractory. The kiln speed is infinitely variable from 7 to 24 revolutions per hour and is normally run at approximately 9 revolutions per hour. The kiln is positioned at a rake of approximately 3 percent (16 inches drop in 45 foot length).

Wastes are held in the kiln for one to two hours. The kiln temperature is normally maintained between 2,200° F to 2400° F. The kiln operates in the slagging mode, discharging molten ash into a water bath for cooling, collection, and disposal. The heat release rate within the kiln is approximately 60 million Btu/hr.

Exhaust gas leaves the kiln and enters a vertical afterburner that is designed for a gas residence time of 2 to 4 seconds at 2400° F. Burners for liquid waste, and a burner used for firing, as required, supplemental natural gas, is located in the initial leg of the afterburner. The downcomer leg is not fired, and serves to provide the required residence time. The afterburner supplemental-fuel burner fires when the chamber temperature falls below 2000° F. If the temperature in the secondary combustion chamber drops below 1650° F, feed to the kiln and afterburner will automatically stop. Provisions for ash removal are provided in the design of the secondary-combustion chamber downcomer leg.

A waste-heat recovery boiler at the afterburner exit is sized to generate approximately 44,000 lb/hr of saturated steam at 550 psig and 660° F under full system load. The waste heat boiler includes a water-wall chamber, followed by convection sections, a superheater, and economizer sections.

From the waste-heat boiler, where the temperature drops to approximately 570° F, the flue gases pass through an electrostatic precipitator, where the majority of entrained particulate is removed. The gas then passes through a series of wet scrubbers, for further cooling and acid gas removal. Caustic soda is used as the neutralizing agent. The wet scrubbing system is also used to separate any residual particulate from the gas stream; to remove any volatile metals which might be present; and to catch any acid mist which might be generated from the incineration process. The scrubber exhaust is discharged at a temperature of 140° F to 160° F to a flue containing a series of heat exchangers which re-heat the exhaust to minimize or eliminate the formation of a plume from the exhaust stack.

LIMAY CENTER, FRANCE

The Limay hazardous-waste disposal facility has been operated by a private firm, SARP Industries, since 1975. Of the 150,000 tons per year waste that it receives, approximately 15,000 tons per year are incinerated. The center includes a sludge-stabilization facility, a chemical/physical treatment plant, and a copper-recovery facility.

The incinerator at Limay is a liquid injection furnace designed for the destruction of hydrocarbons, solvents, and other non-corrosive liquid organic wastes.

MITRY-COMPANS CENTER, FRANCE

This hazardous-waste disposal facility is operated by GEREP, a private firm. It has been in operation since 1977. It includes a chemical/physical treatment plant as well as a rotary-kiln incinerator.

The incinerator has a capacity of 18,000 tons per year. It is designed specifically for the incineration of chlorinated or sulfonated liquid organics and for the destruction of unique solid wastes, such as powdered plastics.

NORRTORP, SWEDEN

The SAKAB industrial/hazardous-waste treatment facility is located in Norrtorp. It consists of an incinerator facility, an oil-recovery plant, a chemical/physical treatment plant, a mercury-recovery facility and a secure landfill. The design capacity of the entire facility is 60,000 tons per year. SAKAB is government owned and operated. It began operation in 1984.

The facility includes a single incinerator line, a rotary-kiln followed by an afterburner, a waste-heat boiler and an exhaust-gas cleaning system. The incinerator has a design capacity of 33,000 tons per year of liquid, solid, and semi-solid organic wastes. Wastes contaminated with PCBs are accepted for incineration.

The incinerator has an external diameter of 15 feet and a length of 40 feet. The kiln temperature is maintained at 1,800° F to 2,400° F. The afterburner is kept at approximately 1,800° F.

The air-emissions-control system includes a dry scrubber, which utilizes lime slurry injection to quench and neutralize the exhaust gas stream. The emissions standards for the stack discharge are 35 mg/NCM for particulate matter and HCl, and 5 mg/NCM for HF.

NYBORG, DENMARK

In 1975, a consortium of Danish municipalities built Kommunekemi, a comprehensive waste-management facility, near Nyborg. These municipalities also operate the facility, which is designed to treat 130,000 tons of industrial/hazardous wastes per year. Kommunekemi consists of two incinerator plants and a chemical/physical treatment plant, with a secure landfill located approximately 15 miles from the facility, and all required support services.

Wastes are sent to Kommunekemi from a network of 21 transfer stations located throughout Denmark. They receive hazardous wastes from industries and from over 300 drop-off or collection stations for households and other small generators. Wastes are shipped by rail to the plant from the transfer stations; the rail cars are owned by Kommunekemi.

The two incinerator plants are designated as FI and FIII. The FI plant, shown schematically in Fig. 9-6, was part of the original installation. It accepts solid and liquid wastes with a chlorine or sulfur content of less than 1 percent by weight. This installation is not equipped with acid-gas scrubbing and the feed must be controlled to ensure that the acid-gas emission limits are not exceeded. For HCl, the maximum allowable stack discharge is 300 mg/NCM; for SO_2 this figure is 750 mg/NCM. Other emissions limits are for hydrofluoric acid and lead; both 5 mg/NCM; arsenic, 0.1 mg/NCM; and particulate matter, 100 mg/NCM. These figures are related to 50 percent excess air.

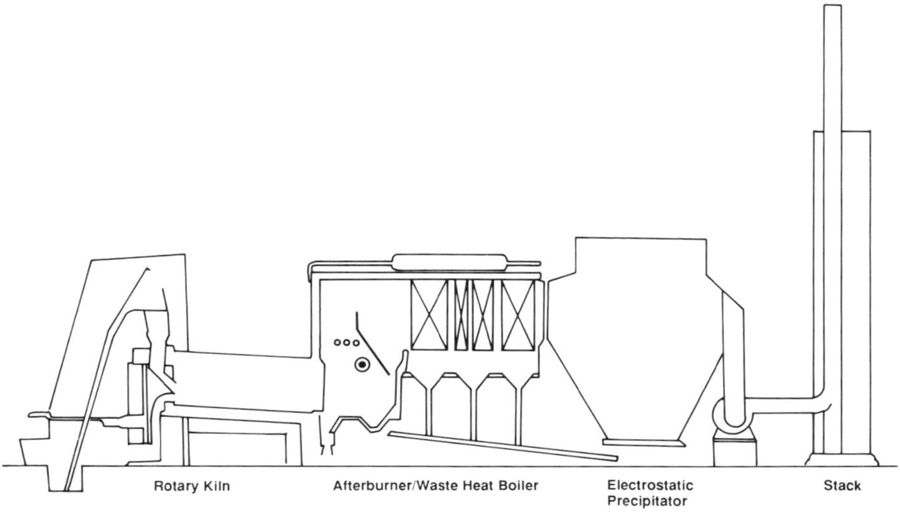

Fig. 9-6. Kommunekemi incinerator line FI.

Solid waste is received and charged to the incinerator without preprocessing. Drums can be loaded directly into the kiln. The kiln exhaust gases discharge to an afterburner (secondary-combustion chamber). Waste-feed lances are located in both the kiln and in the afterburner for the injection of aqueous and other liquid wastes. Liquid wastes are blended in mixing tanks to provide uniformity of feed, viscosity, and heating value. Sludges can be fed to the kiln through feed nozzles at the kiln face.

Air is injected in both the kiln and the afterburner to control temperatures and to ensure adequate waste combustion. From the afterburner, hot gases enter a waste-heat recovery boiler which generates steam for in-plant use and for sale to the City of Nyborg. Approximately 20 percent of the generated steam is used internally.

The FI system utilizes an ESP for particulate control. The discharge from the ESP is collected and combined with the fly-ash discharge from the afterburner and waste-heat boiler hoppers for removal by truck to the Kommunekemi landfill.

The FIII plant, shown in Fig. 9-7, is a newer line, similar to FI, but with acid-gas control on the exhaust. As shown schematically in Fig. 9-8, exhaust gases exiting the waste-heat boiler pass through a dry-scrubbing system. Lime slurry is atomized within this chamber by a high-speed rotary atomizer. The acid-gas component of the exhaust stream is neutralized by the lime and both the spent lime and particulate matter are captured in an electrostatic precipitator located immediately downstream of the dry scrubber.

The ESP discharge, which includes calcium chloride ($CaCl_2$) and calcium hydrogen sulfite ($CaHSO_3$) from acid-gas neutralization, lime, and particulate matter, is collected and disposed of with other particulate in the dedicated Kommunekemi landfill. The water injected with the slurry into the furnace evaporates, reducing the temperature of the gas stream to below 190° F.

Except for the exhaust-gas cleaning system, both FI and FIII are similar. They each utilize a rotary kiln 14 feet in outside diameter and 40 feet in length. The kiln/afterburner can incinerate 5.5 tons per hour of solids and sludges plus 2.2 tons per hour of contaminated water, which is equivalent to a total heat release of 75 million Btu per hour.

Fig. 9-7. Kommunekemi line FIII. Source: J. McClure, Jr.

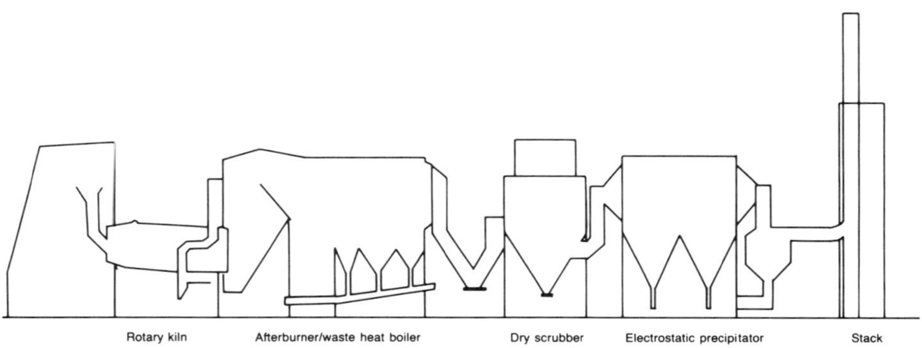

Fig. 9-8. Kommunekemi incinerator line FIII.

The kilns are designed to operate in the slagging mode, at a kiln temperature of 2600° F. At this temperature the steel drums, salts, and many of the other inorganics present collect in a molten phase at the bottom of the kiln. This slag is discharged continuously from the kiln into a wet pit, collected, and trucked to the dedicated landfill.

The afterburner is maintained at a temperature less than 1800° F to reduce accretion on the boiler tubes. The organic destruction at this temperature has been found to be in excess of 99.90 percent.

At full load, each kiln system can generate approximately 50,000 lb/hr of dry or saturated steam at 175 psig. Gases exit the waste-heat boiler section at a temperature of approximately 500° F.

RIIHIMAKI, FINLAND

Construction of a central industrial/hazardous waste treatment and disposal facility was completed in 1985. It includes an incinerator, a chemical/physical treatment plant, wastewater treatment, and a secure landfill.

The incineration system system utilizes a single rotary kiln with an afterburner, followed by a waste-heat boiler. The exhaust-gas cleaning system includes a dry scrubber fed by a lime-slurry system followed by a fabric filter (baghouse). It is designed to remove acid gases from the exhaust. Solids, liquids, and sludges can be incinerated. Those with a significant chloride fraction are not excluded.

Solids and sludges are mixed before being fed to the kiln. A nitrogen atmosphere is maintained above the mixing area to help prevent explosions. The spent nitrogen, along with other vapors generated in the preparation area, are fed to the kiln.

This unit has a design capacity of 35,000 tons per year. The waste-heat boiler generates steam for in-plant use with the excess sold to the City of Riihimaki for district heating.

Stack emissions limits are 35 mg/NCM for particulate and HCl and 5 mg/NCM for HF, at 50 percent excess air.

ROTTERDAM, THE NETHERLANDS

A hazardous-waste disposal facility, AVR, was built by a consortium of private and public interests. The national government owns 10 percent of AVR, the City of Rotterdam owns 45 percent and the remaining 45 percent of facility ownership is shared by eight large industrial firms.

There is a rotary-kiln incinerator at this facility and a unique moving-grid furnace. They are designed to incinerate approximately 100,000 tons per year of non-chlorinated organic wastes. A new rotary kiln, which will be capable of incinerating up to 40,000 tons per year of chlorinated organic wastes, is under construction.

SAINT VULBAS CENTER, FRANCE

This is an integrated hazardous-waste disposal facility that has been operated by the private firm TREDI since 1975. The incinerator is a rotary kiln which is followed by an afterburner. It includes a wet scrubber in its air emissions control train. This system is capable of incinerating solid and semisolid wastes, such as contaminated packing material, chlorinated plastics, and paint wastes. The incinerator has a design capacity of 20,000 tons per year.

SANDOUVILLE, FRANCE

Sandouville, near LeHavre, is the location of a regional hazardous-waste disposal facility that was designed and built by the consortium SERPE (the Society of Regional Study for the Protection of the Environment). SERPE was formed in the mid-1970's by 20 firms, including Ashland Oil, Esso Chemical, Esso Standard, Firestone, Goodyear, Mobil, and Shell. SIDIBEX, a division of Compagne Generale Des Eaux, is operator of the facility.

The facility can accept 72,000 tons per year waste, of which 55,000 tons per year are incinerated. The wastes entering the plant are divided into eleven categories for management and pricing, as indicated in Table 9-3. Seven of these categories are organic materials, which are to be incinerated. The other four are sent to the chemical/physical plant for treatment.

The incinerator system consists of a rotary kiln followed by an afterburner, a waste heat boiler, and an electrostatic precipitator (ESP) for

Table 9-3. Sandouville Waste Categories.

Incineration Categories
1. Barrels of soldified wastes requiring manual handling; compressed blocks of wastes requiring cutting
2. Barrels of sludge or other semi-liquid wastes
3. Loose solids, unusual sludges
4. Bulk solids, sludges
5. Liquids in barrels
6. Bulk low aqueous liquids
7. Bulk organic liquids
Chemical/Physical Treatment
8. Acids
9. Neutral wastes or weak bases
10. High aqueous wastes
11. Strongly alkaline wastes

air emissions control. The kiln dimensions are 13 feet outside diameter by 40 feet long. It is designed for a capacity of 6,600 lb/hr organic waste. The afterburner is rectangular, 16 feet long × 43 feet high × 13 feet deep. It is lined with 18 inches of refractory. Liquid waste can be injected into the afterburner at a rate of 3,100 lb/hr (from 450 to 550 gallons per hour).

Gas exits the kiln at 1650° F; the afterburner at approximately 1750° F. Steam is generated at 590 psig and 610° F, at a rate of 55,000 lb/hr when the incinerator is operating at full load. The gas temperature at the boiler exit averages approximately 530° F.

SCHWABACH, WEST GERMANY

This facility, the Sondermullplatze Mittelfranken (SMM) hazardous-waste disposal facility, owned by the ZVSMM, a governmental organization, is the oldest operating hazardous-waste disposal facility in the world. It went on line in 1967, with a secure landfill. In 1968 an oil/water separation plant was built, followed by a physical/chemical separation plant and an incinerator system in 1970. The plant is located in Bavaria.

The incinerator is a rotary kiln followed by an afterburner, an electrostatic precipitator (ESP) and a wet scrubber. It does not incinerate PCBs or other organics with a significant chloride content. The present system is not equipped for acid-gas removal and has no provisions for waste-heat recovery.

Incinerator design capacity is 20,000 tons per year. Its external diameter is 8 feet and its length is 28 feet. The kiln temperature is maintained at approximately 1650° F and the temperature of the afterburner is normally 1560° F.

A new scrubbing system is being installed to allow the incinerator to accept wastes with up to 10 percent chlorine content. The new scrubber is designed to comply with new regulations which include heavy-metal restrictions as well as gas-phase restrictions. These standards are listed in Table 9-4.

SMM is responsible for its own environmental control monitoring. SMM laboratory and technical personnel test all stack emissions and plant effluent, and monitor groundwater on a regular basis. Livestock on adjacent pasture lands is also monitored periodically by plant personnel.

SLEMMESTAD, NORWAY

The Norcem Cement Company operated a cement kiln at Slemmestad through 1984, when economic conditions forced the plant to shut down. From 1981 through the time that the plant was closed, a demonstration program was implemented to incinerate hazardous wastes. Pumpable hazardous wastes were injected into the kiln. Toxic organic materials that

Table 9-4. Emissions Standards at Schwabach, West Germany.

Contaminant	Standard, mg/NCM
HCl	50
HF	2
Particulate	50
Sulfur Dioxide	200 as SO_2
Nitrogen Oxides	500 as NO_2
Organic Carbon	20
CO	100
Hg, As, Cd	0.02 in sum
Cr, Ni, Th	2.0 in sum
Vd, Pb, Cu, other metals	5.0 in sum

Note: The standard is corrected to 50% excess air.

were not pumpable were mixed with water to form a slurry and were then fed into the cement kiln.

The destruction of organics within the kiln was found to be at least equal to the level of destruction that was found in conventional kiln incineration processes. The cement kiln was 16 feet in outside diameter and 560 feet long. The temperature in the kiln is maintained at 2550° F, as required for the production of cement. Residence time is measured in minutes, rather than seconds as in more conventional waste-incineration systems. An additional feature of the cement process kiln is the neutralizing effect of the cement clinker which makes acid gas scrubbing unnecessary.

A full-scale hazardous-waste feed system is being built for another Norcem cement kiln based on the success of the previous operations.

VIENNA, AUSTRIA

A private firm, Entsorgungsbeitriebe Simmering Gmbh. (EBS) built an industrial/hazardous-waste treatment facility, including an incinerator, outside Vienna in 1980. In 1982 the City of Vienna obtained a controlling interest in the plant.

EBS consists of a hazardous-waste incinerator facility, sewage-sludge incinerators, a chemical/physical treatment plant, and a wastewater treatment facility. The hazardous-waste incineration plant has a capacity of 100,000 tons per year of organic solids, liquids, and drums. It includes two rotary-kiln incinerators (Fig. 9-9 shows one of these kilns), each followed by an afterburner, a waste-heat recovery system, and an air emissions control system. Drums of waste can be accepted by the kilns.

The two sewage-sludge incinerators are fluid-bed furnaces with a combined capacity of 740 tons per day wet sludge feed.

Fig. 9-9. EBS rotary-kiln installation. Source: J. McClure, Jr.

REFERENCES

1. J. McClure, Jr., "Western Europe Hazardous Waste Management Systems," *Engineering Bulletin,* Number 55, March 1981.

2. B. Piasecki and G. Davis, "A Grand Tour of Europe's Hazardous Waste Facilities," *Technology Review,*, July 1984.

10

Materials Handling

Major functions of an incineration facility include receiving, handling, storing, blending, and feeding waste. Materials-handling systems are specific to waste type (solid, liquid, sludge, etc.), and to the mode of generation of that stream.

INITIAL PROCEDURES

When an incinerator facility is designed and permitted for wastes generated on-site, the types of wastes and waste streams brought to it are usually known. RCRA regulations require that the incinerator feed be maintained within the envelope defined by permit conditions; that envelope of waste quality and feed rate is generally narrow. Contract waste disposers and other installations where waste is not generated on site are faced with a different set of circumstances. The waste stream is unknown and the permitted operating envelope is much broader than for on-site disposal.

Wastes must be evaluated prior to acceptance at a central or contract disposal facility. That facility may have restrictions on the types of wastes that it will accept, such as restrictions against dioxin-containing materials, PCBs, no inorganics in excess of 5 percent, etc. Another consideration is disposal cost, particularly for contract disposal facilities. Waste must be evaluated to establish a disposal cost before a contract for waste disposal can be executed.

The facility illustrated in Fig. 10-1 is an industrial waste incineration facility in Cincinnati which primarily contracts for liquids and sludge disposal. It has an on-site laboratory.

Fig. 10-1. Waste receipt at incineration facility. Copyright © C. R. Brunner, 1979.

Before a truck is brought to the facility, a waste sample is shipped by the generator. The sample is evaluated for unacceptable materials (radioactivity, PCBs, excessive inorganics, etc.), and if found acceptable, additional analyses are performed to establish the disposal cost. This cost considers the waste heating value, the neutralization required (if any) and the ash residual of the stream. As part of the waste-sample evaluation process an infra-red scan is made, which essentially identifies the waste's "fingerprint".

Based on this laboratory work, a contract is established for disposal of the waste. Only then is the waste allowed into the facility. Before the truck is unloaded, a sample of the waste is extracted and an infrared scan is taken. If this scan matches that of the initial pre-shipped sample for which the disposal contract was drawn, the waste is unloaded from the truck. Truck turn-around is an important factor in the operation of the facility and, in Cincinnati, the truck can be sampled and unloaded within four hours.

THE LIABILITY ISSUE

Under RCRA regulations, the generator of waste is responsible for the ultimate disposition of that waste. Through a manifest procedure establishing a chain-of-custody, the disposition of a hazardous waste is always known. This procedure was instituted to protect the public safety by identifying the waste flow path. It does not relieve the generator of his responsibility for proper disposal of the waste.

Until the waste is certified destroyed, the generator is responsible for any damages as a result of any of the following procedures, as applicable:

- generation and initial storage;
- interim storage;
- storage in transport vehicle (or drum);
- transportation;
- transportation to final storage;
- final storage;
- waste preparation (blending, shredding, etc.); and
- transportation to incinerator.

In the contract disposal of waste, where more than one generator is disposing of his waste, waste mixing or intermingling can be a potential problem. A contract disposer may, after determining that there are no compatibility problems, place incoming waste into a tank with waste from other generators. This tank can be used for incinerator feeding and will be emptied only for cleaning and service, perhaps once or twice per year. In this situation, a particular generator cannot know when his waste was destroyed. Waste will always be present in the tank, and the waste cannot be identified as generated from a specific source.

As long as there is any waste in a mixed tank, any of the generators contributing waste to that tank are subject to liability under a claim. Claims would likely be made against all generators if, for instance, the tank caught fire and caused damage; the tank leaked and created a soil contamination problem; etc.

BULK LIQUIDS UNLOADING

Typically, liquids and other pumpable materials are delivered to an incineration facility in bulk by tank trucks or rail tank cars. For on-site disposal, liquid wastes are usually pumped to the facility. Drums are normally discouraged as a method of liquids transfer because drums, containing 40 to 55 gallons, require significantly more handling than tankers, which are usually in excess of 1000 gallon capacity. Each handling operation creates a potential safety and spillage problem.

Liquids from bulk systems are normally transferred through a piping system by pumps, gravity flow, or compressed gas systems. Pumping systems, using positive displacement or centrifugal pumps, are most commonly used for moving liquid wastes. Positive displacement pumps are generally preferred over centrifugal pumps because their construction prevents siphoning when not in use.

Gravity unloading may be preferred when liquids have a relatively high vapor pressure. High vapor pressure can result in vapor-lock to conventional pumps, rendering them non-functional and creating a potential safety hazard. Unlike pumps, which only pressurize the system during demand, gravity unloading systems operate at constant pressure, which is maintained even when transfer is not occurring. In addition, quick automatic or manual shut-off is more difficult with gravity systems than with pumped systems.

Compressed-gas unloading normally utilizes an inert gas, such as nitrogen or carbon dioxide. As with gravity systems, the system is always under pressure, and flow shut-off is more difficult than with pumped systems. Storage tanks must be built to withstand the pressure of the compressed gas. The compressed-gas unloading system, however, is very effective with high-vapor-pressure liquid wastes.

Figure 10-2 illustrates a typical bulk-liquid unloading facility. A connection is made between the discharge of the truck or railcar unloading pump and the flange shown center left. Facility pumps, which are fixed-displacement type to avoid potential problems with low-vapor-pressure liquids, discharge liquids through a meter system and strainers into storage tanks within the facility.

CONTAINER UNLOADING

Containers, such as steel or fiber drums, barrels, and special bulk units loaded on rail box cars or semi-trailer trucks, are used to transport

Fig. 10-2. Bulk-liquid unloading facility. Copyright © C. R. Brunner, 1981.

waste to an incineration facility. The containers can be unloaded from the trailers and rail cars manually with special drum-handling equipment. Forklift trucks are normally used to unload drums on pallets.

Once a container has been unloaded it can be placed in storage, and the contents can be transferred to other storage. For instance, liquids can be dumped to a stationary storage tank and bulk solids can be dumped into a storage pit or hopper.

Figure 10-3 shows a drummed-liquid unloading facility. The hopper directs the liquid to a sump. The liquid waste is pumped from the sump to a storage tank. An explosion-retardation system is provided alongside the hopper to reduce the intensity of any conflagration that may occur.

BULK SOLIDS UNLOADING

Three mechanisms are used to unload bulk solids transported by truck or rail: gravity, pressure differential, and fluidized systems.

Gravity systems are typically used with discharge pits for unloading. They are also used with mechanical conveyors (screw conveyor, belt conveyor, bucket elevator, etc.) to transfer the solid waste to storage or directly to the incinerator.

Pressure differential (pneumatic) systems are commonly used to transfer dry powdered materials or granular solids up to approximately ¼-inch mean particle size. Pneumatic conveyors require that the materials must move through piping as well as auxiliary equipment (valves) without clogging, degradation, or segregation, and must be easily separated from the

Fig. 10-3. Drummed-liquid disposal facility. Copyright © C. R. Brunner, 1979.

conveying air stream. The air discharge from this system may contain volatile organics from the waste stream, and treatment may be necessary before discharge to the atmosphere.

Fluidizing systems wash the solid waste with air to create a smoothly flowing material, which can be discharged from the transport container. It is most commonly used with rail cars.

LIQUID-STORAGE SYSTEMS

Three types of tanks are commonly used for liquid-waste storage. Temporary holding tanks provide initial storage of liquid wastes, allowing the transport vehicle to leave the facility. Batching tanks allow preparation of wastes prior to feeding to the incinerator. Preparation might include blending for viscosity control or mixing to reduce the net choloride content of the waste feed. Main storage tanks are used to store wastes that have been accepted by the facility. There may be a number of storage tanks in a single facility to hold wastes segregated by heating content, moisture content, viscosity, reactivity, etc.

Both horizontal and vertical tanks are used to store liquids; these tanks are often coated with corrosion-resistant materials compatible with the stored liquids. Vertical tanks are more commonly used because they are generally less expensive to install and require less area.

Although liquid-waste storage tanks can be installed either above or below ground, above-ground tanks are recommended. They can be readily inspected; leaks can be detected and repaired. Above ground tanks are also less expensive to install than underground tanks.

BULK-SOLIDS STORAGE

Bulk solids received at an incineration facility can be stored in enclosed bins or silos; concrete pits or below-grade stockpiles; and stockpiles on grade. Materials with potential toxicity problems or wastes with explosive, flammable or corresive properties, are generally stored in totally enclosed units, such as single-outlet bins, multiple-outlet silos, and portable bins. Concrete hoppers are normally not used for storage of hazardous materials.

Generally, stockpiles are used for storage of relatively small quantities of waste materials when the waste has a low moisture content. Stockpiles can be located inside a building or maintained outside, on a concrete pad and under appropriate cover.

Characteristics of bulk solids that should be considered in the selection and design of bulk-storage systems include the following:

- bulk density;
- moisture content;
- particle size;
- angle of repose;
- angle of slide;
- temperature requirements;
- abrasiveness;
- cohesiveness;
- material melting point; and
- hygroscopicity.

STORAGE IN CONTAINERS AND TANK CARS

Metal or fiber containers that are used to hold waste materials are normally stored on a concrete pad or other relatively impervious base. The storage area is isolated with a dike to contain any spills and protect the pathway to ground water.

Containers may be stacked above ground level to prevent damage from vehicular movement. The entire area may be covered to minimize weathering and subsequent damage to containers.

Tank trucks and rail tank cars used to deliver waste materials to an incineration facility can also serve as temporary storage for the waste.

LIQUID BLENDING AND FEEDING

Usually, the blending or mixing of liquid wastes is carried out as part of the overall liquid-feed system. The feed system usually consists of a feed pump, a mixing vessel with associated recirculation pipes, and a pipe network to delivery the liquid waste to the incinerator, as shown in Fig. 10-4.

Vessels used for mixing can be either closed top or open top. For hazardous organic wastes, closed tanks are used. They are required to prevent splashing and control vapor emissions into the atmosphere. In addition, inert-gas blanketing of the mixing vessel may be required to reduce the possibility of fire in the tank.

The liquid storage and feed system typically includes recirculation loops to insure good mixing of the waste and to keep any solids in suspension.

The two main reasons for blending are viscosity control and feed-quality control. The viscosity of a liquid should not exceed 10,000 SSU, to ensure that it can be pumped and handled in pipelines. Generally a viscosity of 750 SSU or less is necessary for atomization or burning in suspension. By mixing different compatible wastes of different viscosity characteristics, an acceptable viscosity can be obtained. Additionally, blend tanks will normally be provided with tank steam coils to aid in viscosity control, and slow-speed mixers to encourage mixing.

Feed-quality control refers to the control of feed properties such as halogen content, solids content, viscosity, moisture, and heating value.

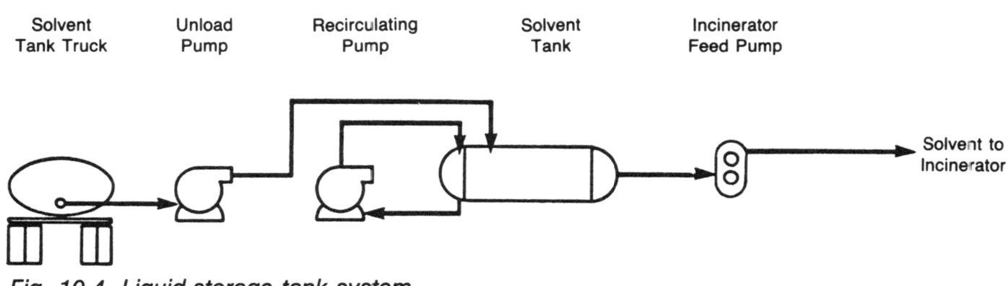

Fig. 10-4. Liquid-storage tank system.

Blending of wastes will also provide a feed that, for each batch, is of uniform consistency. Uniformity of incinerator feed prevents swings in temperature and extreme variations in supplemental fuel consumption which, in turn, would result in poor incinerator performance and excessive maintenance.

SOLIDS PROCESSING AND FEEDING

Available systems for charging solid wastes into an incinerator make use of pneumatic, mechanical, or gravity techniques. Heterogeneous solid wastes are generally subject to some form of size reduction (shredding or pulverizing, for instance) to meet feed system requirements and to facilitate proper injection, distribution, and combustion within the incinerator.

Conveyors are often used for solid-waste transport. The three most common types of conveying systems are the belt conveyor, the apron conveyor, and the flight conveyor.

A belt conveyor, shown in Fig. 10-5, utilizes a neoprene, synthetic, or metal belt to move materials. It is usually limited in incline to no more than 15° from the horizontal. Variations of this conveyor have intermediate flights or ribs, which allow higher belt inclination.

The apron conveyor, illustrated in Fig. 10-6, consists of a series of metal pans moving along a track. These pans move more slowly than the belt in a belt conveyor; however, they can withstand greater impact from dropped waste than can belt systems. Figure 10-7 shows an apron conveyor feeding a rotary-kiln hazardous-waste incineration system.

A drag (or flight) conveyor system is used for ash disposal, and, occasionally, for solid waste transfer. Figure 10-8 shows a drag conveyor, which is essentially a series of flights pulled across a stationary pan by

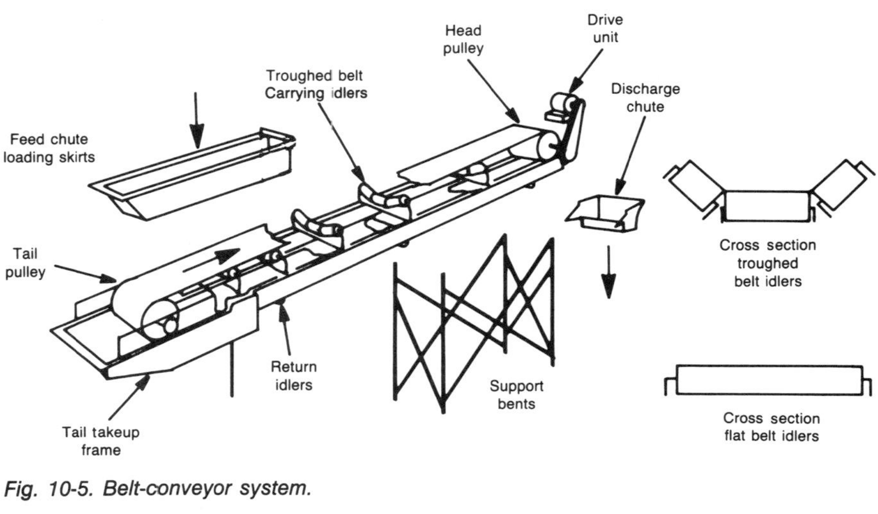

Fig. 10-5. Belt-conveyor system.

Fig. 10-6. Apron conveyor.

Fig. 10-7. Rotary-kiln feed with apron conveyor. Copyright © C. R. Brunner, 1979.

a chain system. A variation of the flight conveyor is the tubular conveyor, shown in Fig. 10-9, which is essentially a series of flights, totally enclosed in a round or square tubular section, pulled by a chain.

Solid-waste feeding equipment may be required to transfer material from rail car to storage bin, from storage bin to conveyor, or from conveyor to incinerator. Four types of feeding systems in use for solids are the rotary feeder, screw feeder, vibratory feeder, and belt feeder.

A rotary feeder is used at the bin discharge to provide an air seal and to regulate the discharge rate of material from the bin or other storage equipment. A typical rotary feeder is shown in Fig. 10-10.

A screw feeder, or conveyor, is shown in Fig. 10-11. It is applicable to granular materials and sludges. Screws can be solid or ribbon (a typical ribbon conveyor is shown in Fig. 10-12), depending on the nature of the

Fig. 10-8. Flight conveyor. Copyright © C. R. Brunner, 1980.

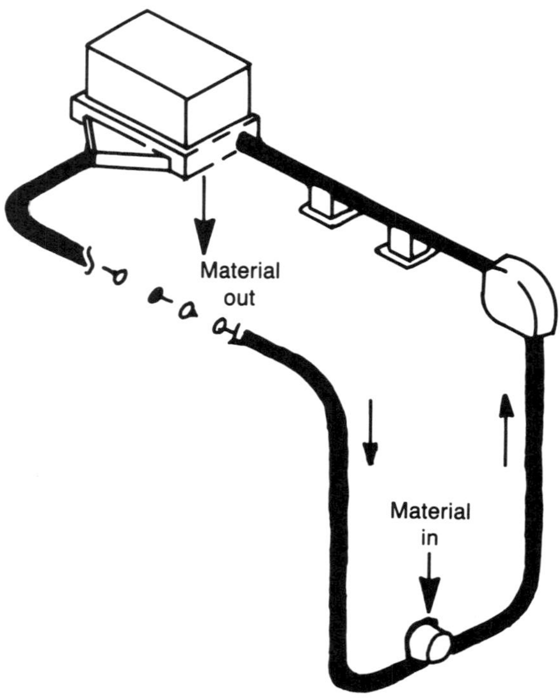

Fig. 10-9. Tubular conveyor system.

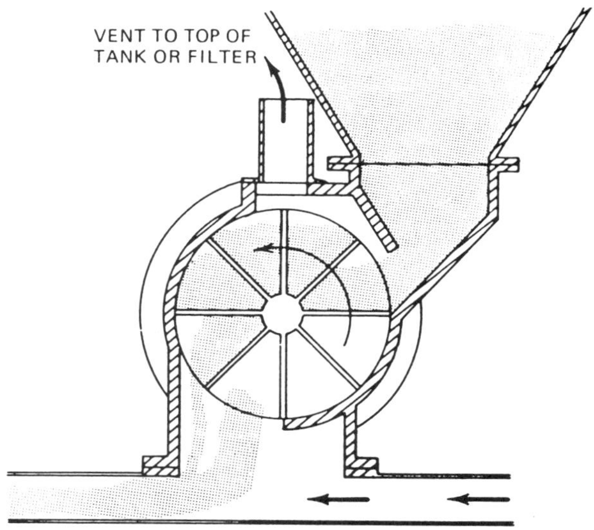

Fig. 10-10. Rotary feeder.

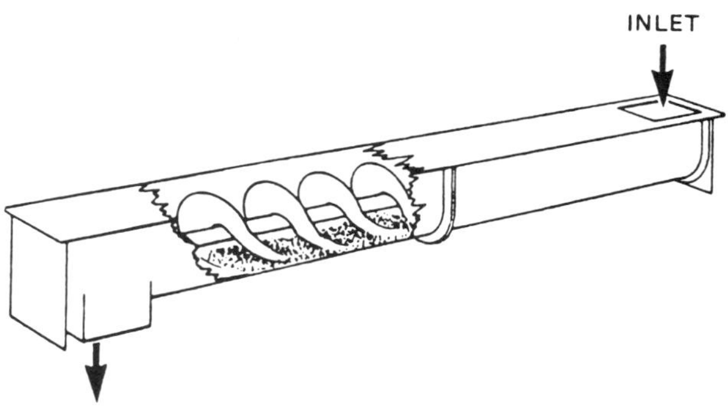

Fig. 10-11. Screw feeder.

feed. Sludges in particular require a thin-ribbon, long-flight screw because they would tend to adhere to a wider and less smooth screw surface.

Vibratory feeders are stationary metal pans or grids fitted with vibratory mechanisms. They can be used to regulate feed and to segregate feed at the same time. When a grid is used, the vibrating action tends to drop those materials which are smaller than the grid openings onto a separate conveyor for alternate storage or processing. In this manner, the feed is sized.

Belt feeders are short-length belt conveyors. Their speed is normally variable and is automatically controlled to feed at a rate compatible with the downstream operation.

Fig. 10-12. Ribbon screw feeder. Copyright © C. R. Brunner, 1980.

Solid-waste charging to the combustion zone of an incinerator can be batch (open charging or air-lock feeders) or continuous (a screw feeder).

CONTAINERS

There are currently a number of automated methods of charging containerized hazardous waste into an incinerator. Most of them require that the top and bottom of the container be removed to prevent gas build-up and explosions. The drum is often quartered to protect against explosions caused by a rapid build-up of gases on the drum inner surface. An emergency discharge stack at the feeding end of the incinerator can contain and direct explosions which may occur.

Hydraulic drum and pack feeding is often used. Wastes are placed in the throat of a charging hopper. A charging ram, either on a timed cycle or operated manually, charges the waste into the incinerator.

Drums may be conveyed to an air-lock section at the kiln entrance. After the drum enters the air lock, the door closes and the incinerator charging door opens. The conveyor feeds the drum through the charging door.

As an alternative to cutting the container, the drum-punching system illustrated in Fig. 10-13 was developed. A container is conveyed to a vestibule, immediately upstream of the primary incinerator chamber, where a punch punctures its top.

SLUDGE HANDLING

Generally, sludges are generated at the incinerator location and extensive storage facilities are not required. Vacuum filters, belt filter

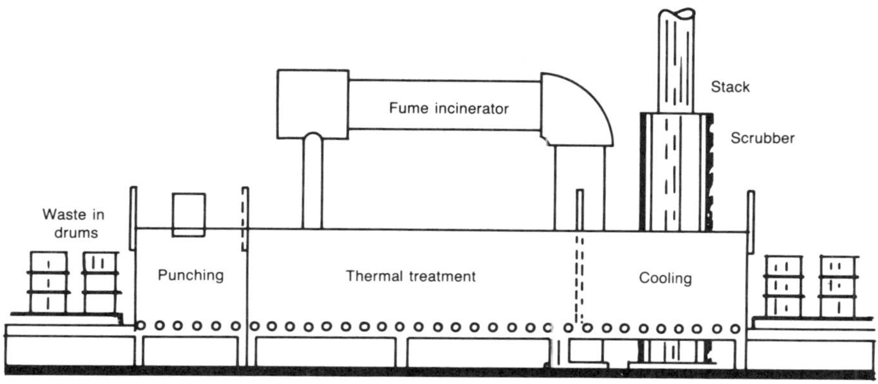

Fig. 10-13. Container-punching feed system.

presses, and centrifuges are common sludge-dewatering devices, and each of these generate a continuous sludge feed to an incinerator.

Another common method of sludge dewatering is the plate-and-frame press, which generates sludge in batches, one 15 to 30 minute batch every 2 or 3 hours. The feed to an incinerator should be constant and consistent. To even out this batch generation, sludge storage must be provided between a plate-and-frame press and an incinerator. This storage can be in a hopper beneath the press, or it can be in a separate silo, as illustrated in Fig. 10-14.

This silo is used for the storage of hazardous-sludge incinerator feed. As indicated in the schematic, Fig. 10-15, sludge is dropped into the silo by gravity. Its discharge is controlled by a variable-speed wiper rotating at the bottom of the silo. The wiper discharges sludge into a screw conveyor, which feeds the furnace.

Sludge handling can be a difficult problem. Sludge can adhere to conveyor and feeder surfaces and the physical sludge characteristics are sensitive to changes in moisture content. Normally, sludges are moved with belt conveyors, screw conveyors, plunger pumps, or progressive-cavity pumps.

Belt and screw conveyors are used when gravity feeding of an incinerator is possible. An incinerator that is under positive internal pressure, such as a fluid bed incinerator, requires that sludge be injected at positive pressure. The plunger pump and progressive-cavity pump will inject sludge into pressurized systems. Figure 10-16 illustrates a progressive-cavity pump, which is essentially a tight-fitting screw-type conveyer.

The distance limit for sludge pumping is normally 50 feet. The longer the pipeline, the greater the load on the pump. By reducing the distance

Fig. 10-14. Sludge storage silo. Copyright © C. R. Brunner, 1986.

to the incinerator, the feed system can be simplified. It is not uncommon for a 150-horsepower motor driving a pump to 300 psig to feed sludge to an incinerator maintained at 1 psi. Reducing this developed pressure, sludge feeding will be more reliable and will require less maintenance.

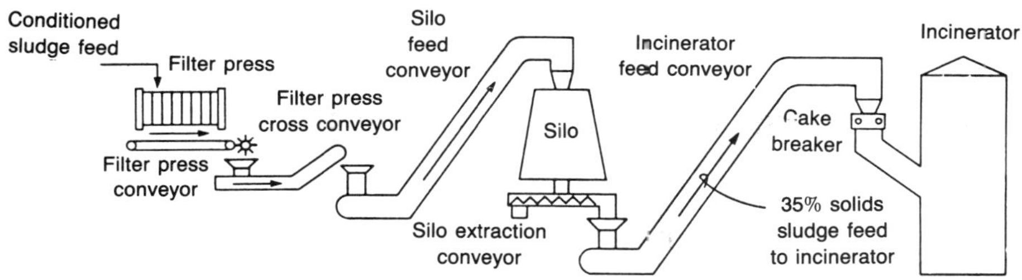

Fig. 10-15. Sludge storage silo schematic.

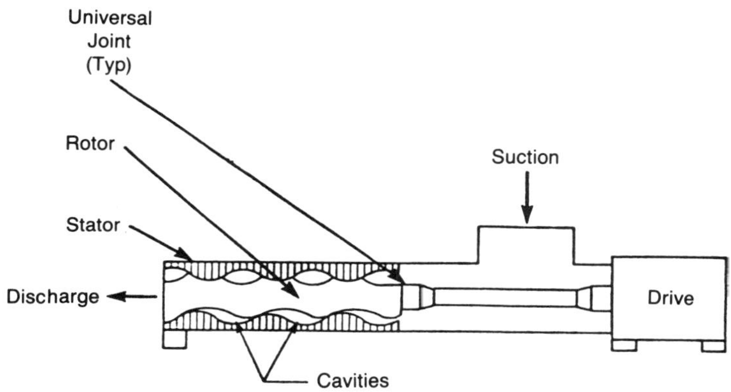

Fig. 10-16. Progressive-cavity pump.

11

Emissions Generation

There are three emissions pathways from an incineration system, namely, air emissions, wastewater discharge, and solid residue, or ash. All three of these discharges are sensitive to the presence of effective combustion within the furnace. With poor combustion, organics will not be effectively destroyed, and intolerable amounts of hazardous constituents could be transferred to the ash or to the scrubber water discharge.

In this chapter, the nature of the emissions discharge is discussed.

THE COMBUSTION PROCESS

Incinerator combustion must be effective to minimize emissions to the gas stream and to the residual (ash) discharge. Poor combustion can be readily identified. Often, smoke is its result. Carbon monoxide or total hydrocarbon levels (THC) in the incinerator exhaust can be good indicators of combustion efficiency. TOSCA requires monitoring of CO, as do many permit writers issuing RCRA permits.

Carbon monoxide and carbon dioxide measurements can be made reliably and continuously. The measurement of THC is more difficult and less certain. For this reason, THC is not normally used as a compliance parameter.

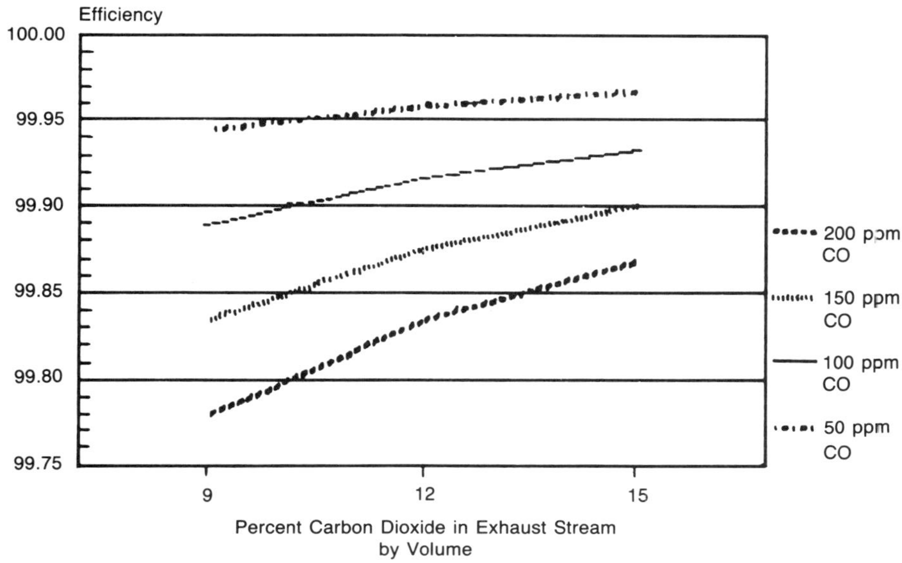

Fig. 11-1. Combustion efficiency.

Combustion efficiency (CE) is defined in terms of the presence of CO in the gas stream. With [CO_2] and [CO] the concentrations of CO_2 and CO in the exhaust gas, respectively, the CE is calculated as follows:

$$CE = 100\% \times [CO_2] / \{[CO_2] + [CO]\}$$

Figure 11-1 indicates the relationship between combustion efficiency and CO level. TOSCA requires a combustion efficiency of 99.9 percent. This value is becoming generally recognized as an indicator of good combustion. From Fig. 11-1, the presence of CO in a concentration in excess of 150 ppm makes achievement of a combustion efficiency of 99.9 percent impossible. (Normal operation precludes CO_2 concentrations greater than 15 percent.)

VISIBLE PARTICULATE DISCHARGE

Particulate matter is both solid and liquid. The term comprises a complex category of materials, also termed *aerosols*, that inhabit the atmosphere. The size range of particulates of interest varies from just over that of large individual molecules, 0.1 μ (microns), to 500 μ in diameter, where 1 million μ equals 1 meter.

Hazardous-waste regulations typically require a maximum particulate discharge of 0.08 grains per dry standard cubic foot. The relationship of this requirement to a visible particulate discharge is not necessarily significant.

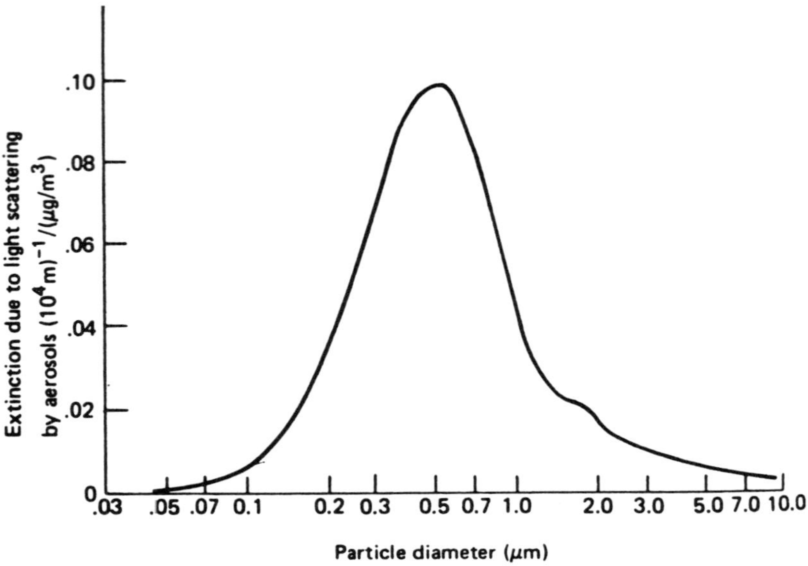

Fig. 11-2. Light scattering as a function of particle diameter. Source: Reference 1, page 43.

Figure 11-2 relates light scattering to particulate diameter. The greater the scatter coefficient, the more visible will be the discharge. Below 0.1 μ, particles are sufficiently small compared to the wavelength of light to obey the same laws of light scattering as do molecules, termed *Raleigh scattering*, a function of the sixth power of particle diameter. This mechanism is relatively inconsequential in its effect upon visibility. Particles much larger than 1 μ are so much larger than the wavelength of visible light that they obey the laws of macroscopic objects, intercepting or scattering light in proportion to their cross-sectional areas.

In summation, the curve in Fig. 11-2 indicates that if a discharge has a relatively high proportion of particles smaller than 0.2 μ (or greater than 1.0 μ) mean diameter, it will have significantly different visible properties than a distribution of particles in the range of 0.2 μ to 1.0 μ mean diameter. Opacity is a function of particle size, in addition to the weight density (grains per cubic foot) of the discharge.

SMOKE

Smoke is a suspension of solid or liquid particulate in a gaseous discharge that produces a visible effect. Generally, the particles range from fractions of a micron to over 50 μ in diameter. The visibility of smoke is a function of the quantity of particles present rather than the weight of particulate matter. The weight of particulate emissions is therefore not necessarily indicative of the optical density of the emission. For instance, a weight density of so many grains of emissions per cubic foot of gas is not directly related to the opacity of the discharge. Neither is the color of a discharge related to opacity, or smoke density. Smoke can be either black or non-black (white smoke).

The formation of white or other opaque, non-black smoke is usually due to insufficient furnace temperatures when burning carbonaceous materials. Hydrocarbons heated to a level at which evaporation or cracking occurs within the furnace produce white smoke. The temperatures are not high enough to produce complete combustion of these hydrocarbons. With stack temperatures in the range of 300° F to 500° F, many hydrocarbons will condense to liquid aerosols and, with the solid particulate present, will appear as non-black smoke.

Increased furnace or stack temperatures and increased turbulence are two methods of controlling white smoke. Turbulence helps ensure thermal uniformity within the off-gas flow.

Excessive air flow may provide excessive cooling, and an evaluation of reducing white smoke discharges includes investigating the air quantity introduced into the furnace. Inorganics in the exit gas may also produce non-black smoke. For instance, sulfur and sulfur compounds appear yellow in a discharge; calcium and silicon oxides appear light to dark brown.

Black smoke is formed when hydrocarbons are burned in an oxygen-deficient atmosphere. Carbon particles are found in the off-gas. Causes of oxygen deficiency are poor atomization, inadequate turbulence (or mixing), and poor air distribution within a furnace chamber. These factors will each generate carbon particulate which, in the off-gas, produces dark, black smoke.

Pyrolysis reactions occur within an oxygen-starved atmosphere. This generates stable, less complex hydrocarbon compounds that form dark, minute particulate, generating black smoke.

A common method of reducing or eliminating black smoke has been steam injection into the furnace. The carbon present is converted to methane and carbon monoxide as follows:

$$3\ C\ (\text{smoke}) + 2\ H_2O \text{ - - - -> } CH_4 + 2\ CO$$

Similar reactions occur with other hydrocarbons present, and the methane and carbon monoxide produced burn clean in the heat of the furnace, eliminating the black carbonaceous smoke that would have been produced without steam injection:

$$CH_4 + 2\ O_2 \text{ - - - -> } CO_2 + 2\ H_2O\ (\text{smokeless})$$

$$2\ CO + O_2 \text{ - - - -> } 2\ CO_2\ (\text{smokeless})$$

Steam injection normally requires from 20 to 80 pounds of steam per pound of flue gas, or 0.15 to 0.50 lb of steam per pound of hydrocarbon in the gas stream. (It should be noted that there is some controversy regarding the effect of steam injection on carbonaceous discharges. Some argue that the steam produces, primarily, good mixing, and that the turbulence, or effective mixing with air, eliminates the smoke discharge as opposed to methane generation.)

METALS EMISSIONS

There is growing concern over the introduction of certain metals into the atmosphere from incineration and other industrial discharges. Five metals which have been of particular concern are lead, nickel, hexavalent chromium, mercury, and cadmium.

Generally, metals will be discharged on (adsorbed onto) solid particulate matter. Their disposition in an incineration system can best be determined by measurement, but estimates of where they may be found can be made.

Table 11-1 lists the disposition of metals in the discharge of a solid-waste incinerator with a wet scrubber. Table 11-2 lists the fate of metals

Table 11-1. Estimate of the Fate of Particulate Emissions, Solid Waste Incinerator with a Scrubber.

Metal	In Bottom Ash	In Scrubber Water	Stack Discharge
Aluminum	57	42	1
Antimony	45	54	1
Arsenic	30	62	8
Barium	39	60	1
Beryllium	40	59	1
Cadmium	31	62	7
Chromium	31	59	10
Cobalt	45	52	3
Copper	47	51	2
Iron	53	46	1
Lead	16	82	2
Magnesium	33	66	1
Manganese	20	78	2
Mercury	0	10	90
Molybdenum	58	2	40
Nickel	30	66	4
Selenium	1	19	80
Titanium	45	54	1
Vanadium	18	79	3
Zinc	20	76	4

Note: All of the above figures are in percent.

Table 11-2. Estimate of the Fate of Particulate Emissions, Solid Waste Incinerator with an Electrostatic Precipitator.

Metal	In Bottom Ash	ESP Capture	Stack Discharge
Aluminum	57	35	8
Antimony	45	28	27
Arsenic	30	47	23
Barium	39	60	1
Beryllium	40	58	2
Cadmium	31	47	22
Chromium	31	57	12
Cobalt	45	53	2
Copper	47	39	14
Hafnium	49	50	1
Iron	53	39	8
Lead	16	76	18
Magnesium	33	66	1
Manganese	20	78	2
Mercury	0	1	99
Molybdenum	58	17	25
Nickel	30	52	18
Potassium	52	47	1
Selenium	1	19	80
Sodium	44	55	1
Tantalum	37	62	1
Titanium	45	52	3
Vanadium	18	78	4
Zinc	20	60	20

Note: All of the above figures are in percent.

in an incinerator system firing solid waste and discharging through an electrostatic precipitator. The metals discharge from a liquid-waste stream fired in an incinerator system with a wet scrubber is listed in Table 11-3.

INORGANIC GAS DISCHARGES

The burning process produces carbon monoxide, oxides of nitrogen, and sulfur oxides when sulfur is present in the feed. When halogen compounds are present, hydrogen halides will be formed, such as hydrogen chloride, hydrogen fluoride, hydrogen iodide, etc.

The quantity of sulfur oxides in the exhaust gas stream is estimated by assuming that all sulfur present in the incinerator feed is oxidized and is discharged in the exhaust gas. The majority of sulfur oxides will be in the form of sulfur dioxide, SO_2. Normally less than 2 percent of the sulfur oxide will be sulfur trioxide, SO_3. Figure 11-3 can be used to determine the generation of SO_3 for a 1 percent sulfur fuel. To use these

Table 11-3. Estimate of the Fate of Particulate Emissions, Liquid Waste Incinerator with a Scrubber, No Bottom Ash.

Metal	In Scrubber Water	Stack Discharge
Arsenic	81	19
Chromium	81	19
Cobalt	88	12
Copper	96	4
Magnesium	91	9
Manganese	87	13
Molybdenum	87	13
Nickel	88	12
Selenium	88	12
Vanadium	72	28
Zinc	72	28

Note: All of the above figures are in percent.

curves, assume that all of the sulfur in the fuel is oxidized. Calculate the amount of SO_3 generated; subtract from the total sulfur present to obtain the amount of SO_2 present in the gas stream. SO_3 is soluble and will readily convert to sulfuric acid whereas SO_2 is relatively insoluble.

If organic halogens are present, assume that all of the halogens are converted to their hydrogen halide and exit in the exhaust gas stream. Inorganic halogens, such as NaCl or $CaCl_2$, can be assumed to be inert and to exit the process in the ash residual.

The amount of carbon monoxide formed is a function of the ratio of carbon to hydrogen in the fuel and the temperature of the process. Figure 11-4 can be used to determine the moles of carbon monoxide generated per mole of stoichiometric oxygen entering the process.

Nitrogen oxides are generated in the form of nitrogen oxide (NO) and nitrogen dioxide (NO_2). The majority of nitrogen oxides generation is NO.

The variation in nitrogen oxides generation as a function of temperature and excess air is shown graphically in Fig. 11-5. Above 2000° F, the amount of NO_x generated starts increasing substantially with temperature. The amount of NO generated can be calculated using Table 11-4, and Table 11-5 lists rates for NO_2 generation. NO is the predominant oxide, with NO_2 generated in extremely small quantities.

ORGANIC-GAS GENERATION

When organic materials are fired, other organic compounds are generated which were not present in the original feed. The general term for

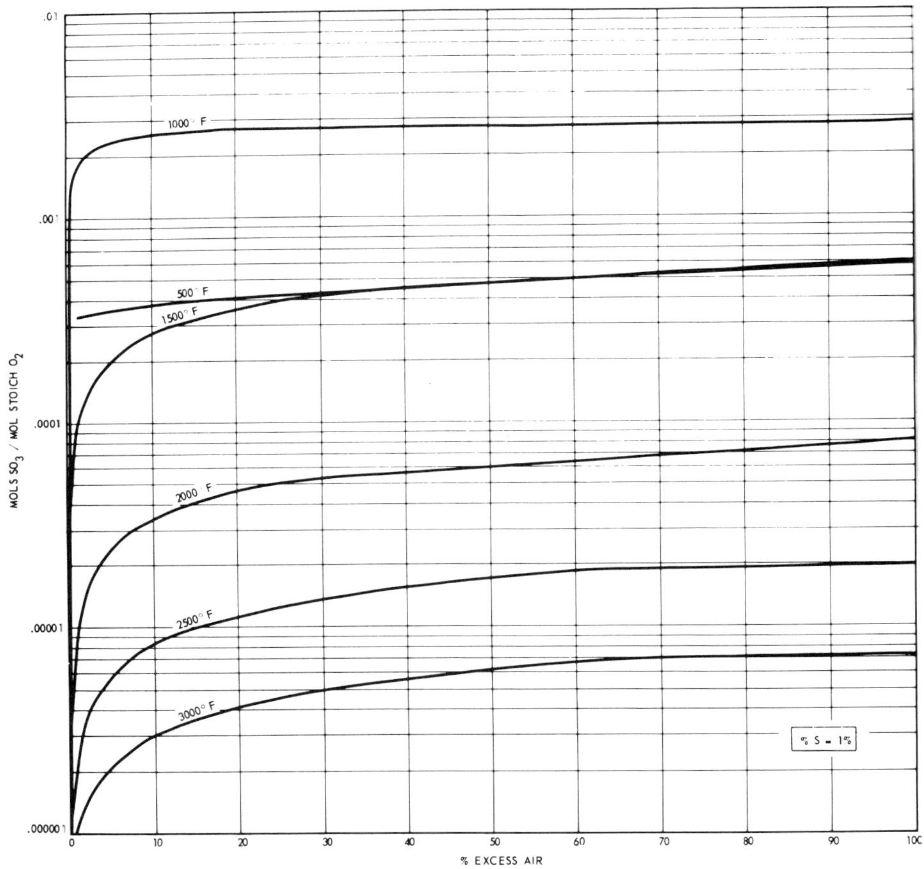

Fig. 11-3. Generation of sulfur trioxide. Source: Reference 2.

these compounds are *PICs* (products of incomplete combustion) and the more notable of these are PCBs, dioxins, and dibenzofurans. Current regulations have not yet addressed the issue of PIC generation and control. Some of these compounds are considered to be dangerous to public health and regulations concerning them can be expected to be promulgated as part of RCRA.

At the current time, there are no reliable techniques to forecast the generation of these compounds. No specific control technology exists for these organics, although it has been found that with good combustion (greater than 99.9 percent combustion efficiency or less than 100 ppm CO in the flue gas stream) the generation of PICs is extremely small.

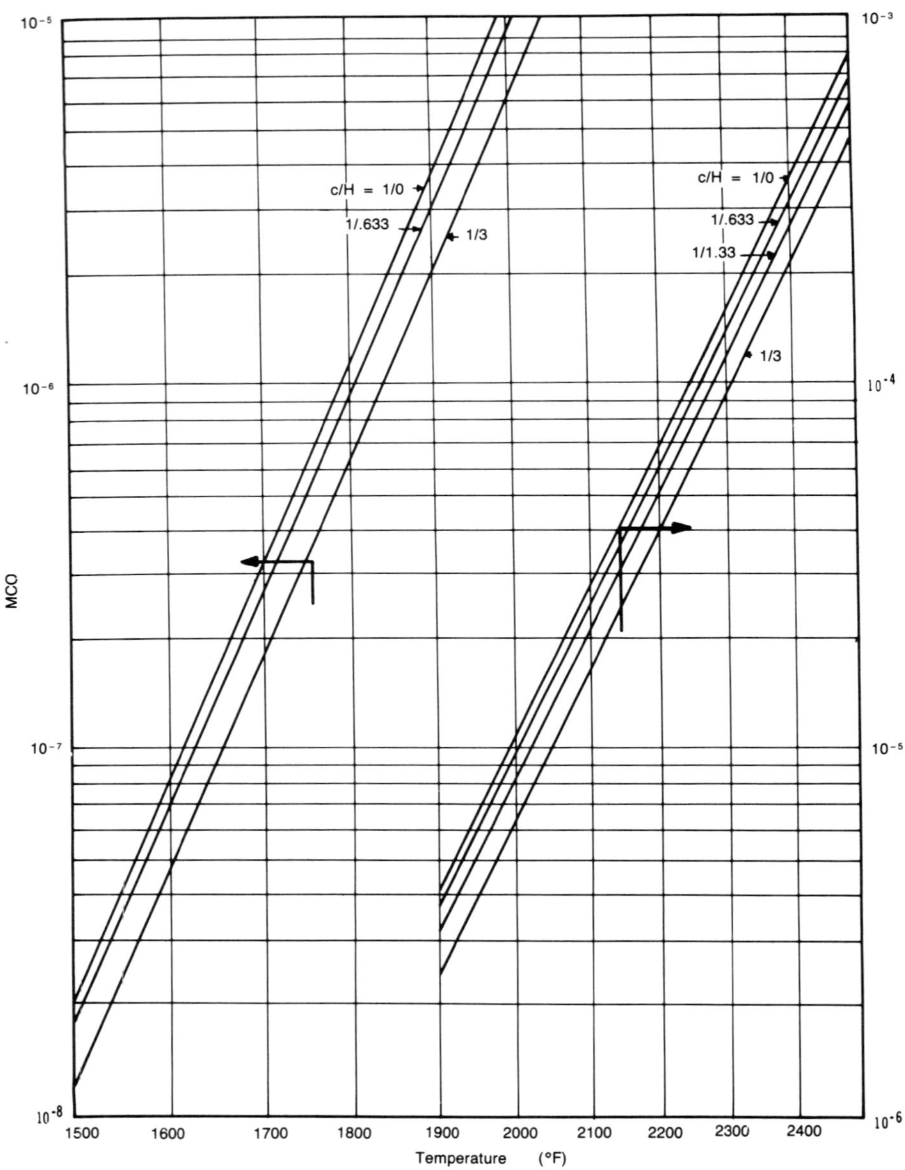

Fig. 11-4. Carbon monoxide generation. Source: Reference 2.

REFERENCES

1. C. R. Brunner, *Hazardous Air Emissions from Incinerators*, Second Edition (New York: Chapman & Hall, 1986).

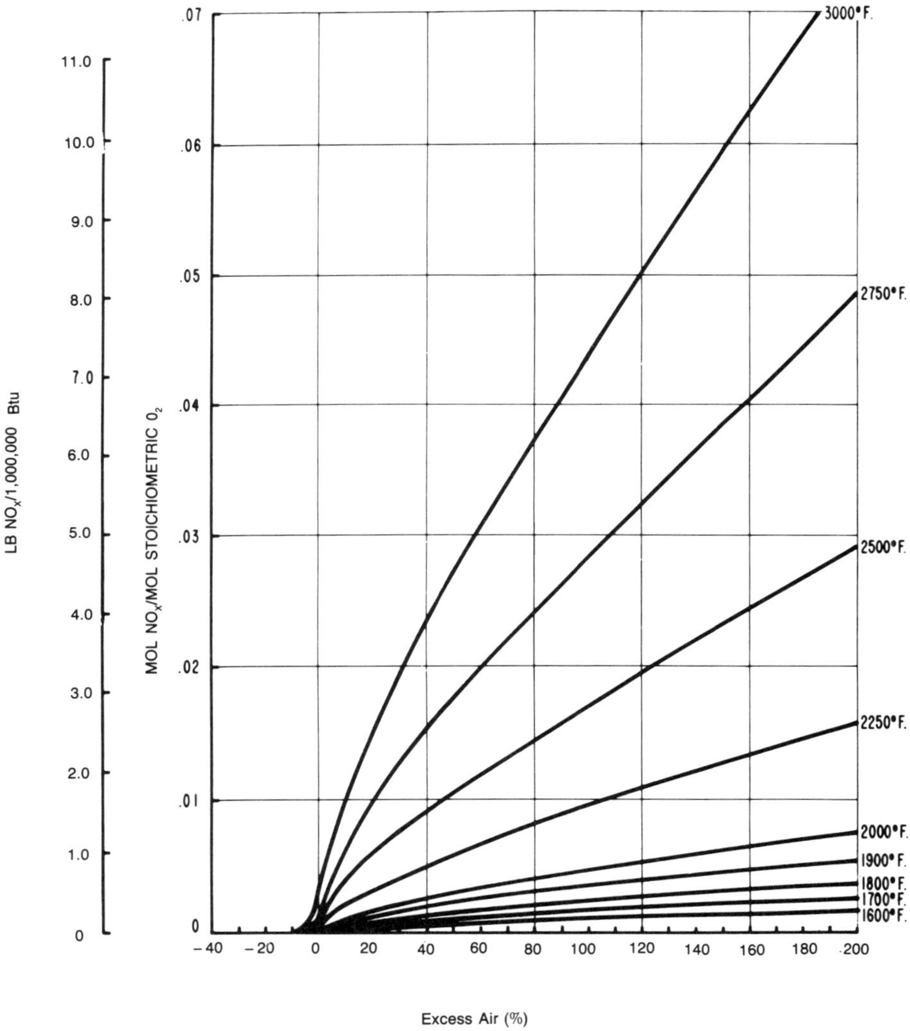

Fig. 11-5. Generation of nitrogen oxides. Source: Reference 2.

2. *Combustion Fundamentals for Waste Incineration* (New York: American Society of Mechanical Engineers, 1974).

3. C. R. Brunner, *Incineration Systems: Selection & Design* (Reston, VA: Incineration Consultants Incorporated, 1988).

Table 11-4. Generation of Nitrogen Oxide.

Temp. °F	0	10%	20%	Excess Air 30%	50%	100%	150%	200%
For C_1H_0 (Carbon)								
1000	—	9.587E-07	1.416E-06	1.805E-06	2.503E-06	4.088E-06	5.599E-06	7.082E-06
1500	—	2.931E-05	4.331E-05	5.523E-05	7.657E-05	1.251E-04	1.712E-04	2.166E-04
1832	1.712E-06	1.260E-04	1.862E-04	2.375E-04	3.293E-04	5.380E-04	7.366E-04	9.316E-04
2192	2.022E-05	4.081E-04	6.033E-04	7.696E-04	1.074E-03	1.744E-03	2.388E-03	3.022E-03
2500	8.198E-05	8.510E-04	1.260E-03	1.607E-03	2.232E-03	3.645E-03	4.993E-03	6.318E-03
3000	5.677E-04	2.204E-03	3.242E-03	4.194E-03	5.760E-03	9.420E-03	1.291E-02	1.633E-02
For C_3H_4 ($C_1H_{1.33}$)								
1000	—	9.587E-07	1.416E-06	1.805E-06	2.503E-06	4.088E-06	5.599E-06	7.082E-06
1500	2.575E-07	2.933E-05	4.331E-05	5.523E-05	7.657E-05	1.251E-04	1.712E-04	2.166E-04
1832	1.932E-06	1.260E-04	1.862E-04	2.375E-04	3.293E-04	5.380E-04	7.366E-04	9.316E-04
2192	1.899E-05	4.079E-04	6.033E-04	7.694E-04	1.068E-03	1.744E-03	2.388E-03	3.022E-03
2500	7.870E-05	8.502E-04	1.259E-03	1.670E-03	2.230E-03	3.645E-03	4.993E-03	6.318E-03
3000	5.341E-04	2.191E-03	3.241E-03	4.138E-03	5.751E-03	9.411E-03	1.290E-02	1.632E-02
For C_0H_2 (Hydrogen)								
1000	—	9.587E-07	1.416E-06	1.805E-06	3.372E-06	4.088E-06	5.599E-06	7.082E-06
1500	2.835E-07	2.933E-05	4.331E-05	5.523E-05	7.657E-05	1.251E-04	1.712E-04	2.166E-04
1832	1.849E-06	1.260E-04	1.862E-04	2.375E-04	3.293E-04	5.380E-04	7.366E-04	9.316E-04
2192	1.792E-05	4.079E-04	6.031E-04	7.694E-04	1.068E-03	1.744E-03	2.388E-03	3.018E-03
2500	5.949E-05	8.493E-04	1.258E-03	1.606E-03	2.230E-03	3.645E-03	4.993E-03	6.315E-03
3000	4.281E-04	2.169E-03	3.222E-03	4.125E-03	5.738E-03	9.398E-03	1.288E-02	1.632E-02

Note: 1.5E03 = .0015

Source: Reference 3.

Table 11-5. Generation of Nitrogen Dioxide.

Temp. °F	0	10%	20%	Excess Air 30%	50%	100%	150%	200%
For C_1H_0 (Carbon)								
1000	—	1.208E-07	2.416E-07	3.622E-07	6.039E-07	1.208E-06	1.812E-06	2.416E-06
1500	—	4.045E-07	8.086E-07	1.213E-06	2.022E-06	4.045E-06	6.065E-06	8.089E-06
1832	—	6.528E-07	1.352E-06	2.029E-06	3.386E-06	6.768E-06	1.015E-05	1.354E-05
2192	6.538E-10	1.018E-06	2.008E-06	3.066E-06	5.113E-06	1.024E-05	1.536E-05	2.049E-05
2500	1.375E-08	1.330E-06	2.671E-06	4.015E-06	6.701E-06	1.343E-05	2.015E-05	2.668E-05
3000	1.359E-07	1.863E-06	3.718E-06	5.589E-06	9.341E-06	1.875E-05	2.816E-05	3.758E-05
For C_3H_4 ($C_1H_{1.33}$)								
1000	—	1.180E-07	2.364E-07	3.552E-07	5.936E-07	1.192E-06	1.793E-06	2.395E-06
1500	—	3.948E-07	7.916E-07	1.189E-06	1.988E-06	3.991E-06	6.006E-06	8.020E-06
1832	—	6.598E-07	1.324E-06	1.989E-06	3.326E-06	6.681E-06	1.005E-05	1.342E-05
2192	6.961E-10	9.944E-07	1.998E-06	3.004E-06	5.027E-06	1.010E-05	1.262E-05	2.031E-05
2500	1.219E-08	1.296E-06	2.611E-06	3.932E-06	6.585E-06	1.325E-05	1.994E-05	2.664E-05
3000	1.172E-07	1.798E-06	3.615E-06	5.453E-06	9.151E-06	1.846E-05	2.782E-05	3.718E-05
For C_0H_2 (Hydrogen)								
1000	—	1.107E-07	2.228E-07	3.362E-07	5.656E-07	1.149E-06	1.740E-06	2.335E-06
1500	—	3.705E-07	7.460E-07	1.126E-06	1.894E-06	3.848E-06	5.826E-06	7.820E-06
1832	—	6.189E-07	1.247E-06	1.883E-06	3.168E-06	6.438E-06	9.751E-06	1.309E-05
2192	—	9.318E-07	1.882E-06	2.843E-06	4.787E-06	9.734E-06	1.475E-05	1.980E-05
2500	1.471E-08	1.213E-06	2.457E-06	3.718E-06	6.269E-06	1.276E-05	1.934E-05	2.597E-05
3000	7.021E-08	1.654E-06	3.376E-06	5.123E-06	8.679E-06	1.775E-05	2.694E-05	3.655E-05

Note: 1.5E-03 = .0015

Source: Reference 3.

12

Air Emissions Control Systems

Air emissions control is critical to an incinerator system. Different incinerator systems and incinerator feeds have different requirements. This chapter presents various control systems with application to hazardous waste incinerator systems.

REMOVAL EFFICIENCY

A control device is characterized by its particle removal efficiency. *Removal efficiency* is the weight of material removed divided by the weight of material entering the device. It can be calculated for a specific particulate size, or can be based upon total particulate loading.

PARTICLE-SIZE ANALYSIS

The efficiency of any particulate-control device is a function of particle size. Particulate matter is normally designated by mean diameter, measured in microns (μ).

The particulate size distribution of an emission cannot be calculated. It can only be inferred from prior data based on measurement. Data on particle-size distribution are rare. It is a function of waste properties, furnace design, and furnace operation, and will vary from test to test. It is virtually impossible to define particle-size distribution with any degree of confidence prior to construction and operation of the incinerator in question.

SETTLING CHAMBERS

Settling chambers are basically long, boxlike structures in which the gas velocity is low enough to allow settling of larger particles. Generally, settling chambers are ineffective for particulate smaller than 40 μ mean particle size.

A settling chamber is not practical for use as a pollution control device, except as part of a larger system, in which it helps remove larger particulate matter from the exhaust, upstream of another control system.

DRY IMPINGEMENT SEPARATORS

Impingement separators are essentially a series of baffles placed in the gas stream. The relatively high inertia of particles within the gas stream maintains their direction of flow while the gaseous component of the stream changes its direction and flows around a baffle. Particulate matter drops and the gas flow continues through the process stream.

This method of particulate removal can be effective for larger particles, above 15 μ, but smaller particulate matter will continue to flow with the gas stream.

DRY CYCLONIC SEPARATORS

The cyclone is an inertial separator. As shown in Fig. 12-1, gas entering the cyclone forms a vortex, which reverses direction, and forms

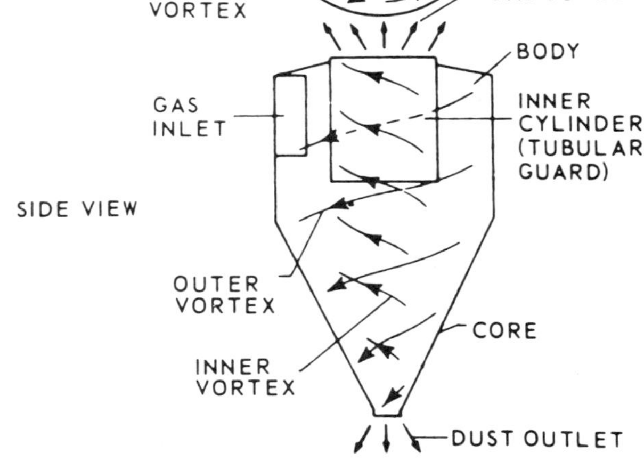

Fig. 12-1. Conventional reverse-flow cyclone.

a second vortex leaving the cyclonic chamber. Particulate matter, because of its inertia, tends to move toward the outside wall. The particles drop from this wall, the sides of the cyclone, to an external receiver for ultimate disposal.

Cyclones will remove larger-sized particulate from the gas stream (greater than 15 μ mean particle size), but will have negligible effect on smaller particles. A cyclone collector is often placed before another control device, such as an electrostatic precipitator or a baghouse. The cyclone removes larger particles from the gas stream, and in many cases these particles are discharged directly to the incinerator, to encourage complete burning through a second cycle of incineration. Removal of these larger particles from the gas stream results in increased efficiency for the downstream equipment, which now have only the smaller-sized particles to contend with.

With D_{pc} the particle cut size, the ratio D_p/D_{pc} is the actual particle size D_p related to the particle cut size. The cut size is the diameter of those particles collected by a particular piece of equipment with 50 percent efficiency. Figure 12-2 shows the relationship between cyclone efficiency and particle cut size ratio. Collection efficiency decreases rapidly as the particle size decreases.

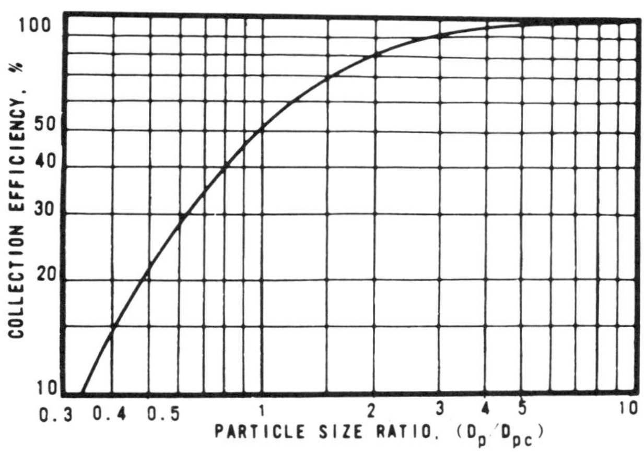

Fig. 12-2. Cyclone efficiency versus particle cut size ratio.

One concern regarding the use of cyclonic separators is the reluctance of collected particulate to drop from the cyclone walls. This condition, *agglomeration*, may occur if the dust is fibrous, sticky, or hygroscopic (water absorbing), or if the gas stream contains too much particulate matter (100 grains per cubic foot is a practical maximum).

Cyclonic collectors are normally designed in a series of 4 to 24 cyclones.

WET COLLECTION

Wet devices used for particulate removal utilize two mechanisms. First, particles are wetted by contact with a liquid droplet. Second, the wetted particles impinge on a collecting surface, from which they are subsequently removed by a liquid flush.

One of the simplest means of wet collection of particles is wetting a collecting baffle. Particulate is continuously washed from the baffle, or wall preventing agglomeration and presenting a continual clean liquid surface to aid in catch and removal.

LIQUID SPRAYING

A spray directed along the path of dust or other particulate matter will impinge upon those particles with an efficiency directly proportional to the number of droplets and to the momentum imparted to the droplets. The smaller the size of the individual droplets the greater their number. However, the smaller the droplet particle size, the less impingement, or momentum, is associated with them. The optimum water droplet size is approximately 100 μ. Above 100 μ there are too few particles and below 100 μ the droplets have insufficient inertia.

The mechanism of diffusion promotes deposition of dust particles on water droplets. *Diffusion*, or *Brownian movement*, is that property of particles of different diameters to intermingle although they may be at rest, much as natural gas diffuses within a contained room although the air within that room is at rest. Diffusion helps the particulate and water droplets come in contact and is inversely proportional to the size of water and solid particulate. The smaller the particle or droplet size, the more rapid the diffusion and the quicker the wetting process.

Spray chambers are at times utilized as a low-cost means of removing heavier particulate from a gas stream. Water is sprayed at rates of 3 to 5 gpm per 100 cfm of gas flow, and the heavier particulate matter in wetted and dries out in the gas stream.

WET CORROSION

Gases exiting an incinerator may contain sulfide and halogen components from the waste feed. These components, in addition to organic residuals, can result in significant acid formation in the gas-scrubbing liquid. The use of non-metallic chambers or non-metallic-coated carbon steel is a growing method of corrosion control. These non-metallic materials, such as flake polyester, fiberglass, neoprene, and epoxy, have good anti-corrosion properties; however, they are temperature sensitive. In most cases these materials cannot withstand temperatures above 160° F to 200° F. Design with such materials must include provisions for emergency quenching to provide cooling if normal cooling water supply is lost.

Table 12-1. Evaluation of Alloys for Incinerator Scrubbers.

Alloy	CORROSION RESULTS Scrubber Solutions	Fan Deposits
Ti6A1-4V	good resistance	good resistance
Hastelloy C	good resistance	good resistance
Inconel 625	good resistance	good resistance
Hastelloy F	good resistance	–
Hastelloy C-276	good resistance	–
Hastelloy G	good resistance	–
Ti75A	good resistance	–
S-816	good resistance	pitted
Carpenter 20	pitted	pitted, SCC
Incoloy 825	pitted	pitted
Incoloy 800	–	pitted
316L	pitted, SCC	pitted, SCC
310	pitted	–
446	pitted	–
Inconel 600	trenches	–
Inconel 601	trenches	–
Armco 22-13-5	pitted	pitted, SCC
USS 18-18-2	pitted	pitted, SCC
304	pitted, SCC	pitted, SCC

SCC: Stress corrosion cracking

Source: Reference 1.

Table 12-1 lists the resistance of several alloys to incinerator scrubber solutions having relatively high acidity. Also listed is the resistance of these alloys to corrosion when used in an induced draft fan downstream of the scrubber. Note that stainless steel (304, 310, 316L, 446 and USS 18-18-2) has poor resistance to acid attack.

MIST ELIMINATORS

Wet gas cleaning equipment floods a gas with scrubbing liquid. Water droplets are usually carried off by the gas stream exiting the scrubber. A mist eliminator is passive equipment that removes most of the entrained water droplets from the gas stream. This is desirable for the removal of collected particulate that is sorbed onto the water particles to reduce the size of the plume exiting the stack (excess water in off-gas will appear as a white plume at the stack discharge) and will reduce the moisture collected in the induced draft fan and other downstream equipment. Decreasing this liquid accumulation also decreases attendant corrosion.

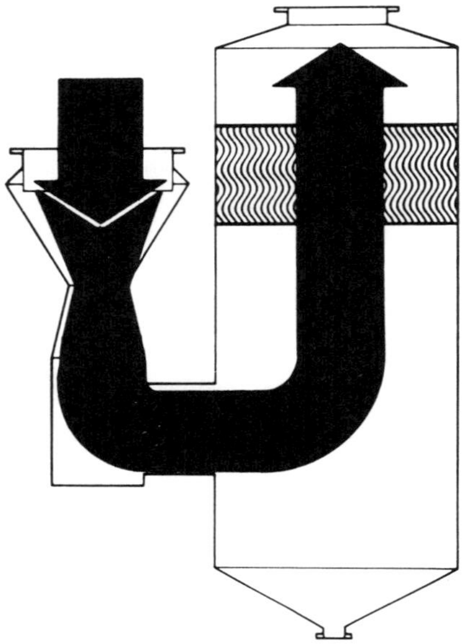

Fig. 12-3. Chevron-type mist eliminator. Source: Peabody-Galion, Princeton, New Jersey.

through a series of baffles. Gas exits the baffles, and the heavier moisture droplets, because of their inertia, hit the chevron-shaped baffles and fall to the bottom of the chamber, leaving the gas stream.

Figure 12-4 is a representation of an impingement-tray tower which functions as a mist eliminator. The wet gas stream passes through a perforated plate and then strikes a series of small impingement plates. As with the chevron mist eliminator, the gas flows around the impingement plates while the entrained moisture falls to the bottom of the tower. The use of impingement plates for gas scrubbing is discussed later in this chapter.

A packed tower used as a mist eliminator is shown in Fig. 12-5. Moisture-laden gas passes through a bed of spherical elements (they can be other shapes) which allows the flow of gas. Entrained moisture collects on the packing and eventually drops to the bottom of the tower.

GAS SCRUBBING

The wet scrubbers previously discussed are essentially washing devices used in conjunction with centrifuges or other inertial collection devices. Scrubbing of a gas stream, however, requires more than inertial effects. There are several mechanisms by which gas scrubbing occurs.

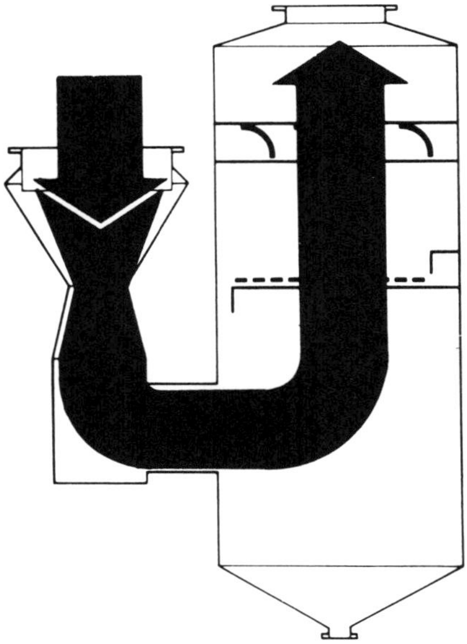

Fig. 12-4. Impingement-type mist eliminator. Source: Peabody-Galion, Princeton, New Jersey.

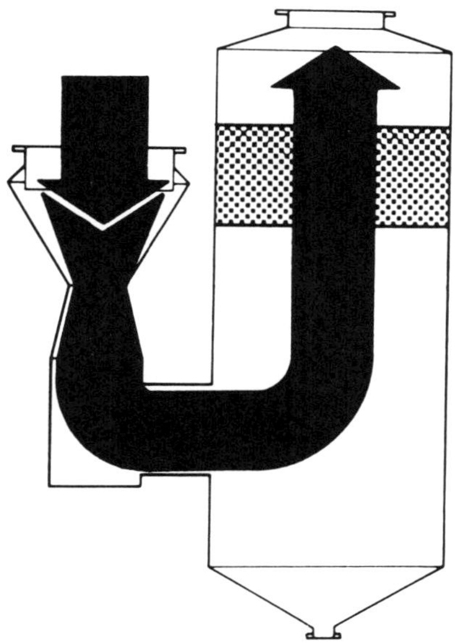

Fig. 12-5. Packed-tower-type mist eliminator. Source: Peabody-Galion, Princeton, New Jersey.

Interception

Interception of a solid particle with a liquid particle occurs when the two particles collide.

Gravitational Force

Gravitational force causes a particle, when passing an obstacle, to fall from its streamline and settle upon the surface of that obstacle.

Impingement

Impingement occurs when an object, placed in the path of a particle-laden gas stream, causes the gas to flow around the obstacle. Larger particles tend to continue in a straight path because of their inertia, and may impinge on the obstacle, fall from the gas stream, and be collected.

Contraction/Expansion

Contraction of a gas stream tends to produce condensation of moisture from the stream. High turbulence in a contracted area results in good contact between solid particulate and liquid droplets. The dust-laden droplets have the same velocity as the rest of the gas stream and, when passing through an area of expansion, will maintain their direction of flow while the balance of the gas stream can be directed to flow in another direction. This process of contraction/expansion produces good separation of particulate matter from the gas stream.

The above mechanisms are normally all present in varying degrees in gas scrubbing equipment at the same time.

The effectiveness of a scrubbing system is usually directly related to the pressure drop across the scrubber. The higher the pressure drop, the greater the turbulence and mixing; therefore, the more effective the scrubbing action. This feature is illustrated by the graph in Fig. 12-6. For a 2 μ diameter particle, for instance, a pressure differential of less than 3 inches WC will provide a removal efficiency of 99%, where as a 20-inch WC differential is necessary for 99% removal of a 0.8 μ particle from the gas stream.

Scrubbing systems are often categorized by system pressure drop. A low-energy system is normally defined as one producing less than 12 inches WC for particulate removal. High-energy systems will have significantly higher pressure drops from 20 to over 60 inches WC. Medium-energy systems operate at 7 to 20 inches WC.

CYCLONIC SCRUBBER

A cyclonic scrubber is shown in Fig. 12-7. The damper (referred to as a *spin damper*) contracts the entering gas stream. Normally, water

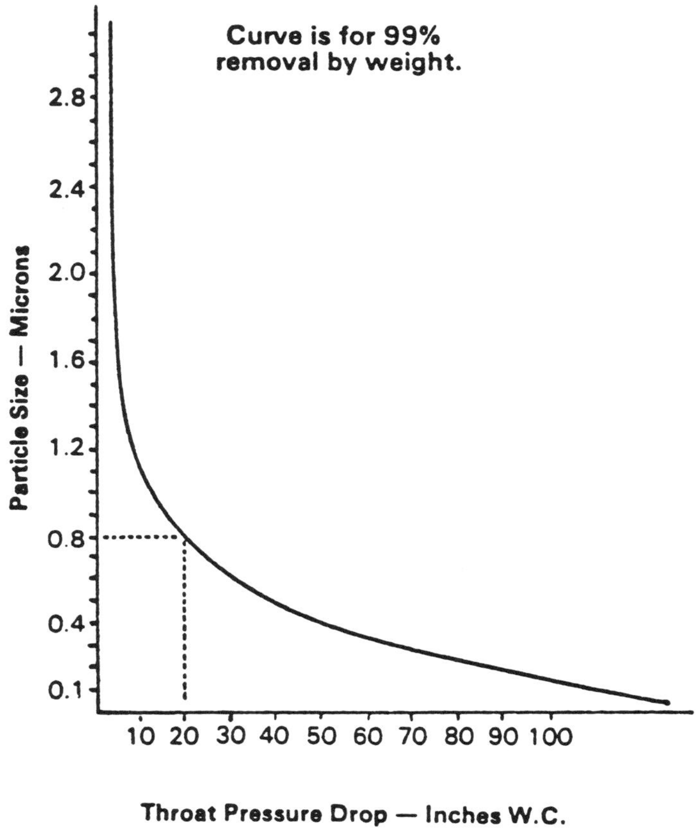

Fig. 12-6. Collection efficiency versus particle size. Source: Reference 2.

is injected immediately upstream of the spin damper. The flow expands within the cyclonic chamber and the contraction/expansion effect combined with inertial effects scrub the gas and separate the dust-laden water from the gas stream.

This is a low-energy device. The pressure drop across the entire unit ranges from 6 to 12 inches WC. The spin damper position can usually be controlled to vary the inlet velocity to the cyclone. This control also provides the necessary adjustment when passing low gas flow versus high or rated flow. The water consumption varies from 5 to 15 gallons per 1000 cubic feet of gas.

VARIABLE-ORIFICE SCRUBBER

A variable-orifice scrubber is little more than a butterfly valve within the gas stream, shown in Fig. 12-8. A nozzle immediately upstream of

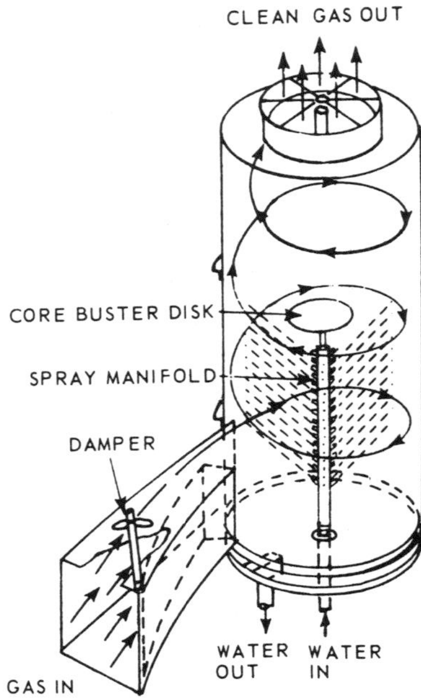

Fig. 12-7. Cyclonic scrubber. Source: Pease Anthony Construction Corp.

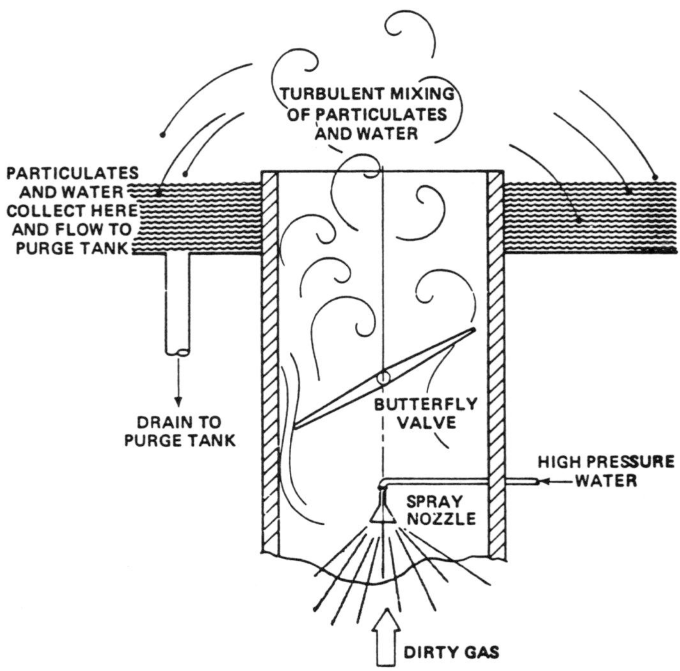

Fig. 12-8. Variable-orifice scrubber. Source: Reference 2.

the butterfly valve injects water into the gas stream. High gas turbulence downstream of the valve causes intimate mixing of water droplets and particulate.

This scrubber is capable of removing submicron particles from the gas stream. It can be readily adjusted to maintain a fixed pressure differential as the gas flow changes. Required water flow is in the range of 8 to 16 gallons per minute per 1000 cfm of gas.

VENTURI SCRUBBER

Venturi scrubbers are widely used (where water is readily available) as high-efficiency, high-energy gas cleaning devices. The heart of the system is a venturi throat, where gases pass through a contracted area, such as that shown in Fig. 12-9, reaching velocities of 200 to 600 feet per second, and then pass through an expansion section. From the expansion section, the gas enters a large chamber for separation of particles or for further scrubbing.

Water is injected at or just upstream of the venturi throat. Figure 12-10 illustrates a venturi scrubber in which water is injected at its throat. For this design, 5 to 7 gallons of water per 1000 cubic feet of gas are required.

A scrubber with an adjustable venturi throat is shown in Fig. 12-11. Two throat flaps are illustrated; however, they are often designed with a single flap. Water is injected into a pre-cooler section immediately before the throat. The throat area is adjustable and is normally controlled (manually or automatically) to maintain a desired pressure drop. Note that

Fig. 12-9. Venturi throat. Copyright © C. R. Brunner, 1980.

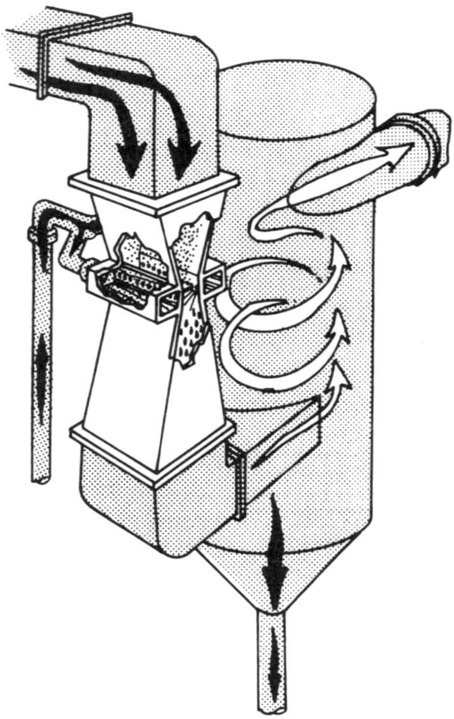

Fig. 12-10. Venturi scrubber. Source: UOP Air Correction Division, Des Plaines, Illinois.

in this illustration a vane or chevron demister (water separator) is included in the outlet chamber to remove entrained water droplets.

A flooded-disk (or plumb-bob) scrubber design is illustrated in Fig. 12-12. The conical plug is positioned to increase or decrease the gas flow area and is similar in function to the throat flap. Water is injected above the venturi section, tangentially to the inlet gas flow. A remote, automated control system is often used to move the plug, as shown in Fig. 12-13.

A serious concern associated with venturi sections is the erosive effect of the gas/liquid mixture passing through the throat. As indicated, the throat velocity is extremely high, creating the potential for wear of the throat surface. In addition, the corrosive effect of the wet, acidic gas flow is heightened by turbulence.

IMPINGEMENT TRAY SCRUBBER

Impingement tray scrubbers are essentially perforated plates with target baffles. Tray scrubbers have no large, gas-directing baffles; they are simply perforated plates within a tower, usually immediately downstream

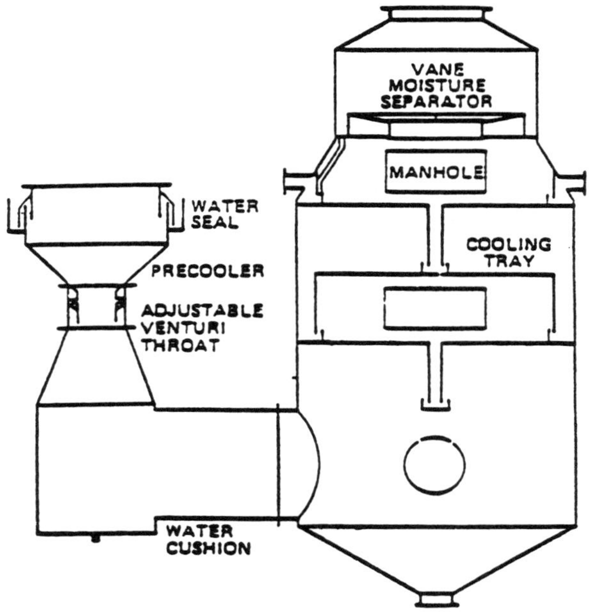

Fig. 12-11. Variable-throat venturi scrubber.

of a venturi. Figure 12-14 illustrates a typical impingement plate. The water level is maintained above the trays (there are usually two or more trays). The specific tray thickness, hole diameter and spacing, and target-plate details, result in a high-efficiency device for the removal of small-size particles less than 5 μ diameter.

As many as 300 openings are provided per square foot of tray area. The openings can be from 1/16 to 3/8 inch in diameter. Gas flows up through the openings, against the static water pressure. Highly effective turbulence with attendant scrubbing of the solid particulate with the water effectively catches the small micron particulate and removes them from the gas stream.

Tray scrubbers have a pressure drop of 2 to 3 inches WC per tray. Impingement plates add another 1/2 to 1 inch WC per tray.

SELF-INDUCED SCRUBBER

A number of scrubbers employ a specialized geometry, utilizing the gas flow to produce scrubbing action. For instance, in the unit shown in Fig. 12-15, the water level controls the scrubbing action of the gas. The higher the water level, the greater and longer the contact between the gas stream and the water. As the water level decreases, however, the gas discharge into the water bath will generate a surface effect, atomizing some of the water. The net effect of this system is to obtain relatively

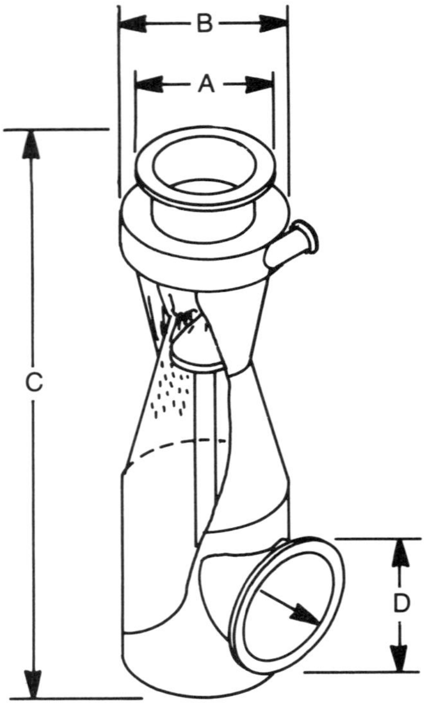

Saturated Gas Volume (ACFM)	A	B	C	D
2,500	1′-0″	3′-6″	6′-9″	1′-1″
5,000	1′-4″	3′-10″	7′-11″	1′-6″
10,000	1′-10″	4′-4″	8′-10	2′-0″
25,000	3′-0″	5′-6″	10′-4″	3′-3″
50,000	4′-2″	7′-2″	13′-7″	4′-5″
75,000	5′-2″	8′-2″	14′-10″	5′-7″
100,000	5′-11″	8′-11″	16′-3″	6′-0″
150,000	7′-3″	10′-3″	18′-3″	7′-6″
350,000	11′-0″	13′-6″	25′-0″	11′-0″

Fig. 12-12. Flooded-disc scrubber. Source: Reference 1.

good efficiency while maintaining relatively low water flows and low gas differential pressure.

ELECTROSTATIC PRECIPITATOR

Electrostatic precipitators (ESPs) are effective devices for the removal of airborne particulate matter. The ESP (a typical unit is shown in cuta-

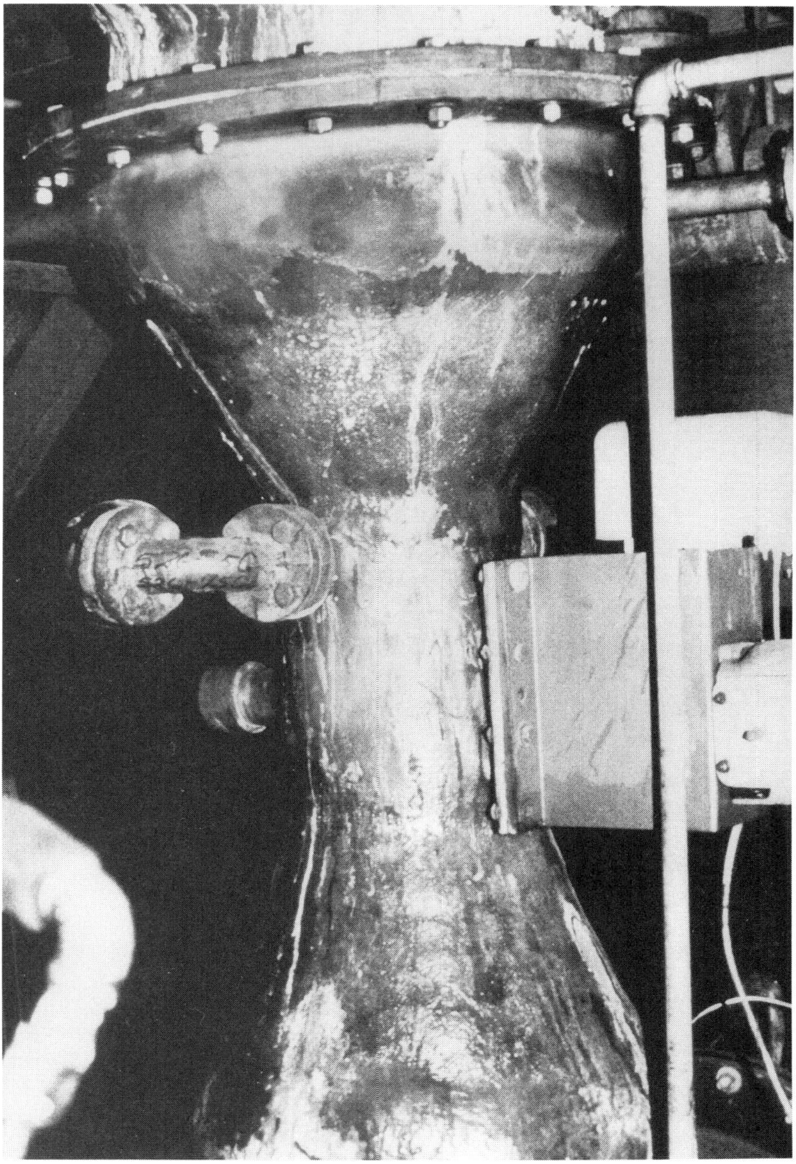

Fig. 12-13. Remote automated venturi scrubber. Copyright © C. R. Brunner, 1982.

way in Fig. 12-16 and an actual ESP installation is shown in Fig. 12-17 operates as follows:

- The gas stream passes through a series of discharge electrodes. These electrodes are negatively charged, at a potential of 1000 to

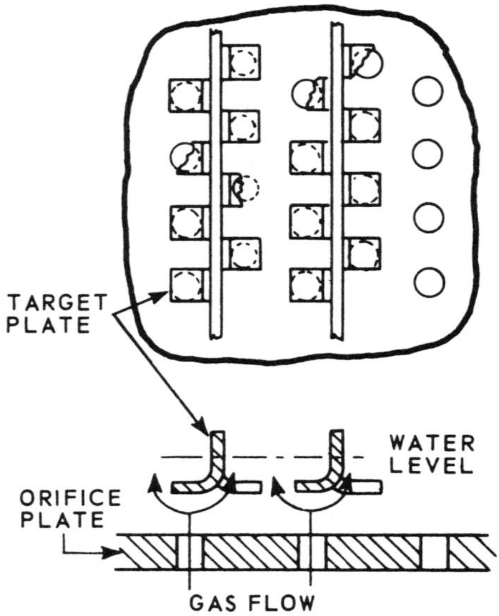

Fig. 12-14. Impingement target plates. Source: Reference 1.

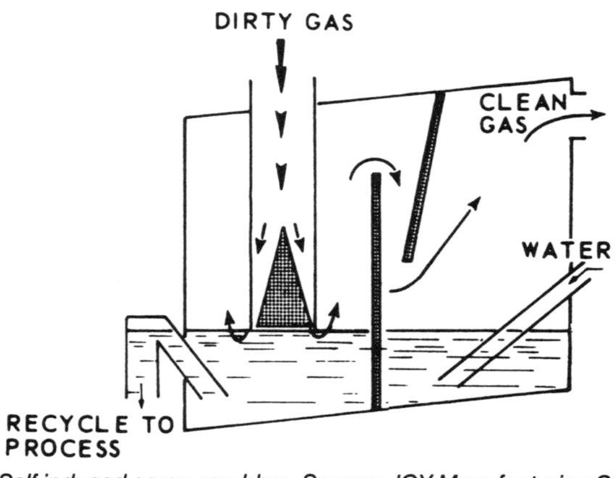

Fig. 12-15. Self induced spray scrubber. Source: JOY Manufacturing Co., Los Angeles, California.

6000 volts. This voltage creates a corona around the electrode. A negative charge is induced in the particulate matter passing through the corona.

- A grounded surface, or collector electrode, surrounds the discharge electrode. Charged particulate collects on the grounded surface, which is usually a plate.
- Particulate matter is removed from the collector surface for ultimate disposal.

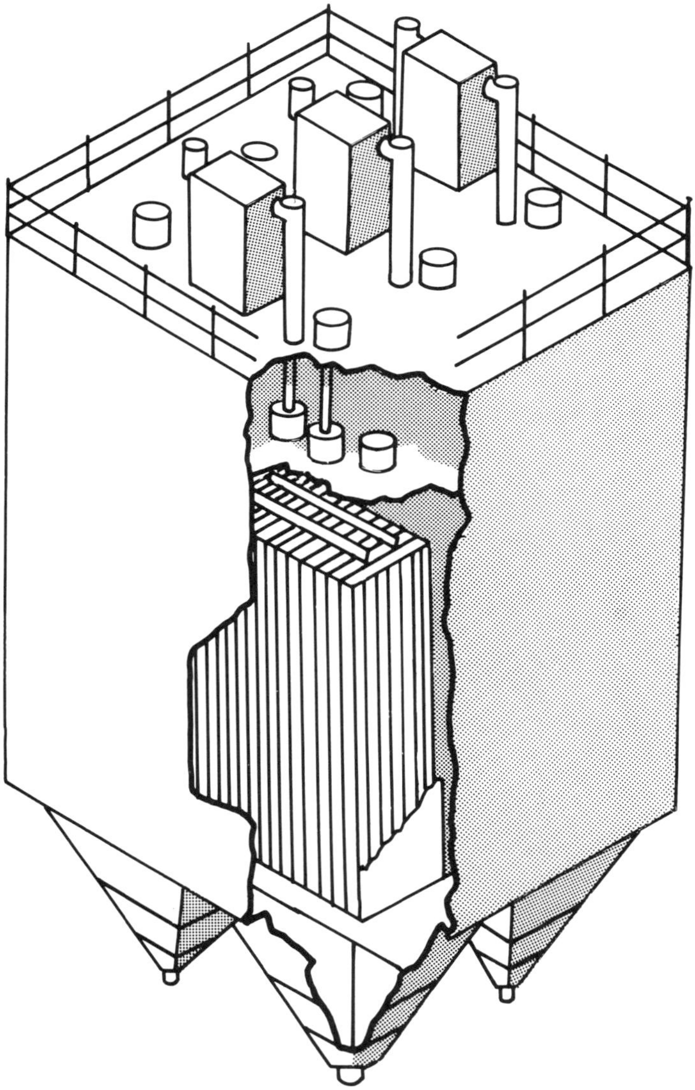

Fig. 12-16. Parallel plate electrostatic precipitator.

Typical discharge electrode and collector plate designs are shown in Fig. 12-18. A variation on these designs is the two-stage ESP, in which the gas passes through a corona discharge prior to entering the collector plate area.

The ESP is very efficient in the collection of small-size particulate, down to the sub-micron range. It can be designed for temperatures as

Fig. 12-17. ESP system. Copyright © C. R. Brunner, 1983.

high as 750° F; however, its efficiency is sensitive to variations in temperature and flue-gas humidity. As shown in Fig. 12-19, ESP efficiency can have substantial variance when operated outside its design envelope.

Removal of accumulated particulate from the collecting surface is key to the success of an ESP installation. If particulate is not removed, it will

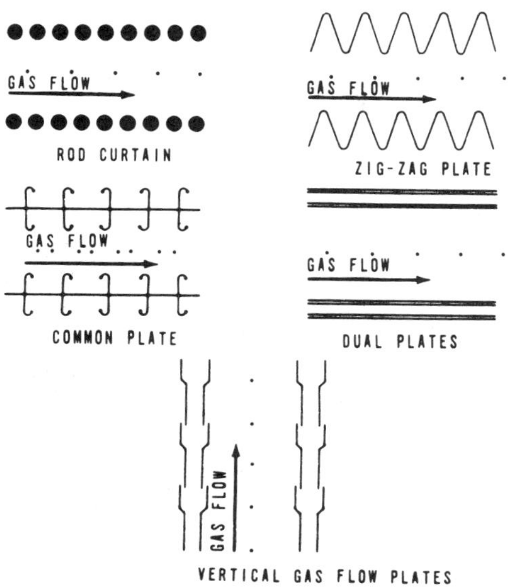

Fig. 12-18. Electrode/collector plate designs.

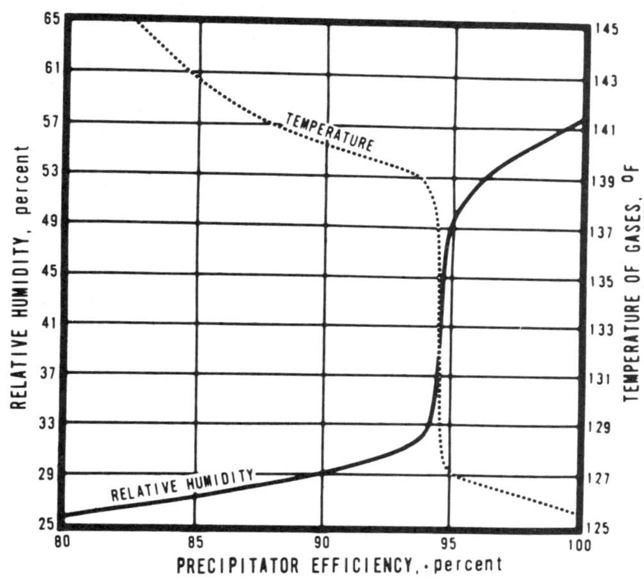

Fig. 12-19. Precipitator efficiency versus relative humidity and temperature. Source: Reference 3.

act as an insulator, preventing the required electrostatic action from occurring.

Various methods have been developed for removing particulate. The most common system is use of "rappers," members that are sequenced to "rap" each plate section at regular intervals. Particulate falls off the plates by this action, collecting in a hopper or series of hoppers below the ESP for eventual final disposal. The ESP shown in Fig. 12-16 has four such collection hoppers.

The resistivity of the particulate matter in the gas stream is a significant parameter in ESP design. If a particle has a high resistivity, it is unable to give up its electric charge to the collecting electrode. The dust will, therefore, build up on the collector, acting as an insulating layer. As this layer increases in depth, its surface will develop a significant electric charge relative to the collector. This charge generates a back discharge, or back corona, which can damage the collecting electrode. Eventually the insulating dust layer prevents the ESP from collecting particulate.

With too small a resistivity, dust will readily relinquish its negative charge to the collector and will assume a positive charge. With the collecting electrode at a positive potential, the particle is repelled back into the gas stream. In the gas stream, which is saturated with negatively charged particles, the dust picks up a negative charge again and eventually returns to the collector plate to be again repelled. The low-resistivity particle, therefore, will not be collected; it will pass through the ESP system.

Electrical resistivity is measured in ohm-centimeters. The optimum resistivity for particulate matter to be effectively collected within an ESP is from 10,000 to 10,000,000,000 ohm-centimeters. The resistivity of most materials varies significantly with temperature, and the use of an ESP may very well mandate the temperature range of collection to that temperature range where the resistivity is within acceptable limits.

With the proper resistivity, a dust particle will relinquish part of its charge to the collecting electrode. The rate at which the charge dissipates increases as the dust layer builds up. When the weight of the collected dust exceeds the electrostatic force available to hold the layer to the collector, the dust particles will fall off under their own weight or will be jarred loose when the collectors are rapped.

Another particulate quality is the tendency to *agglomerate*, or form a hard or tar-like mass impossible to remove from the collector electrodes by rapping. The tendency to agglomerate is influenced by the quantity of moisture in the gas stream and by the temperature of the gas stream.

Collection is a function of gas velocity as well as the other factors noted. The velocity through the collector plates normally ranges from 2 to 4 feet per second.

WET ELECTROSTATIC PRECIPITATOR

Another method of cleaning plates from an ESP is to wet them down. Wet electrostatic precipitators have been developed wherein sprays wet

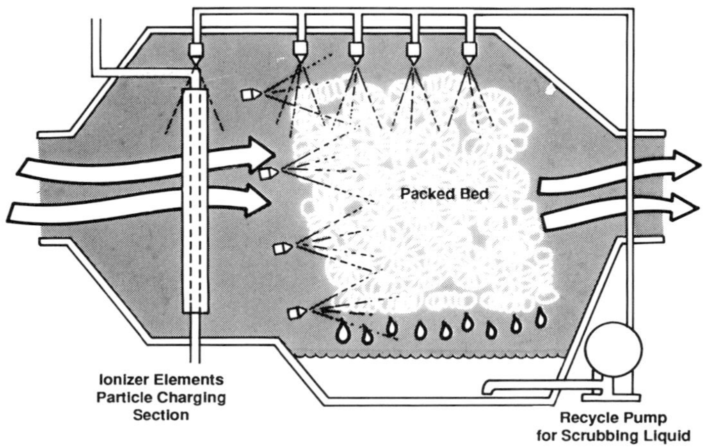

Fig. 12-20. Ionized wet scrubber. Source: Reference 4.

the incoming flue-gas stream to a saturated or super-saturated condition. The electric charge is transferred to the liquid droplets and the liquid charges, collects, and washes away particulate from the gas stream.

The ionized wet scrubber is related to the wet ESP. It consists of a high-voltage ionizer followed by a packing section, shown in Fig. 12-20. The ionizer is continuously washed with water to prevent particulate build-up. A high-voltage dc source negatively charges particulate passing into the unit. These charged particles flow through the packing section, which is also continuously washed with water. In general, particulate matter above 3 μ will be removed from the gas stream by striking the packing. Smaller particles are removed by *image force attraction.*

Figure 12-21 is an illustration of the effect known as image force attraction. When charged particles pass close to the surface of the packing (within 1 millimeter), they induce a charge of opposite polarity. This charge, or attraction, can be compared to the attraction that would exist between a charged particle and its mirror image of equal but opposite polarity at an equal distance behind a neutral surface, i.e., the packing surface. The charged particles are then attracted to the surface of the water, or the packing. Collected particles are washed from the packing surface with water.

EJECTOR SCRUBBING SYSTEM

The ejector-type scrubbing system has been developed as a prime mover as well as a gas-cleaning device. The system is shown in Fig. 12-22 and includes a quench chamber to reduce the gas volume to its adiabatic

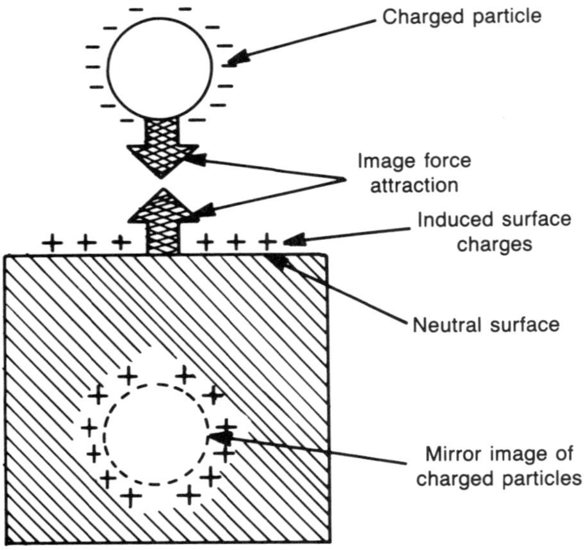

Fig. 12-21. Image force attraction. Source: Reference 4.

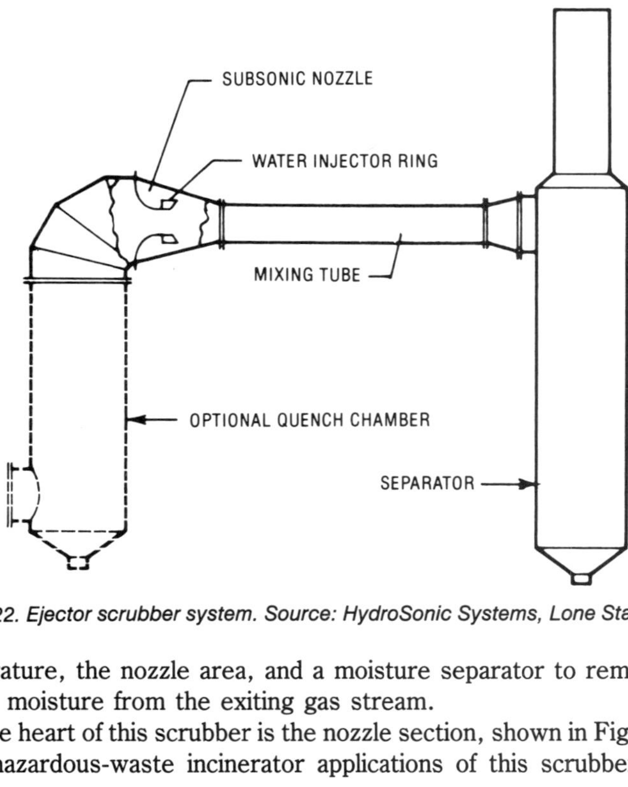

Fig. 12-22. Ejector scrubber system. Source: HydroSonic Systems, Lone Star, Texas.

temperature, the nozzle area, and a moisture separator to remove entrained moisture from the exiting gas stream.

The heart of this scrubber is the nozzle section, shown in Fig. 12-23. Most hazardous-waste incinerator applications of this scrubber utilize

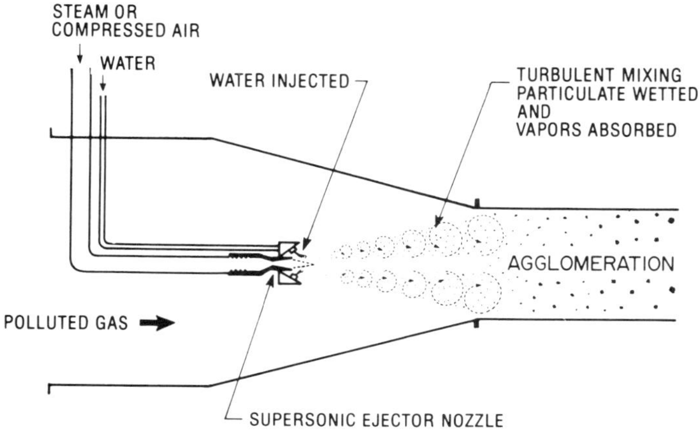

Fig. 12-23. Ejector scrubber nozzle section. Source: HydroSonic Systems, Lone Star, Texas.

steam. A high-pressure (approximately 50 psig) steam nozzle is surrounded by a series of water injection nozzles. Figure 12-24 indicates the required steam rate for a range of particulate loading.

Water is atomized to a size slightly greater than the size of the larger particulate matter. Turbulent contact is encouraged by the steam flow and results in the growth of dust/water particle that can be removed by the cyclonic separator.

The action of the high-velocity gas, steam, and water stream passing through the nozzle creates a strong negative driving pressure, which is sufficient to move gases through a primary combustion chamber (a rotary kiln, for instance), a secondary chamber, a boiler (to generate the required steam flow), a quench chamber, the nozzle itself, and then direct the gases through a mist eliminator and out the discharge stack.

An ejector scrubber and target mist eliminator are shown in Fig. 12-25.

FABRIC FILTERS

Fabric filters, or baghouses, are prevalent in many industrial applications. They are essentially a series of permeable bags which allow the passage of gas, but not particulate matter. They are effective for particle sizes down to the sub-micron range. A typical baghouse on an incinerator is shown in Fig. 12-26.

Filter fabrics are usually woven with relatively large spaces, in excess of 50 μ diameter. The filtration process, therefore, is not just simple sieving, since particles less than 1 μ are caught. Filtration occurs as a result of the combined effects of impact, diffusion, gravitational attraction, and electrostatic forces generated by inter-particle friction. The dust layer itself acts as a filter medium.

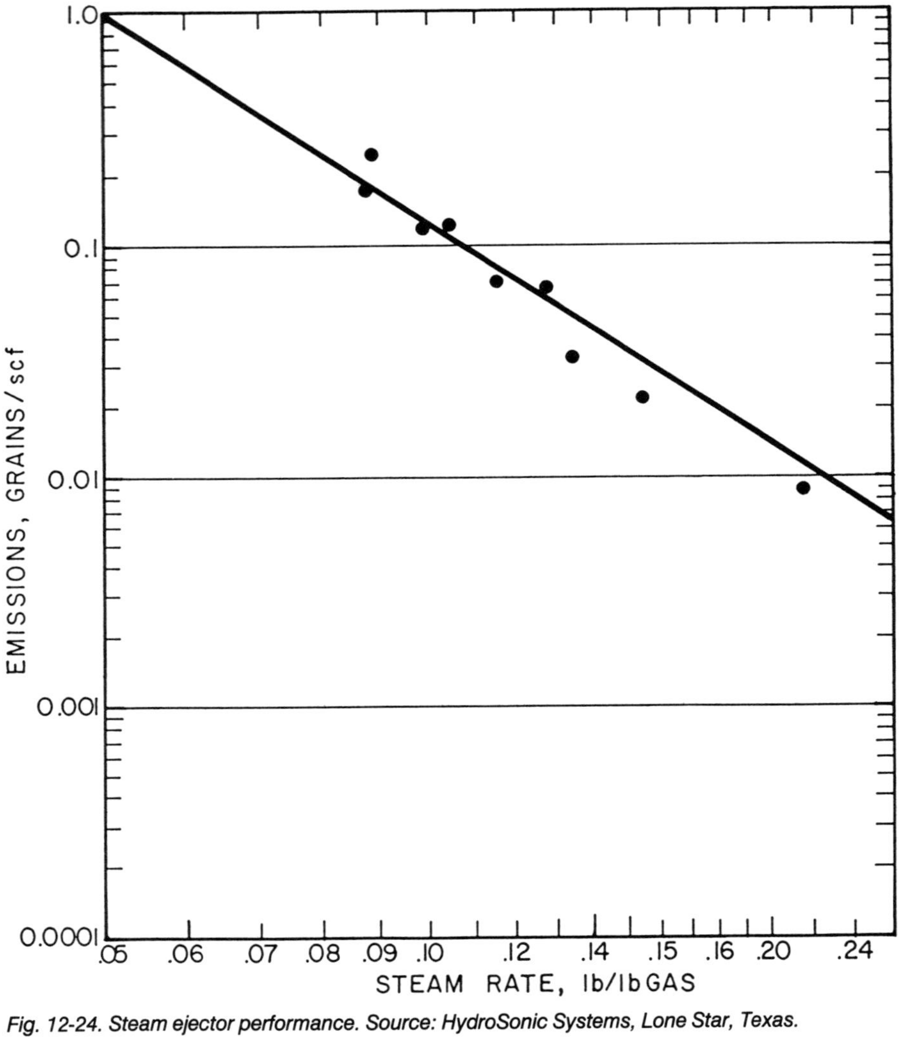

Fig. 12-24. Steam ejector performance. Source: HydroSonic Systems, Lone Star, Texas.

As dust collects on the fabric filter surface the resistance of the filter increases. In addition to this dust-mat resistance there is a clean-cloth resistance which is a function of the type of cloth fiber and its weave. At a particular build-up the filter system reaches its capacity and it must be cleaned, i.e., collected particulate must be discharged. A number of different methods have been developed for particulate removal.

Shaker Mechanism

An eccentric rod physically shakes a bag section, and the falling particles drop to the bottom of a silo, by gravity, for eventual disposal. The

Fig. 12-25. Ejector scrubber and target mist eliminator. Copyright © C. R. Brunner, 1985.

Fig. 12-26. Baghouse installation. Copyright © C. R. Brunner, 1983.

shaker motor is sequenced to operate in conjunction with operation of the fresh air dampers. Fresh air is admitted to that damper section whose bags are shaken, or agitated. The fresh, clean air aids in discharging the accumulated dust (Fig. 12-27).

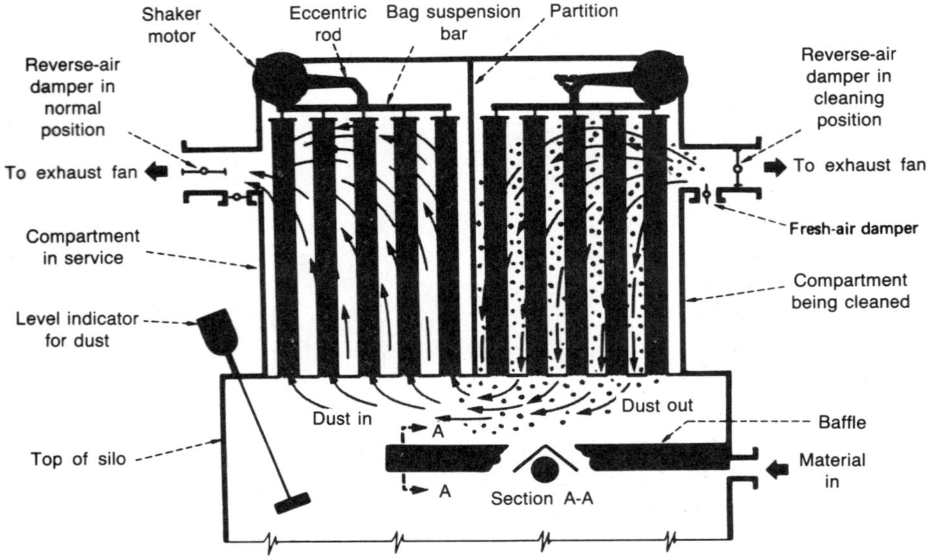

Fig. 12-27. Baghouse with shaker mechanism. Source: Reference 5.

Compressed Air

A blast of compressed air is directed into the inside of each bag discharging the dust accumulated on the external surface of the bags. Wire retainers are provided to help re-inforce the bags against the abrupt action of the air blast (Fig. 12-28).

Re-pressurization

The filter sections are independent of each other. Through a series of inlet and exhaust valves, shown in Fig. 12-29, the dirty gas flow passes through the inside of the bags. On a timed sequence, the flow is reversed to pressurize the outside of each bag. Under external pressure, the internal dust loading will fall from the bag surfaces to a collection hopper for disposal.

Sonic Cleaning

A source of intense sound tuned to the resonant frequency of the bags creates sympathetic vibrations in the bags. Under this vibration, the dust falls from the bags for collection and eventual removal.

Related to a baghouse, an HEPA (High Efficiency Particulate Air) filter has been developed to control extremely small particulate discharges. They can remove over 99.97 percent of particles 0.3 μ diameter and larger.

A typical HEPA filter, shown in Fig. 12-30, is constructed of a glass-fiber mat, pleated to increase its unit surface area. The filter is mounted

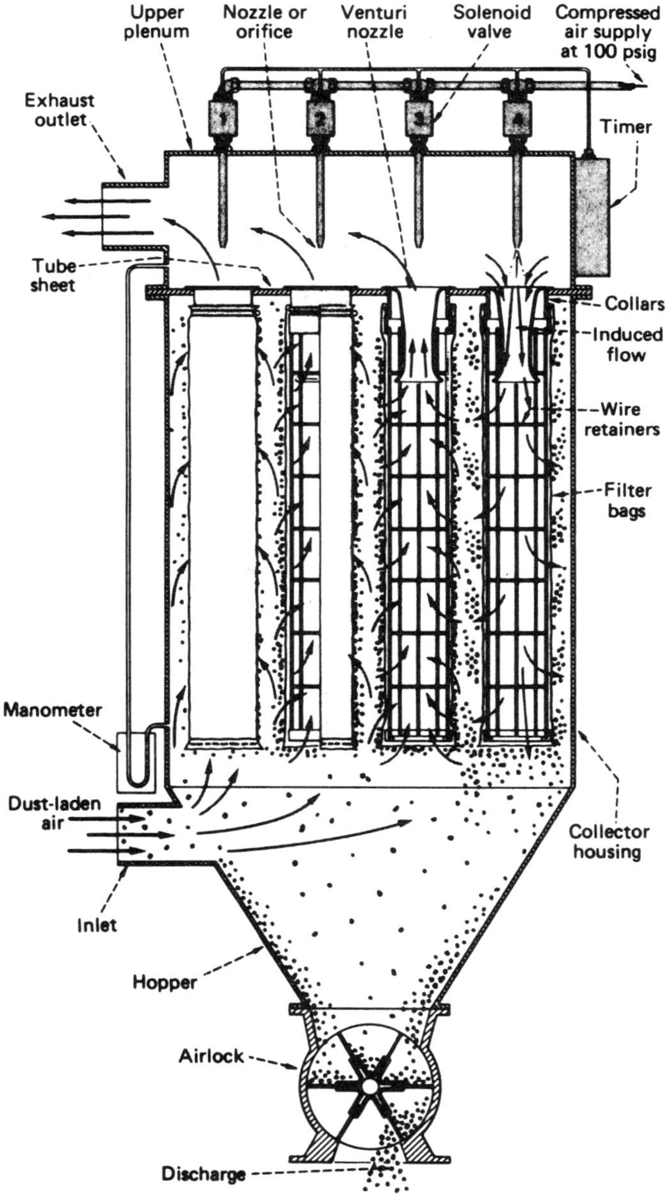

Fig. 12-28. Compressed air utilization in a baghouse. Source: Reference 5.

in a frame and a series of frames are mounted in a filter band to provide the required flow capacity.

HEPA filters are normally used with other control devices. They should be placed downstream of a device that removes larger particulate.

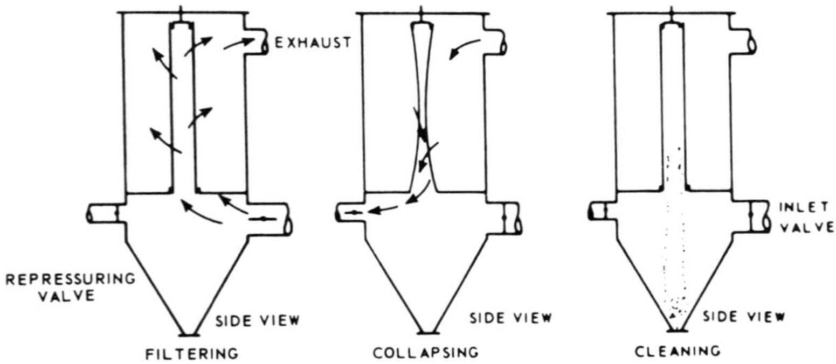

Fig. 12-29. Reverse-air flexing of a baghouse. Source: Reference 6.

Placing an HEPA filter in a stream containing larger particulate matter will result in clogging of the filter, and severely shortens its useful life.

REFERENCES

1. C. R. Brunner, *Incineration Systems: Selection & Design* (Reston, VA: Incinerator Consultants Incorporated, 1988).

2. T. D. Ellis, R.B. Diemer, Jr., and C.R. Brunner, *Industrial Hazardous Waste Course Notes*, AIChE short course text, 1987.

3. R. Coulter, ''Smoke, Dust, Fumes Closely Controlled in Electric Furnaces, '' *Iron Age*, January 1954.

4. S. V. Sheppard, ''Ionizing Wet Scrubbers Control Plant Emissions,'' *Mechanical Engineering*, July 1985.

5. M. Kraus, ''Baghouses,'' *Chemical Engineering*, April 9, 1979.

6. *Control Techniques for Particulate Pollutants*, National Air Pollution Control Administration, AP-51, September 1981.

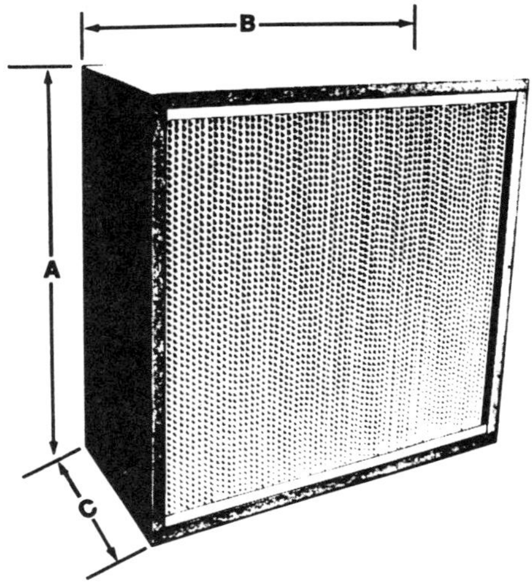

Rated SCFM at 1" W.G.	Actual Dimensions (Less Gaskets) A (in.)	B (in.)	C (in.)
900	24	36	5 7/8
1230	24	48	5 7/8
1550	24	60	5 7/8
1900	24	72	5 7/8
750	30	24	5 7/8
925	30	30	5 7/8
1150	30	36	5 7/8
1550	30	48	5 7/8
1975	30	60	5 7/8
2350	30	72	5 7/8
900	36	24	5 7/8
1150	36	30	5 7/8
1400	36	36	5 7/8
1900	36	48	5 7/8
2350	36	60	5 7/8
2850	36	72	5 7/8

Source: **American Air Filter Company, Louisville, KY.**

Fig. 12-30. HEPA Filter. Source: American Air Filter Co., Louisville, Kentucky.

13

Acid-Gas Control

A significant proportion of hazardous-waste streams derive from industrial processes in which chlorine is a major constituent of the product. This is particularly true in the plastics industry. To a lesser extent, sulfur is found in many wastes. Both of these constituents generate a potential acid discharge in the incinerator exhaust.

REGULATORY REQUIREMENTS FOR ACID REMOVAL

TSCA requires that all incinerators permitted under its requirements include acid-gas scrubbing. The RCRA incinerator regulations mandate acid-gas scrubbing of any incinerator discharge with more than 4 pounds hydrogen chloride. In this chapter, techniques of acid-gas control are discussed.

WET SCRUBBING

The wet-scrubbing process reduces the temperature of the gas stream and removes the acidic component in one system. This system normally consists of a venturi scrubber, for particulate control, followed by a tray tower or a packed tower. An alkaline solution is circulated in the tower to wash acidic components from the gas stream.

Generally, caustic soda (NaOH) or lime is used for neutralization of incinerator exhaust gas. Removal of acidic gas components is described by the following equations:

Caustic Scrubbing

64.06		16.00		80.00		142.04		18.02
SO_2	+	½ O_2	+	2 NaOH	---->	Na_2SO_4	+	H_2O
1.00				1.25		2.22		0.28

With the atomic weights of each component on the upper line of the above equation, the line below the equation represents each component relative to the weight of SO_2 present. One pound of SO_2 requires 1.25 pounds of NaOH for neutralization, and produces 2.22 pounds of Na_2SO_4 and 0.28 pound of H_2O.

Likewise, for HCl scrubbing with caustic soda:

36.46		40.00		58.44		18.02
HCl	+	NaOH	---->	NaCl	+	H_2O
1.00		1.10		1.61		0.49

One pound of HCl requires 1.10 pounds of NaOH for complete neutralization, producing 1.61 pounds of NaCl and 0.49 pound of H_2O.

Lime Scrubbing

The use of lime (CaO) requires the slaking of lime into a concentrated liquid, calcium hydroxide, $Ca(OH)_2$, per this equation:

56.08		18.02		74.10
CaO	+	H_2O	---->	$Ca(OH)_2$
1.00		0.32		1.32

The $Ca(OH)_2$ produced combines with the acidic component of the gas stream as follows:

64.06		74.10		120.14		18.02
SO_2	+	$Ca(OH)_2$	---->	$CaSO_3$	+	H_2O
1.00		1.16		1.88		0.28

From this equation, 1.16 pounds of $Ca(OH)_2$ is required to neutralize one pound of SO_2. With one pound of lime required for 1.32 pounds of $Ca(OH)_2$, 1.53 pounds of lime is required for this reaction. $CaSO_3$ is generated from this reaction at the rate of 1.88 pounds per pound of SO_2 and 0.28 of H_2O is also produced.

The neutralization of HCl proceeds as follows, using lime as the reagent:

72.92		74.10		110.98		36.04
2 HCl	+	$CA(OH)_2$	---->	$CaCl_2$	+	2 H_2O)
1.00		1.01		1.52		0.49

Neutralization of one pound of HCl requires 1.01 pounds of $Ca(OH)_2$, or 0.77 pounds of lime, CaO. For each pound of HCl neutralized, 1.52 pounds of $CaCl_2$ and 0.49 pounds of H_2O are generated.

To a lesser degree, the use of lime as a reagent results in the absorption if carbon dioxide, as follows:

$$CO_2 + Ca(OH)_2 \text{ - - - -> } CaCO_3 + H_2O$$

This carbonation of lime occurs due to the high carbon dioxide content of the flue gas.

The above equations are for the stoichiometric quantities of alkali. The actual use of alkali will be in excess of these figures, representing the fact that alkali will also scrub carbon dioxide out of the gas stream, that mixing, or scrubbing/contact of reagent and solution is never ideal, and that the distribution of acid gases in the exhaust is not uniform.

In the selection of lime versus caustic for scrubbing, the cost of each and the solubility of the salts generated by each should be considered. Sodium hydroxide costs over ten times more than calcium hydroxide (lime). On the other hand, calcium salts are much less soluble than those of sodium. For instance, at 200° F, the solubility of sodium sulfate is 43 grams per 100 milliliters while the solubility of calcium sulfate is only 0.2 grams per 100 milliliters. When calcium hydroxide is used for acid-gas cleaning, a white calcium sulfate precipitate is formed, which, if unchecked, settles out in tanks, idle pipes, and treatment ponds.

DRY-SCRUBBING SYSTEMS

Dry-scrubbing systems for acid-gas removal operate as sorbent systems rather than washing systems. Wet systems operate with relatively high pressure requirements, utilizing high-horsepower fans. Their ability to remove small particulate matter from the gas stream below μ is limited, and carbonaceous aerosols are hydrophobic and may not be effectively captured by a wet system. The major problem inherent in a wet scrubbing system, however, is the consumption of relatively large fresh-water supplies and the generation of a wastewater.

There are a number of dry-scrubbing systems in use in hazardous-waste incineration facilities. These systems have been relatively common in Europe during the past five years and recently have been adapted to United States installations.

DRY-SCRUBBER OPERATION

Only lime is used to neutralize acidic gases in a dry scrubber; either lime or caustic can be used in a wet scrubber. Another difference is that in a wet scrubber collection and removal occurs in the same device. A dry scrubber requires a number of major sub-systems; neutralization, collection, and removal require separate units.

A generic system is shown schematically in Fig. 13-1. Incinerator off-gas is normally reduced in temperature to the range of 400° F to 600° F, often by passing the hot gas stream through a waste-heat boiler. This

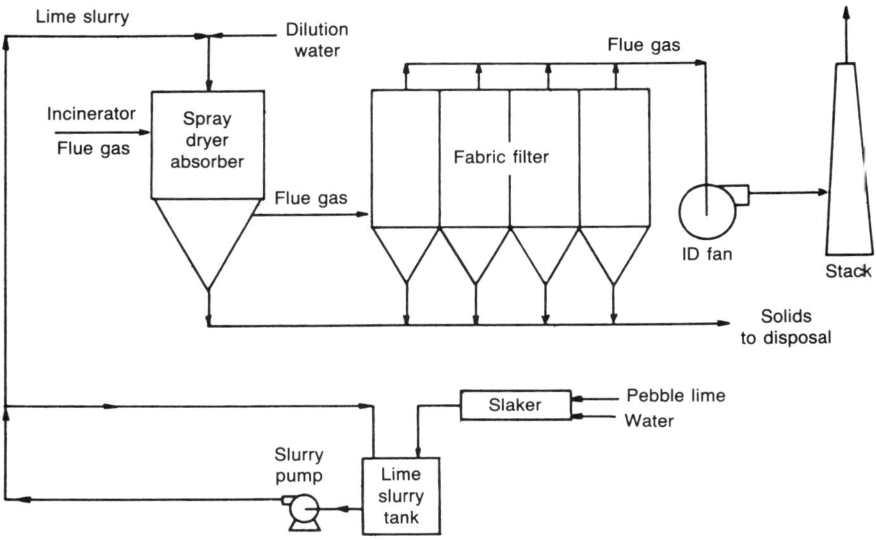

Fig. 13-1. Spray dryer system. Source: Reference 1.

temperature range allows the use of conventional materials (carbon steel, etc.) within the absorber.

Lime is injected as a calcium hydroxide slurry into the absorber in the range of 5 to 50 percent calcium hydroxide to water. The reactions are basically those described in equations 4 and 5 for the removal of SO_2 and HCl from the gas stream. The slurry strength is controlled through the dilution-water line.

The absorbent (lime slurry) is injected into the absorber (spray dryer) as a finely atomized spray, producing droplets in the range of 30 to 100 μ diameter. The absorber is sized to provide from 10 to 122 seconds gas residence time. With this retention time and with appropriate absorber geometry, the lime slurry evaporates and does not come in contact with the reactor walls. The temperature of the gas exiting the absorber is normally maintained at 250° F to 300° F.

Flue gases exiting the absorber pass through a collection device, either an electrostatic precipitator, or, as shown in the illustration, a baghouse. Solids in the gas stream form a cake on the surface of the bags, and the growth of these solids provides additional opportunity for acid removal in the system. Use of an electrostatic precipitator requires that all acid be removed in the absorber, and the absorber has to be designed appropriately for this increased loading.

Some designs utilize a dry injection system. Water is injected upstream of the absorber, with dry lime (not lime slurry) injected into the absorber. A wet-dry (lime slurry) system has been found to be more difficult to maintain than a dry injection system. Lime tends to cake in pipes, nozzles, and on surfaces exposed to the slurry. The major advantage of such a system, however, is that it requires less equipment than a dry system, which requires a separate quench chamber, or cooling tower.

FLAKT

Flakt has developed a number of dry scrubbing systems for use on hazardous-waste incinerators for acid-gas control. Two of their designs are the dry system and the wet-dry system. Both of these utilize a baghouse for particulate lime collection and disposal.

The dry system, shown in Fig. 13-2, accepts flue gas from the incinerator at a temperature of 490° F to 520° F. Quench water is atomized by nozzles in the gas cooler to provide a uniform reduction of temperature of the gas stream to the range of 230° F to 280° F. The injected water is completely vaporized in this section.

From the gas cooler, the stream enters the bottom of a cyclonic absorber tower, or dry scrubber. Larger particles are removed from the gas stream by inertial effects. Above the point of gas entrance, dry lime powder is injected into the scrubber with compressed air. The lime nozzle is directed downwards, counter to the upwards direction of the exhaust gas stream. The gas, now cooled and conditioned with lime, and containing dry lime reaction products, is directed to a baghouse (fabric filter) for collection, further growth and acid removal on the bag surfaces, and eventual disposal.

In the wet-dry system, shown in Fig. 13-3, hot off-gases exiting the incinerator enter a scrubber directly. Lime is injected into the scrubber as a slurry, through a nozzle arrangement that produces a fine, well-atomized spray. As with the dry system, water is evaporated completely. No entrained moisture is present in the gas stream exiting the scrubber.

From the wet-dry scrubber, the gas stream passes through a fabric filter for particulate collection and removal.

In general, baghouses used in these systems are designed to withstand a temperature of 480° F, and the gas entering the baghouse is limited to a maximum sustained temperature of 280° F. Bags are fabricated of chemically inert materials, such as Teflon conglomerates. Collection occurs on the outside of the bags. The Flakt system uses compressed air to clean the bags; it is pulsed to provide continual cleaning during use.

These systems can operate with fabric filter "ash" recirculation. This "ash" contains unreacted hydrated lime which is mixed with fresh lime in either slurry or dry form before re-injection into its scrubber.

Table 13-1 demonstrates the effectiveness of dry scrubbing in the removal of hydrogen chloride from the gas stream. Removal efficiency is affected by temperature and differences between wet-dry and dry systems may not be significant.

Sulfur dioxide removal is presented in Table 13-2. It appears that temperature has a greater effect on SO_2 removal than on the removal of HCl,

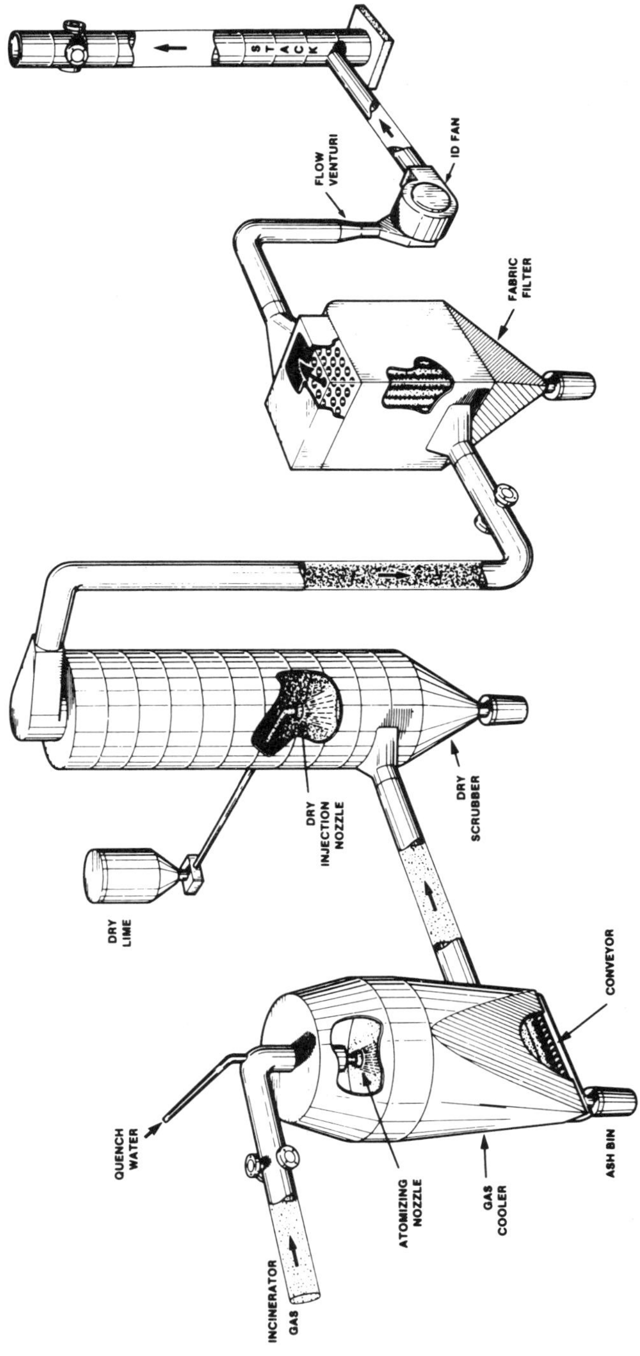

Fig. 13-2. Flakt's dry system. Source: Reference 7.

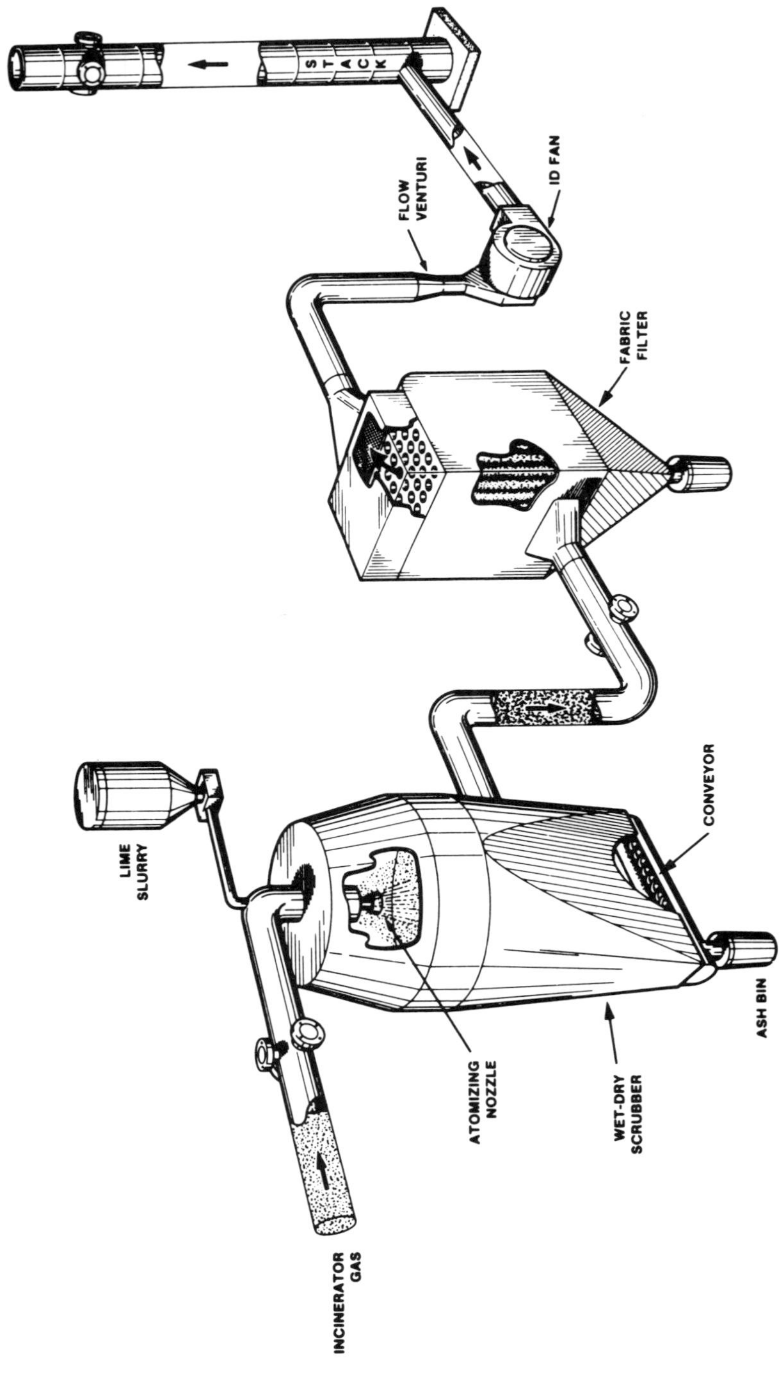

Fig. 13-3. Flakt's wet-dry system. Source: Reference 7.

and at 285° F there is no difference in performance of the wet versus the wet-dry system.

In removal of either HCl and SO_2, recycle appears to have no effect on performance.

The removal of dioxins and dibenzofurans is demonstrated in Table 13-3 and Table 13-4, respectively, based on a series of tests on the Flakt

Table 13-1. Hydrogen Chloride Concentrations and Removal Efficiencies Corrected to 8% Oxygen in the Gas Stream.

	Dry System				Wet-Dry System	
Flue Gas Temp. @ Baghouse Inlet, °F	230	260	285	>400	285	285 + recycle
Inlet (ppmw)	423	464	475	392	366	470
Outlet (ppmw)	7	9	29	91	29	42
Efficiency, %	98	98	94	77	92	91

Note: ppmw = parts HOl per million parts of gas by weight

Source: Reference 7.

Table 13-2. Sulfur Dioxide Concentrations and Removal Efficiencies Corrected to 8% Oxygen in the Gas Stream.

	Dry System				Wet-Dry System	
Flue Gas Temp. @ Baghouse Inlet, °F	230	260	285	>400	285	285 + recycle
Inlet (ppmw)	119	118	99	117	106	106
Outlet (ppmw)	4	10	41	83	35	43
Efficiency, %	96	92	58	29	67	60

Note: ppmw = parts HCl per mission parts of gas by weight

Source: Reference 7.

Table 13-3. Polychlorinated Dibenzo-p-dioxin Concentrations and Removal Efficiencies Corrected to 8% Oxygen in the Gas Stream.

	Dry System				Wet-Dry System	
Flue Gas Temp. @ Baghouse Inlet, °F	230	260	285	>400	285	285 + recycle
Inlet (ng/Nm^3)	580	1400	1300	1030	1100	1300
Outlet (ng/Nm^3)	0.2	ND	ND	6.1	ND	0.4
Efficiency, %	>99.9	>99.9	>99.9	99.4	>99.9	>99.9

Note: ND = not detected

Source: Reference 7.

Table 13-4. Polychlorinated Dibenzofuran Concentrations and Removal Efficiencies Corrected to 8% Oxygen in the Gas Stream.

	Dry System				Wet-Dry System	
Flue Gas Temp. @ Baghouse Inlet, °F	230	260	285	>400	285	285 + recycle
Inlet (ng/Nm^3)	300	940	1000	560	660	850
Outlet (ng/Nm^3)	2.3	ND	1.0	1.2	ND	0.9
Efficiency, %	99.3	>99.9	99.9	99.8	>99.9	99.9

Note: ND = not detected

Source: Reference 7.

scrubber. Efficiencies at all test points are extremely high. In many cases, the level of dioxins or dibenzofurans was undetectable.

TELLER SYSTEM

The Teller dry-scrubber system uses lime for acid neutralization and a separate absorbent. Hot incinerator exhaust gas enters the inlet of a quench reactor, as shown in Fig. 13-4. This is a cyclone, where larger particles drop out of the gas stream through inertial action. Rising through the reactor, an alkali solution is sprayed into the swirling gas stream and fully wets it. The alkali (a lime slurry or sodium carbonate solution) neutralizes the acid component of the gas stream and quenches any sparklers within the gas flow. The reactor is designed to neutralize the acidic gas components within one second by formation of an alkaline mist within the reactor. The gas-residence time within the reactor is nominally seven seconds.

The neutralized gas stream leaves the quench reactor at its adiabatic temperature (normally 150° F to 180° F) and passes through the dry venturi, Fig. 13-5. A highly crystalline, inert material, from 3-μ to 20-μ diameter particle size, is injected into the gas stream through the venturi. Talc or waste product fines from several manufacturing industries can be used.

Within the venturi, the generated turbulence tends to complete the neutralization process. In addition, the injected powder will adsorb particulate matter within the gas stream and will also act as a catalist in promoting agglomeration of particulate matter. Particles down to the submicron range have been found to agglomerate to particles of 10 μ and greater diameter.

A baghouse downstream of the dry venturi catches the particles in the gas stream. From 95 to 99 percent of the particulate matter exiting the incinerator is removed, in addition to over 90 percent of the HCl and over 65 percent of the SO_2 in the exhaust gas. Teller estimates that use of an ESP in lieu of a baghouse would result in a drop of particulate removal

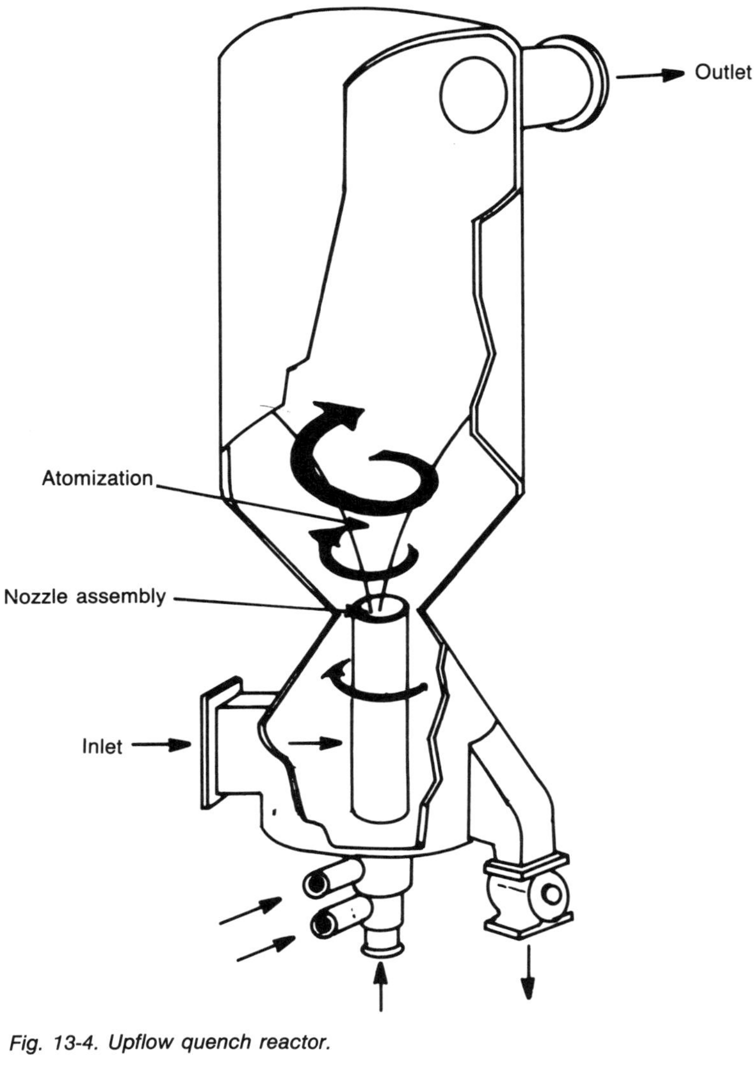

Fig. 13-4. Upflow quench reactor.

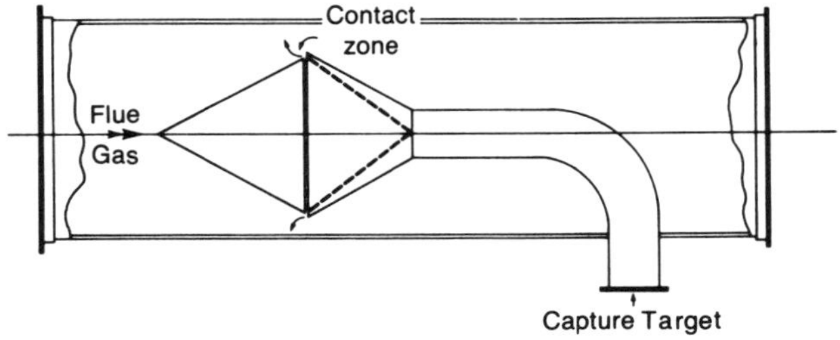

Fig. 13-5. Teller dry scrubber.

Fig. 13-6. Teller dry scrubber installation. Copyright © C. R. Brunner, 1975.

efficiency to approximately 90 percent. This is due to the growth and residence of particles on fabric filter bags, which does not occur in an ESP. A Teller dry scrubber applied to an incinerator is illustrated in Fig. 13-6.

REFERENCES

1. P. Kroll and P. Williamson, "Application of Dry Flue Gas Scrubbing to Hazardous Waste Incineration," *Journal of the Air Pollution Control Association,* Volume 36, Number 11, November 1986.

2. K. Feindler, "Long Term Results of Operating TA Luft Acid Gas Scrubbing Systems," *Proceedings of the 1986 National Waste Processing Conference,* American Society of Mechanical Engineers Solid Waste Processing Division.

3. B. Hurst and C. White, "Thermal DeNox: A Commercial Selective Noncatalytic Nox Reduction Process for Waste-to-Energy Applications," *Proceedings of the 1986 National Waste Processing Conference,* American Society of Mechanical Engineers Solid Waste Processing Division.

4. B. Flynn, H. Hsieh, and D. Curtis, "Effect of Product Recycle on Dry Acid Gas Emission Control," *Proceedings of the 1984 National Waste Processing Conference,* American Society of Mechanical Engineers Solid Waste Processing Division.

5. Y. Kiang, "The Formation of Nitrogen Oxides in Hazardous Waste Incineration," *Proceedings of the 1982 National Waste Processing Conference,* American Society of Mechanical Engineers Solid Waste Processing Division.

6. T. D. Ellis, R. B. Diemer, Jr., and C. R. Brunner, *Industrial Hazardous Waste Incineration,* AIChE short course text, 1987.

7. "The National Incinerator Testing and Evaluation Program: Air Pollution Control Technology," Report EPS 3/UP/2, *Environment Canada,* September 1986.

8. C. R. Brunner, *Hazardous Air Emissions from Incinerators,* Second Edition (New York: Chapman & Hall, 1986).

14

Heating Value Determination

Incineration is a thermal process dependent upon the heating value of the waste. Wastes will either have insufficient heating value to support their own combustion or will have a high enough heat content to maintain autogenous combustion at a specific temperature. Incineration system calculations must start with a determination of heating value, which is covered in this chapter.

WASTE QUALITY

The most important characteristic of any waste stream is its heterogeneity. Almost by definition, waste quality is inconsistent from batch to batch. If a waste's quality were predictable, it would not be considered a waste, and would have utility.

A determination of waste characteristics must always be considered approximate. One waste batch may include constituents not present in another batch; process aberrations may result in more of one component in one batch than in another. If taking a sample of waste, attention must be directed to making the sample as representative of the day-to-day waste stream as possible. Analytical samples are necessarily a very small fraction of the waste stream, and obtaining a representative sample is often the most crucial step in the accurate determination of waste parameters.

HAZARDOUS-COMPONENT HEATING VALUE

In many instances, the hazardous component of a waste stream is relatively small. Many solid and sludge wastes, for instance, have very low concentrations of hazardous waste. With low hazardous-waste concentrations, waste heating value is less a function of the hazardous-waste constituent than of the supporting matrix.

Heating values for hazardous-waste constituents are listed in Table 2-11 and Table 2-12. Methods of determining the heating value of other materials are described in the following sections.

PROXIMATE ANALYSIS

This procedure is a relatively quick and inexpensive laboratory determination of moisture, volatile matter, fixed carbon, and ash contents. This analytic procedure is as follows:

1. Heat a sample for one hour at 220° F - 230° F (105° C - 110° C). Report the weight loss fraction as percent moisture.
2. Raise the temperature of the dried sample, in a covered crucible, to 1340° F (725° C) and hold it at this temperature for 7 minutes. Report the sample weight loss fraction as volatile matter percentage.
3. Ignite the remaining sample in an open crucible at 1740° F (950° C) and allow it to burn to a constant weight. Report the sample weight loss as percent fixed carbon.
4. The sample residual is to be reported as percent ash. The sum of moisture, volatiles, fixed carbon, and ash should equal 100 percent.

Proximate analysis is described in detail in ASTM Standard D 3172.

ULTIMATE ANALYSIS

An *ultimate analysis* is a standard procedure used for the determination of the quantities of elemental constituents present in a sample. It is required to determine the products of combustion of a material, its combustion air requirement, and the nature of the off-gas or combustion products.

In this procedure the following element fractions are usually determined:

- Carbon
- Hydrogen
- Sulfur
- Oxygen
- Nitrogen
- Halogens (chlorine, fluorine, etc.)

- Heavy metals (mercury, lead, etc.)
- Other elements that can effect the combustion process

In addition to the components, analyses may also be performed, under the heading of ultimate analysis for the presence of certain compounds that might be in the waste, such as hazardous constituents, PCBs, dioxins, and other organics.

ASTM Standard D 3176 describes the standard method for performing an ultimate analysis.

Ultimate analyses are performed by specialty laboratories with special equipment developed specifically for elemental or molecular analyses, such as the gas chromatograph, infrared scanners, or mass spectrometers.

MEASUREMENT OF HEATING VALUE

The most common equipment used for determining heating value of a sample of material is the oxygen bomb calorimeter. In this instrument, a measured sample, usually one gram, is ignited in an enclosed atmosphere of pure oxygen by an electric wire. The heat of combustion of the sample heats a water bath surrounding the bomb. The temperature rise of the water is measured, and the heat of combustion is calculated from this temperature increase.

The standard test method for determining heating value with a bomb calorimeter is described in ASTM Standard D 3286.

When dealing with a waste stream, this method of analysis is often impractical. A sample as small as one gram will not be a representative sample of sludge from an industrial wastewater treatment plant collecting a varied effluent from an industrial facility. Neither will it necessarily reflect the properties of a heterogeneous mixture of mixed, solid hazardous waste.

TABULATED VALUES

Table 14-1 lists heating values and other combustion characteristics of common chemicals. Note that the high (or gross) heating value assumes that the moisture of combustion condenses within the combustion systems, i.e., the exit temperature of the flue gas is less than 212° F at atmospheric pressure. The lower heating value (net) considers water present in the vapor phase, i.e., if the exit temperature from the combustion systems is greater than 212° F.

Characteristics of common substances are listed in Table 14-2, including proximate analyses and heating value.

HEAT OF FORMATION

The heat of formation of a compound is that amount of heat absorbed when it is formed from its prime elements. If the formation reaction gener-

Table 14-1. Combustion Constants.

No.	Substance	Formula	Molecular Weight	Lb per Cu Ft	Cu Ft per Lb	Sp Gr Air 1.0000	Heat of Combustion: Btu per Cu Ft Gross (High)	Btu per Cu Ft Net (Low)	Btu per Lb Gross (High)	Btu per Lb Net (Low)	For 100% Total Air, Moles per mole of Combustible or Cu Ft per Cu Ft of Combustible: Required for Combustion O_2	Required for Combustion N_2	Required for Combustion Air	Flue Products CO_2	Flue Products H_2O	Flue Products N_2	For 100% Total Air, Lb per Lb of Combustible: Required for Combustion O_2	Required for Combustion N_2	Required for Combustion Air	Flue Products CO_2	Flue Products H_2O	Flue Products N_2
1	Carbon*	C	12.01	...	...	...	...	...	14,093	14,093	1.0	3.76	4.76	1.0	...	3.76	2.66	8.86	11.53	3.66	...	8.86
2	Hydrogen	H_2	2.016	0.0053	187.723	0.0696	325	275	61,100	51,623	0.5	1.88	2.38	...	1.0	1.88	7.94	26.41	34.34	...	8.94	26.41
3	Oxygen	O_2	32.000	0.0846	11.819	1.1053	...	...	...	...	...	...	...	...	...	...	...	...	...	...	...	...
4	Nitrogen (atm)	N_2	28.016	0.0744	13.443	0.9718	...	...	...	...	...	...	...	...	...	...	...	...	...	...	...	...
5	Carbon monoxide	CO	28.01	0.0740	13.506	0.9672	322	322	4,347	4,347	0.5	1.88	2.38	1.0	...	1.88	0.57	1.90	2.47	1.57	...	1.90
6	Carbon dioxide	CO_2	44.01	0.1170	8.548	1.5282	...	...	...	...	...	...	...	...	...	...	...	...	...	...	...	...
Paraffin series																						
7	Methane	CH_4	16.041	0.0424	23.565	0.5543	1013	913	23,879	21,520	2.0	7.53	9.53	1.0	2.0	7.53	3.99	13.28	17.27	2.74	2.25	13.28
8	Ethane	C_2H_6	30.067	0.0803	12.455	1.0488	1792	1641	22,320	20,432	3.5	13.18	16.68	2.0	3.0	13.18	3.73	12.39	16.12	2.93	1.80	12.39
9	Propane	C_3H_8	44.092	0.1196	8.365	1.5617	2590	2385	21,661	19,944	5.0	18.82	23.82	3.0	4.0	18.82	3.63	12.07	15.70	2.99	1.63	12.07
10	n-Butane	C_4H_{10}	58.118	0.1582	6.321	2.0665	3370	3113	21,308	19,680	6.5	24.47	30.97	4.0	5.0	24.47	3.58	11.91	15.49	3.03	1.55	11.91
11	Isobutane	C_4H_{10}	58.118	0.1582	6.321	2.0665	3363	3105	21,257	19,629	6.5	24.47	30.97	4.0	5.0	24.47	3.58	11.91	15.49	3.03	1.55	11.91
12	n-Pentane	C_5H_{12}	72.144	0.1904	5.252	2.4872	4016	3709	21,091	19,517	8.0	30.11	38.11	5.0	6.0	30.11	3.55	11.81	15.35	3.05	1.50	11.81
13	Isopentane	C_5H_{12}	72.144	0.1904	5.252	2.4872	4008	3716	21,052	19,478	8.0	30.11	38.11	5.0	6.0	30.11	3.55	11.81	15.35	3.05	1.50	11.81
14	Neopentane	C_5H_{12}	72.144	0.1904	5.252	2.4872	3993	3693	20,970	19,396	8.0	30.11	38.11	5.0	6.0	30.11	3.55	11.81	15.35	3.05	1.50	11.81
15	n-Hexane	C_6H_{14}	86.169	0.2274	4.398	2.9704	4762	4412	20,940	19,403	9.5	35.76	45.26	6.0	7.0	35.76	3.53	11.74	15.27	3.06	1.46	11.74
Olefin series																						
16	Ethylene	C_2H_4	28.051	0.0746	13.412	0.9740	1614	1513	21,644	20,295	3.0	11.29	14.29	2.0	2.0	11.29	3.42	11.39	14.81	3.14	1.29	11.39
17	Propylene	C_3H_6	42.077	0.1110	9.007	1.4504	2336	2186	21,041	19,691	4.5	16.94	21.44	3.0	3.0	16.94	3.42	11.39	14.81	3.14	1.29	11.39
18	n-Butene	C_4H_8	56.102	0.1480	6.756	1.9336	3084	2885	20,840	19,496	6.0	22.59	28.59	4.0	4.0	22.59	3.42	11.39	14.81	3.14	1.29	11.39
19	Isobutene	C_4H_8	56.102	0.1480	6.756	1.9336	3068	2869	20,730	19,382	6.0	22.59	28.59	4.0	4.0	22.59	3.42	11.39	14.81	3.14	1.29	11.39
20	n-Pentene	C_5H_{10}	70.128	0.1852	5.400	2.4190	3836	3586	20,712	19,363	7.5	28.23	35.73	5.0	5.0	28.23	3.42	11.39	14.81	3.14	1.29	11.39
Aromatic series																						
21	Benzene	C_6H_6	78.107	0.2060	4.852	2.6920	3751	3601	18,210	17,480	7.5	28.23	35.73	6.0	3.0	28.23	3.07	10.22	13.30	3.38	0.69	10.22
22	Toluene	C_7H_8	92.132	0.2431	4.113	3.1760	4484	4284	18,440	17,620	9.0	33.88	42.88	7.0	4.0	33.88	3.13	10.40	13.53	3.34	0.78	10.40
23	Xylene	C_8H_{10}	106.158	0.2803	3.567	3.6618	5230	4980	18,650	17,760	10.5	39.52	50.02	8.0	5.0	39.52	3.17	10.53	13.70	3.32	0.85	10.53
Miscellaneous gases																						
24	Acetylene	C_2H_2	26.036	0.0697	14.344	0.9107	1499	1448	21,500	20,776	2.5	9.41	11.91	2.0	1.0	9.41	3.07	10.22	13.30	3.38	0.69	10.22
25	Naphthalene	$C_{10}H_8$	128.162	0.3384	2.955	4.4208	5854	5654	17,298	16,708	12.0	45.17	57.17	10.0	4.0	45.17	3.00	9.97	12.96	3.43	0.56	9.97
26	Methyl alcohol	CH_3OH	32.041	0.0846	11.820	1.1052	868	768	10,259	9,078	1.5	5.65	7.15	1.0	2.0	5.65	1.50	4.98	6.48	1.37	1.13	4.98
27	Ethyl alcohol	C_2H_5OH	46.067	0.1216	8.221	1.5890	1600	1451	13,161	11,929	3.0	11.29	14.29	2.0	3.0	11.29	2.08	6.93	9.02	1.92	1.17	6.93
28	Ammonia	NH_3	17.031	0.0456	21.914	0.5961	441	365	9,668	8,001	0.75	2.82	3.57	...	1.5	3.32	1.41	4.69	6.10	...	1.59	5.51
29	Sulfur*	S	32.06	...	...	...	...	...	3,983	3,983	1.0	3.76	4.76	SO_2 1.0	...	3.76	1.00	3.29	4.29	SO_2 2.00	...	3.29
30	Hydrogen sulfide	H_2S	34.076	0.0911	10.979	1.1898	647	596	7,100	6,545	1.5	5.65	7.15	1.0	1.0	5.65	1.41	4.69	6.10	1.88	0.53	4.69
31	Sulfur dioxide	SO_2	64.06	0.1733	5.770	2.2640	...	...	...	...	...	...	...	...	...	...	...	...	...	...	...	...
32	Water vapor	H_2O	18.016	0.0476	21.017	0.6215	...	...	...	...	...	...	...	...	...	...	...	...	...	...	...	...
33	Air	...	28.9	0.0766	13.063	1.0000	...	...	...	...	...	...	...	...	...	...	...	...	...	...	...	...

*Carbon and sulfur are considered as gases for molal calculations only.

Note: This table is reprinted from *Fuel Flue Gases*, 1941 Edition, courtesy of American Gas Association. All gas volumes corrected to 60°F and 30 in Hg dry.

Table 14-2. Characteristics of Selected Materials.

Waste	B.T.U. value/lb. as fired	Wt. in lbs. per cu. ft. (loose)	Wt. in lbs. per cu. ft.	Content by weight in percentage ASH	Content by weight in percentage MOISTURE
Type 0 Waste	8,500	10		5	10
Type 1 Waste	6,500	10		10	25
Type 2 Waste	4,300	20		7	50
Type 3 Waste	2,500	35		5	70
Type 4 Waste	1,000	55		5	85
Kerosene	18,900		50	.5	0
Benzene	18,210		55	.5	0
Toluene	18,440		52	.5	0
Hydrogen	61,000		.0053	0	0
Acetic acid	6,280		65.8	.5	0
Methyl alcohol	10,250		49.6	0	0
Ethyl alcohol	13,325		49.3	0	0
Turpentine	17,000		53.6	0	0
Naphtha	15,000		41.6	0	0
Newspaper	7,975	7		1.5	6
Brown paper	7,250	7		1.0	6
Magazines	5,250	35		22.5	5
Corrugated paper	7,040	7		5.0	5
Plastic coated paper	7,340	7		2.6	5
Coated milk cartons	11,330	5		1.0	3.5
Citrus rinds	1,700	40		.75	75
Shoe Leather	7,240	20		21.0	7.5
Butyl sole composition	10,900	25		30.0	1
Polyethylene	20,000	40-60	60	0	0
Polyurethane (foamed)	13,000	2	2	0	0
Latex	10,000	15	45	0	0
Rubber waste	9,000-11,000	62-125		20-30	
Carbon	14,093		138	0	0
Wax paraffin	18,621		54-57	0	0
1/3 wax-2/3 paper	11,500	7-10		3	1
Tar or asphalt	17,000	60		1	0
1/3 tar-2/3 paper	11,000	10-20		2	1
Wood sawdust (pine)	9,600	10-12		3	10
Wood sawdust	7,800-8,500	10-12		3	10
Wood bark (fir)	9,500	12-20		3	10
Wood bark	8,000-9,000	12-20		3	10
Corn cobs	8,000	10-15		3	5
Rags (silk or wool)	8,400-8,900	10-15		2	5
Rags (linen or cotton)	7,200	10-15		2	5
Animal fats	17,000	50-60			0
Cotton seed hulls	8,600	25-30		2	10
Coffee grounds	10,000	25-30		2	20
Linoleum scrap	11,000	70-100		20-30	1

The above chart shows the various BTU values of materials commonly encountered in incinerator designs. The values given are approximate and may vary based on their exact characteristics or moisture content. The BTU value is the higher heating value.

Source: Ref. 8-4.

ates heat, the heat of formation (ΔH_f) has a negative sign, and the reaction is *exothermic*. If the reaction absorbs heat (*endothermic*), ΔH_f is positive.

Using values for the heat of formation, the heat of combustion can be calculated. The heat of combustion of a substance is equal to that amount of heat released when the substance is completely burned.

The heat of formation of selected organic compounds is listed in Table 14-3. Table 14-4 lists the heats of formation of solid inorganic oxides and Table 14-5 lists the heats of formation of miscellaneous materials. The units used are those in which this parameter is normally stated, kcal/mole and gram/mole. The conversion to English units is as follows:

$$(\text{kcal/mole}) \bullet (\text{BTU}/0.252 \text{ kcal}) \div (\text{gram/mole}) \bullet (\text{lb}/454 \text{ gram})$$
$$1 \text{ kcal/gram} = 1802 \text{ BTU/lb}$$

The calculation of heating value using the heat of combustion can be illustrated by the burning of benzene, C_6H_6, as follows:

$$\begin{array}{llllll} & & & & -57.8 \text{ (g)} \\ +11.7 & & & -94.1 & -68.3 \text{ (1)} \\ C_6H_6 + O_2 & & \text{--->} & 6\ CO_2 & +\ 3\ H_2O \\ 78.12 & & & & \end{array}$$

The quantities 78.12 and 11.7 are the molecular weight and heat of formation of liquid benzene, from Table 14-3. The heat of formation of carbon dioxide (−94.1), gaseous water/steam (−57.8) and liquid water (−68.3) is from Table 14-5.

Assuming that all water produced remains steam, the sum of the heat of formation ΔH_f is calculated:

$$\begin{array}{rcl} \Delta H_f\ H_2O\ (g) : & -57.8 \times 3 = & -173.4 \\ \Delta H_f\ CO_2 : & -94.1 \times 6 = & -564.6 \\ \Delta H_f\ C_6H_6 : & (-)11.7 \times 1 = & -\ \ 11.7 \end{array}$$

$$\Sigma\ \Delta H_f = -749.7 \text{ kcal/mole}$$

Relating $\Sigma\ \Delta H_f$ to the weight of benzene present:

$$1/(78.12 \text{ gram/mole}) \bullet (-749.7) \text{ kcal/mole} \bullet 1802 = -17293 \text{ BTU/lb}$$

If the products of combustion exit the reaction at a temperature below 212° F, the higher heating value of benzene is developed, and the heat of combustion would be calculated utilizing −68.3 kcal/mole of water, as follows on page 294:

Table 14-3. Heat of Formation of Selected Organic Compounds.

Formula	Weight, g/mole	Name	ΔH_f, kcal/mole
CBr_3	251.71	Carbon tribromide (g)	+ 42.0
CBr_4	331.61	Carbon tetrabromide (g)	+ 19.0
CBr_4	331.61	Carbon tetrabromide (c)	+ 4.5
CCl_3	118.36	Carbon trichloride (g)	+ 14.0
CCl_4	153.81	Carbon tetrachloride (g)	- 24.6
CCl_4	153.81	Carbon tetrachloride (liq)	- 32.4
CF_2O	66.01	Carbonyl difluoride (g)	- 151.7
CF_3	69.01	Carbon trifluoride (g)	- 114.0
CF_4	88.01	Carbon tetrafluoride (g)	- 223.0
CO_3Fe	115.86	Iron carbonate (c)	- 177.0
$CHBr_3$	252.72	Bromoform (g)	+ 4.0
$CHBr_3$	252.72	Bromoform (liq)	- 6.8
$CHCl_3$	119.37	Chloroform (g)	- 24.7
$CHCl_3$	119.37	Chloroform (liq)	32.1
CHF_3	70.02	Fluoroform (g)	- 164.5
CHI_3	393.72	Iodoform (c)	33.7
CH_2N_4	70.07	Tetrazole (c)	+ 56.7
CH_2N_4O	86.07	5-hydroxytetrazole (c)	+ 1.5
CH_2O	30.03	Formaldehyde (g)	- 26.0
CH_2O_2	46.03	Formic acid (liq)	- 101.5
CH_3Br	94.94	Methyl bromide (g)	- 8.4
CH_3Cl	50.49	Methyl chloride (g)	- 19.3
CH_3Hg	215.63	Methyl mercury (g)	+ 40.0
CH_3I	141.94	Methyl iodide (g)	+ 3.1
CH_3I	141.94	Methyl iodide (liq)	- 3.7
CH_4	16.05	Methane (g)	- 17.9
CH_4N_2O	60.07	Urea (c)	- 79.7
CH_4O	32.05	Methanol (g)	- 48.0
CH_4O	32.05	Methanol (liq)	- 57.0
$CMnO_3$	114.95	Manganese carbonate (c)	- 213.7
CO_3Zn	125.38	Zinc carbonate (c)	- 194.3
C_2F_4	100.02	Tetrafluoroethylene (g)	- 155.5
C_2F_6	138.02	Hexaflouroethane (g)	- 310.0
$C_2H_2Cl_4$	167.84	1,1,2,2-tetrachloroethane (g)	- 35.7
$C_2H_2CoO_4$	148.97	Cobaltous formate (c)	- 208.7
$C_2H_2CuO_4$	153.59	Copper formate (c)	- 186.7
$C_2H_2MnO_4$	144.98	Manganese formate (c)	- 249.7
$C_2H_2NiO_4$	148.75	Nickel formate (c)	- 208.4
$C_2H_2O_2$	58.04	Glyoxal (g)	- 50.7
$C_2H_2O_4$	90.04	Oxalic acid (c)	- 197.7
$C_2H_2O_4Pb$	297.23	Lead formate (c)	- 210.0
$C_2H_2O_4Zn$	155.41	Zinc formate (c)	- 235.8
$C_2H_3AgO_2$	166.92	Silver acetate (c)	- 95.3
C_2H_3Br	106.95	Vinyl bromide (g)	+ 18.7
C_2H_3BrO	122.95	Acetyl bromide (liq)	- 53.4
$C_2H_3Br_3O_2$	298.75	Bromal hydrate (c)	- 112.0

Table 14-3. Continued.

Formula	Weight, g/mole	Name	ΔH_f, kcal/mole
C_2H_3Cl	62.50	Vinyl chloride (g)	+ 8.5
C_2H_3Cl	62.50	Vinyl chloride (liq)	+ 3.5
C_2H_3ClO	78.50	Acetyl chloride (g)	- 58.2
C_2H_3ClO	78.50	Acetyl chloride (liq)	- 65.4
$C_2H_3ClO_2$	94.50	Chloral hydrate (c)	- 137.7
$C_2H_3ClO_2$	94.50	Chloral hydrate (g)	- 107.2
C_2H_4	28.06	Ethylene (g)	+ 12.5
$C_2H_4O_2$	60.06	Acetic acid (liq)	- 115.7
$C_2H_4O_2$	60.06	Methyl formate (g)	- 83.7
$C_2H_4O_2$	60.06	Methyl formate (liq)	- 90.6
$C_2H_4O_3$	76.06	Gylcolic acid (c)	- 158.6
$C_2H_4O_4$	92.06	Glyoxylic acid (c)	- 199.7
C_2H_5Br	108.97	Ethyl bromide (g)	- 15.4
C_2H_5Br	108.97	Ethyl bromide (liq)	- 22.0
C_2H_5Cl	64.52	Ethyl chloride (g)	- 26.8
C_2H_5Cl	64.52	Ethyl chloride (liq)	- 32.6
C_2H_5ClO	80.52	Ethylene chlorohydrin (liq)	- 70.6
C_2H_5I	143.89	Ethyl iodide (g)	- 1.8
C_2H_5I	143.89	Ethyl iodide (liq)	- 9.6
C_2H_5N	43.08	Ethylenimine (liq)	+ 21.9
C_2H_6	30.08	Ethane (g)	- 20.2
C_2H_6O	46.08	Ethanol (g)	- 56.2
C_2H_6O	46.08	Ethanol (liq)	- 66.4
$C_3H_4O_3$	78.07	Ethylene carbonate (c)	- 138.9
C_3H_6	42.09	Propylene (g)	+ 4.9
C_3H_6O	58.09	Propanone (g)	- 51.8
C_3H_6O	58.09	Propanone (liq)	- 59.2
$C_3H_6O_2$	74.09	Methane acetate (liq)	- 106.4
C_3H_8	44.11	Propane (g)	- 24.8
$C_3H_9N_5O_4$	179.17	Acetamideguanidine nitrate (c)	- 119.1
C_3H_9P	76.09	Trimethylphomphine (liq)	- 30.0
$C_3H_{10}N_2$	74.15	1,2-propanediamine (liq)	- 23.4
$C_3H_{10}N_2O_3$	122.15	Trimethylamine nitrate (c)	- 74.1
$C_3H_{10}O_3Si$	122.22	Trimethexymilane (liq)	- 199.0
$C_3H_{10}Sn$	164.82	Trimethyl tin (g)	+ 5.0
$C_3H_{10}Sn$	164.82	Trimethyl tin (liq)	- 2.1
$C_3H_{12}N_6O_3$	180.21	Guanidine carbonate (c)	- 232.1
$C_4H_2N_2S$	110.14	4-cyanothiazole (c)	+ 52.6
$C_4H_4N_2$	80.10	Pyraxine (c)	+ 33.4
$C_4H_4N_2$	80.10	Pyraxine (g)	+ 46.9
C_4H_4NS	99.16	4-Methylthiazole (liq)	+ 16.3
C_4H_6	54.1	Butadiene (g)	+ 38.8
C_4H_6	54.0	1-butyne (g)	+ 39.5
$C_4H_6O_4Pb$	325.29	Lead acetate (c)	- 230.4
$C_4H_6O_4Zn$	183.47	Zinc acetate (c)	- 257.8
$C_4H_6O_6$	150.10	Tartaric acid (c)	- 308.5

Table 14-3. Continued.

Formula	Weight, g/mole	Name	ΔH_f, kcal/mole
C_4H_8	56.12	Cyclobutane (liq)	+ 0.8
C_4H_8	56.12	Butene (g)	- 2.7
C_4H_8O	72.12	Butanone (g)	- 59.3
C_4H_8O	72.12	Butanone (liq)	- 65.3
C_4H_9N	71.14	Pyrrolidine (g)	- 0.9
C_4H_9N	71.14	Pyrrolidine (liq)	- 9.9
C_4H_{10}	58.14	Butane (g)	- 30.1
$C_4H_{10}O$	74.14	Butanol (liq)	- 78.2
$C_4H_{10}O$	74.14	Diethyl ether (g)	- 60.3
$C_4H_{11}N$	73.16	Butylamine (liq)	- 30.5
$C_5H_4N_4O$	136.13	Hypoxanthine (c)	- 26.2
$C_5H_4N_4O_2$	152.13	Xanthine (c)	- 90.5
$C_5H_4N_4O_3$	168.13	Uric acid (c)	- 147.7
$C_5H_4O_2$	96.09	Furfural (liq)	- 47.8
C_5H_5N	79.11	Pyridine (g)	+ 33.6
C_5H_5N	79.11	Pyridine (liq)	+ 24.0
$C_5H_6N_4O_4$	186.15	Pseudouric acid (c)	- 221.7
$C_5H_6O_2$	98.11	Furfuryl alcohol (liq)	- 66.0
C_5H_8	68.13	Pentadine (g)	+ 33.6
$C_5H_9NO_2$	115.15	DL-proline (c)	- 125.7
$C_5H_9NO_3$	131.15	L-hydroxyproline (c)	- 158.1
C_5H_{10}	70.15	Cyclopentane (liq)	- 25.3
C_5H_{10}	70.15	Pentene (g)	- 5.0
$C_5H_{10}O$	86.15	2-pentanone (liq)	- 71.1
C_5H_{12}	72.17	Pentane (g)	- 35.0
C_5H_{12}	88.17	Pentane (liq)	- 41.4
$C_5H_{12}O$	88.17	Pentanol (liq)	- 64.3
$C_5H_{12}O_5$	152.17	Xylitol (c)	- 267.3
C_6H_6	78.12	Benzene (g)	+ 19.6
C_6H_6	78.12	Benzene (liq)	+ 11.7
C_6H_6O	94.12	Phenol (c)	- 39.5
C_6H_7N	93.14	Aniline (liq)	+ 7.6
C_6H_7N	93.14	Methylpyridine (g)	+ 24.1
C_6H_7N	93.14	Methylpyridine (liq)	+ 13.8
$C_6H_8O_2$	112.14	Sorbic acid (c)	- 93.4
$C_6H_8O_6$	176.14	Ascorbic acid (c)	- 278.3
$C_6H_8O_7$	192.14	Citric acid, anhydrous (c)	- 269.0
C_6H_{10}	82.16	Hexyne (g)	+ 29.6
$C_6H_{10}O_2$	114.16	Hydrosorbic acid (liq)	- 110.2
$C_6H_{10}O_5$	162.16	Saccharinic acid lactone (c)	- 249.6
$C_6H_{10}O_8$	210.16	Citric acid monohydrate (c)	- 439.4
$C_6H_{10}O_8$	210.16	Mucic acid (c)	- 423.0
C_6H_{12}	84.18	Cyclohexane (liq)	- 37.3
C_6H_{12}	84.18	Hexene (g)	- 9.9
$C_6H_{12}O_2$	116.18	Caproic acid (liq)	- 139.7
$C_6H_{12}O_3$	132.18	Acetone glycerol (liq)	- 163.0

Table 14-3. Continued.

Formula	Weight, g/mole	Name	ΔH_f, kcal/mole
$C_6H_{12}O_5$	164.18	Fucose (c)	− 262.7
$C_6H_{12}O_6$	180.18	Fructose (c)	− 302.2
$C_6H_{12}O_6$	180.18	Glucose (c)	− 203.8
$C_6H_{12}O_7$	196.18	Gluconic acid (c)	− 379.3
C_6H_{14}	86.20	Dimethylbutane (liq)	− 51.0
C_6H_{14}	86.20	Hexane (liq)	− 47.5
$C_6H_{14}O$	102.20	Hexanol (liq)	− 90.7
$C_6H_{14}O_6$	182.20	Dulcitol (c)	− 321.8
$C_6H_{14}O_7$	198.20	Glucose hydrate (c)	− 375.0
$C_6H_{15}N$	101.22	Triethylamine (liq)	− 32.1
C_7H_6O	106.13	Benzaldehyde (liq)	− 20.1
$C_7H_6O_2$	122.13	Benzoic acid (c)	− 92.0
$C_7H_6O_3$	138.13	Salicylic acid (c)	− 140.9
C_7H_7N	105.15	Vinylpyridine (liq)	+ 37.2
C_7H_7NO	121.15	Benzamide (c)	− 48.4
C_7H_8	92.15	Toluene (liq)	+ 2.9
C_7H_8O	108.15	Cresol (c)	− 48.9
C_7H_8S	124.21	Benzyl mercaptan (liq)	+ 10.5
C_7H_9N	107.17	Ethylpyridine (liq)	− 1.2
C_7H_9N	107.17	Methylaniline (liq)	+ 7.7
C_7H_{12}	96.19	Heptyne (g)	+ 24.6
C_7H_{14}	98.21	Cycloheptane (liq)	− 37.5
C_7H_{14}	98.21	Heptene (g)	− 14.9
C_7H_{14}	98.21	Methylcyclohexane (liq)	− 45.4
C_7H_{16}	100.23	Dimethylpentane (liq)	− 57.0
C_7H_{16}	100.23	Heptane (liq)	− 53.6
$C_8H_6O_4$	166.14	Phthalic acid (c)	− 191.9
C_8H_7N	117.16	Indole (c)	+ 29.8
C_8H_7NO	133.16	Oxindole (c)	− 41.2
$C_8H_7NO_2$	149.16	Dioxindole (c)	− 76.9
C_8H_8	104.16	Ethenylbenzene (liq)	+ 24.8
$C_8H_8N_2O_2$	164.18	Phthalamide (c)	− 104.4
$C_8H_8O_2$	136.16	Methyl benzoate (liq)	− 79.8
C_8H_9NO	135.18	Acetanilide (c)	− 50.3
C_8H_{10}	106.18	Ethylbenzene (liq)	− 3.0
$C_8H_{10}N_4O_2$	194.22	Caffeine (c)	− 76.2
$C_8H_{10}O$	122.18	Ethylphenol (c)	− 49.9
$C_8H_{11}N$	121.20	Dimethylaniline (liq)	+ 8.2
C_8H_{14}	110.22	Octyne (g)	+ 19.7
C_8H_{16}	112.24	Cyclooctane (liq)	− 40.6
C_8H_{16}	112.24	Dimethylcyclohexane (liq)	− 53.3
C_8H_{16}	112.24	Dimethylcyclohexane (liq)	− 51.5
C_8H_{16}	112.24	Dimethylcyclohexane (liq)	− 51.5
C_8H_{16}	112.24	Dimethylcyclohexane (liq)	− 53.1
C_8H_{16}	112.24	Ethylcyclohexane (liq)	− 50.7
C_8H_{16}	112.24	Propylcyclopentane (liq)	− 45.2

Table 14-3. Continued.

Formula	Weight, g/mole	Name	ΔH_f, kcal/mole
$C_8H_{16}N_2O_3$	188.26	DL-leucylglycine (c)	− 205.1
$C_8H_{16}O_2$	144.24	Caprylic acid (liq)	− 151.9
$C_8H_{17}N$	127.26	Couline (liq)	− 57.6
C_8H_{18}	114.26	Dimethylhexane (liq)	− 62.6
C_8H_{18}	114.26	Ethylhexane (liq)	− 59.9
C_8H_{18}	114.26	Octane (liq)	− 59.7
$C_8H_{20}Pb$	323.47	Tetraethyl lead (g)	− 26.2
$C_8H_{20}Pb$	323.47	Tetraethyl lead (liq)	+ 12.6
C_9H_9N	131.19	Skatole (c)	+ 16.3
$C_9H_{10}N_2$	146.21	Dipyrrylmethane (c)	+ 31.4
C_9H_{12}	120.21	Isopropylbenzene (liq)	− 9.8
$C_9H_{14}O_6$	218.23	Glyceryl triacetate (liq)	− 318.3
$C_9H_{14}O_6$	218.23	Mannitol triformal (c)	− 242.0
$C_9H_{15}N$	137.25	Phyllopyrrole (c)	− 20.4
C_9H_{16}	124.25	Nonyne (g)	+ 14.8
C_9H_{18}	126.27	Nonene (g)	− 24.7
C_9H_{18}	126.27	Propylcyclohexane (liq)	− 57.0
$C_9H_{18}O_2$	158.27	Methyl caprylate (liq)	− 141.1
C_9H_{20}	128.29	Nonane (liq)	− 65.8
$C_9H_{21}N$	143.31	Tri-*n*-propylamine (liq)	− 49.5
$C_{10}H_8$	128.18	Naphthalene (c)	+ 18.0
$C_{10}H_8$	128.18	Naphthalene (g)	+ 35.6
$C_{10}H_9N$	143.20	Phenylpyrrole (c)	+ 34.5
$C_{10}H_9N$	143.20	Quinaldene (c)	+ 39.3
$C_{10}H_{10}O_4$	194.20	Dimethyl phthalate (c)	− 171.0
$C_{10}H_{11}NO_3$	193.22	Benzoyl sarcosine (c)	− 135.7
$C_{10}H_{11}NO_4$	209.22	Animoyl glycine (c)	− 180.9
$C_{10}H_{12}O_4$	196.22	Glyceryl benzoate (c)	− 185.8
$C_{10}H_{13}NO_2$	179.24	Phenacetin (c)	− 101.1
$C_{10}H_{14}N_2$	162.26	Nicotine (liq)	+ 9.4
$C_{10}H_{16}O_2$	168.26	Dehydrocampholenolactone (c)	− 130.0
$C_{10}H_{16}O_3$	184.26	Methyl ethyl heptane lactone (c)	− 183.5
$C_{10}H_{18}$	136.26	Decyne (g)	+ 9.9
$C_{10}H_{20}$	140.30	Decene (g)	− 29.6
$C_{10}H_{20}O_2$	172.30	Capric acid (c)	− 170.6
$C_{10}H_{20}O_2$	172.30	Capric acid (liq)	− 163.6
$C_{10}H_{20}O_2$	172.30	Methyl pelargonate (liq)	− 147.3
$C_{10}H_{22}$	142.32	Decane (liq)	− 71.9
$C_{10}H_{22}O_7$	254.32	Dipentaerythritot (c)	− 376.9
$C_{11}H_{12}N_2O_2$	204.25	Tryptophane (c)	− 99.8
$C_{11}H_{14}N_2O_3$	222.27	Glycylphenylalanine (c)	− 163.9
$C_{11}H_{20}$	152.31	1-undecyne (g)	+ 5.0
$C_{11}H_{20}N_2O_2$	212.33	Valylleucyl anhydride (c)	− 150.1
$C_{11}H_{22}$	154.33	1-undecene (g)	− 34.6
$C_{11}H_{22}O_2$	186.33	Methyl caprate (liq)	− 153.1
$C_{11}H_{24}$	156.35	Undecane (liq)	− 78.0

Table 14-3. Continued.

Formula	Weight, g/mole	Name	ΔH_f, kcal/mole
$C_{12}H_8N_2$	180.22	Phenazine (c)	+ 56.4
$C_{12}H_9N$	167.22	Carbazole (c)	+ 30.3
$C_{12}H_{10}S_2$	218.34	Diphenyl disulfide (c)	+ 35.8
$C_{12}H_{14}N_2O_2$	218.28	Alanylphenylalanyl anhydride (c)	- 89.3
$C_{12}H_{14}H_4O_6$	310.30	Desoxyamalic acid (c)	- 285.7
$C_{12}H_{14}N_4O_8$	342.30	Amalic acid (c)	- 367.0
$C_{12}H_{15}NO_4$	237.28	DL-phenylalanine-*N*-carboxylic acid dimethyl ester (c)	- 184.3
$C_{12}H_{16}N_2O_3$	236.30	Alanylphenylalanine (c)	- 170.2
$C_{12}H_{16}O_8$	288.28	Levoglucosan triacetate (c)	- 371.3
$C_{12}H_{22}$	166.34	Dodecyne (g)	+ 0.1
$C_{12}H_{22}N_2O_2$	226.36	Leucine anhydride (c)	- 160.0
$C_{12}H_{22}O_6$	262.34	Diacetonemannitol (c)	- 350.0
$C_{12}H_{22}O_{11}$	342.34	Cellobiose (c)	- 532.5
$C_{12}H_{22}O_{11}$	342.34	Lactose (c)	- 530.1
$C_{12}H_{22}O_{11}$	342.34	Maltose (c)	- 530.8
$C_{12}H_{22}O_{11}$	342.34	Sucrose (c)	- 531.9
$C_{12}H_{24}$	168.36	Dodecene (g)	- 39.5
$C_{12}H_{24}O_2$	200.36	Lauric acid (c)	- 185.1
$C_{12}H_{24}O_2$	200.36	Lauric acid (liq)	- 176.4
$C_{12}H_{24}O_2$	200.36	Methyl undecylate (liq)	- 159.0
$C_{12}H_{24}O_{12}$	360.36	Lactose monohydrate (c)	- 602.0
$C_{12}H_{26}$	170.38	Dodecane (liq)	- 84.1
$C_{12}H_{26}O_{13}$	378.38	Trehalose dihydrate (c)	- 676.1
$C_{13}H_9N$	179.23	Acridine (c)	+ 44.8
$C_{13}H_{10}O$	182.23	Benzophenone (c)	- 8.0
$C_{13}H_{11}NO$	197.25	Benzanilide (c)	- 22.3
$C_{13}H_{24}$	180.37	Tridecyne (g)	- 4.9
$C_{13}H_{26}$	182.39	Tridecene (g)	- 44.4
$C_{13}H_{26}O_2$	214.39	Methyl laurate (liq)	- 165.6
$C_{13}H_{28}$	184.41	Tridecane (liq)	- 90.2
$C_{14}H_{10}$	178.24	Anthracene (c)	+ 29.0
$C_{14}H_{10}$	178.24	Anthracene (g)	+ 53.7
$C_{14}H_{10}$	178.24	Phenanthrene (c)	+ 27.3
$C_{14}H_{10}$	178.24	Phenanthrene (g)	+ 48.4
$C_{14}H_{18}N_2O_2$	246.34	Valylphenylalanyl anhydride (c)	- 94.3
$C_{14}H_{19}N_3O_4$	293.36	Glycylalanyl phenylalanine (c)	- 222.0
$C_{14}H_{10}N_2O_3$	254.26	Valylphenylalanine (c)	- 183.5
$C_{14}H_{20}O_9$	332.34	Rhamnose triacetate (c)	- 455.4
$C_{14}H_{23}N_3O_{10}$	393.40	Diethylenetriaminepentaacetic acid (c)	- 531.8
$C_{14}H_{26}$	194.40	Tetradecyne (g)	- 9.8
$C_{14}H_{28}$	196.42	Tetradecene (g)	- 49.3
$C_{14}H_{28}O_2$	228.42	Myristic acid (c)	- 199.2
$C_{14}H_{28}O_2$	228.42	Myristic acid (liq)	- 188.5
$C_{14}H_{30}$	198.44	Tetradecane (liq)	- 96.3
$C_{15}H_{20}O_6$	296.35	Tricyclobutyrin (liq)	- 247.0

Table 14-3. Continued.

Formula	Weight, g/mole	Name	ΔH_f, kcal/mole
$C_{15}H_{28}$	208.43	Pentadecyne (g)	− 14.7
$C_{15}H_{30}$	210.45	Pentadecene (g)	− 54.3
$C_{15}H_{30}O_2$	242.45	Methyl myristate (liq)	− 177.8
$C_{15}H_{32}$	212.47	Pentadecane (liq)	− 102.4
$C_{16}H_{10}$	202.26	Fluoranthene (c)	+ 45.8
$C_{16}H_{10}$	202.26	Pyrene (c)	+ 27.4
$C_{16}H_{10}N_2O_2$	262.28	Indigotin (c)	− 32.0
$C_{16}H_{12}N_2O_4$	296.30	Insatide (c)	− 139.0
$C_{16}H_{15}NO_3$	269.32	Benzoylphenylalanine (c)	− 129.6
$C_{16}H_{22}O_{11}$	390.38	Galactose pentaacetate (c)	− 532.8
$C_{16}H_{22}O_{11}$	390.38	Glucose pentaacetate (c)	− 532.1
$C_{16}H_{30}$	222.46	Hexadecyne (g)	− 19.6
$C_{16}H_{32}$	224.48	Hexadecene (g)	− 59.1
$C_{16}H_{32}O_2$	256.48	Palmitic acid (c)	− 213.1
$C_{16}H_{32}O_2$	256.48	Palmitic acid (liq)	− 200.4
$C_{16}H_{34}$	226.50	Hexadecene (liq)	− 108.5
$C_{17}H_{21}NO_4$	303.39	Morphine monohydrate (c)	− 170.1
$C_{17}H_{32}$	236.49	Heptadecyne (g)	− 24.6
$C_{17}H_{34}$	238.51	Heptadecene (g)	− 64.1
$C_{17}H_{36}$	240.53	Heptadecane (liq)	− 114.6
$C_{18}H_{12}$	228.30	Naphthacene (c)	+ 38.3
$C_{18}H_{12}$	228.30	Triphenylene (c)	+ 33.7
$C_{18}H_{18}N_2O_2$	294.38	Phenylalanyl anhydride (c)	− 69.3
$C_{18}H_{23}NO_4$	317.42	Codeine monohydrate (c)	− 151.2
$C_{18}H_{26}O_6$	338.44	Tricyclovalerin (liq)	− 270.0
$C_{18}H_{32}O_{16}$	504.50	Melezitose (c)	− 815.0
$C_{18}H_{32}O_{16}$	504.50	Raffinose (c)	− 761.0
$C_{18}H_{34}$	250.52	Octadecyne (g)	− 29.5
$C_{18}H_{34}O_2$	282.52	Oleic acid (c)	− 187.2
$C_{18}H_{34}O_2$	282.52	Oleic acid (liq)	− 178.9
$C_{18}H_{36}$	252.54	Octadecene (g)	− 69.0
$C_{18}H_{36}O_2$	284.54	Stearic acid (c)	− 226.5
$C_{18}H_{36}O_2$	284.54	Stearic acid (liq)	− 212.5
$C_{18}H_{38}$	254.56	*n*-octadeacne (liq)	− 120.7
$C_{18}H_{42}O_{21}$	594.60	Raffinose penthydrate (c)	−1122.0
$C_{19}H_{21}NO_3$	311.41	Thebaine (c)	+ 63.0
$C_{19}H_{22}N_2O$	294.43	Cinchonine (c)	+ 7.4
$C_{19}H_{24}N_2O$	296.45	Cinchonamine (c)	− 10.4
$C_{19}H_{25}N_3O_4$	359.47	Cinchonamine nitrate (c)	− 79.9
$C_{19}H_{32}$	260.51	Androstane (c)	− 75.0
$C_{19}H_{36}$	264.55	Nonadecyne (g)	− 34.4
$C_{19}H_{36}O_2$	296.55	Methyl elaidate (liq)	− 175.8
$C_{19}H_{36}O_2$	296.55	Methyl oleate (liq)	− 174.2
$C_{19}H_{38}$	266.57	1-nonadecene (g)	− 73.9
$C_{19}H_{40}$	268.59	Nonadecane (liq)	− 126.8
$C_{20}H_{12}$	252.32	Perylene (c)	+ 43.7
$C_{20}H_{21}NO_4$	339.42	Papaverine (c)	− 120.2
$C_{20}H_{24}N_2O_2$	324.26	Quinidine (c)	− 38.3
$C_{20}H_{24}N_2O_2$	324.26	Quinine (c)	− 37.1

Table 14-3. Continued.

Formula	Weight, g/mole	Name	ΔH_f, kcal/mole
$C_{20}H_{27}NO_{11}$	457.48	Amygdalin (c)	− 455.0
$C_{20}H_{38}$	278.58	Elcosyne (g)	− 39.4
$C_{20}H_{40}$	280.60	Elcosene (g)	− 78.9
$C_{20}H_{40}O_2$	312.60	Arachidic acid (c)	− 241.8
$C_{20}H_{40}O_2$	312.60	Arachidic acid (liq)	− 224.6
$C_{20}H_{42}$	282.62	Eicosane (liq)	− 132.9
$C_{21}H_{16}N_2$	296.39	Lophine (c)	+ 65.0
$C_{21}H_{18}N_2$	298.41	Amarine (c)	+ 63.0
$C_{21}H_{19}N_2O_5$	379.42	Amarine heminhydrate (c)	+ 29.0
$C_{21}H_{22}N_2O_2$	334.45	Strychnine (c)	− 41.0
$C_{21}H_{42}O_4$	358.63	Glyceryl ntearate (c)	− 315.8
$C_{22}H_{23}NO_7$	413.46	Narcotine (c)	− 210.9
$C_{22}H_{42}O_2$	338.64	Erucic acid (c)	− 207.0
$C_{22}H_{44}O_2$	340.66	Behenic acid (c)	− 235.0
$C_{22}H_{44}O_4$	372.66	Dihydroxybehenic acid (c)	− 337.0
$C_{23}H_{26}N_2O_4$	394.51	Brucine (c)	− 118.6
$C_{23}H_{31}NO_{10}$	481.55	Narceine dihydrate (c)	− 421.2
$C_{24}H_{20}O_6$	404.44	Glyceryl tribenzoate (c)	− 214.0
$C_{24}H_{24}N_2O_3$	388.50	Anisine (c)	− 51.0
$C_{24}H_{40}O_{20}$	648.64	Diamylose (c)	− 850.0
$C_{24}H_{42}O_{21}$	666.66	Stachyose (c)	− 987.0
$C_{32}H_{36}N_4O_2$	508.72	Pyrroporphyrin monomethyl ester (c)	+ 88.6
$C_{32}N_{38}N_4$	478.74	Aetioporphyrin (c)	− 1.8
$C_{34}H_{34}N_4O_4$	562.72	Protoporphyrin (c)	− 120.6
$C_{34}H_{36}N_4O_3$	548.74	Phylloerythrin monomethyl ester (c)	− 83.2
$C_{36}H_{36}N_4O_6$	620.76	Methyl pheophorbide b (c)	− 200.7
$C_{36}H_{38}N_4O_4$	590.78	Protoporphyrin dimethyl ester (c)	− 122.1
$C_{36}H_{38}N_4O_5$	606.78	Methyl pheophorbide a (c)	− 156.1
$C_{36}H_{38}N_4O_5$	606.78	Pheoporphyrin a_5 dimethyl ester (c)	− 164.3
$C_{36}H_{40}N_4O_6$	624.80	Chlorin p_6 trimethyl ester (c)	− 292.0
$C_{36}H_{42}N_4O_4$	594.82	Mesoporphyrin (1x) dimethyl ester (c)	− 196.4
$C_{36}H_{46}N_4$	534.86	Octaethylporphyrin (c)	− 39.9
$C_{36}H_{60}O_{30}$	972.96	Tetamylose (c)	−1360.0
$C_{37}H_{40}N_4O_7$	652.81	Dimethyl pheopurpurin 7 (c)	− 245.8
$C_{39}H_{74}O_6$	639.13	Glyceryl trilaurate (c)	− 489.0
$C_{40}H_{46}N_4O_8$	710.90	Coproporphyrin (I) tetramethyl ester (c)	− 348.3
$C_{45}H_{86}O_6$	723.31	Glyceryl trimyristate (c)	− 520.3
$C_{47}H_{88}O_5$	733.35	Glyceryl dibrassidate (c)	− 472.0
$C_{47}H_{88}O_5$	733.35	Glyceryl dierucate (c)	− 447.0
$C_{48}H_{54}N_4O_{16}$	943.06	Isouroporphyrian octamethyl ester (c)	− 620.1
$C_{48}H_{80}O_{40}$	1297.28	Hexamylose (c)	−1853.0
$C_{69}N_{128}O_6$	1053.97	Glyceryl tribrassidate (c)	− 625.0
$C_{69}H_{128}O_6$	1053.97	Glyceryl trierucate (c)	− 596.0

(g) Gaseous.
(c) Crystalline.
(liq) Liquid.

Compiled from References 1, 2, and 3.

Table 14-4. Heat of Formation of Inorganic Oxides, Solid State.

Formula	Weight, g/mole	Name	ΔH_f, kcal/mole
Al_2O_3	101.96	Aluminum oxide	−390.0
B_2O_3	69.62	Boric oxide	−299.8
BaO	153.34	Barium oxide	−132.3
BeO	25.01	Beryllium oxide	−145.7
BrO_2	111.90	Bromine dioxide	11.6
CaO	56.08	Calcium oxide	−151.8
CdO	128.40	Cadmium oxide	−61.7
Co_3O_4	240.79	Cobalt oxide	−213.0
CrO_2	84.00	Chromium dioxide	−143.0
CrO_3	100.00	Chromium trioxide	−140.9
Cr_3O_4	220.00	Chromium oxide	−366.0
CuO	79.55	Cupric oxide	−37.6
Cu_2O	143.10	Cuprous oxide	−40.3
FeO	71.85	Ferrous oxide	−65.0
Fe_2O_3	159.70	Ferric oxide	−197.0
HgO	216.59	Mercuric oxide	−21.4
K_2O	94.20	Potassium oxide	−86.8
Li_2O	29.88	Lithium oxide	−143.1
MgO	40.31	Magnesium oxide	−142.9
MnO	70.94	Manganese oxide	−92.1
MnO_2	86.94	Manganese dioxide	−124.3
MoO_2	127.94	Molybdenium dioxide	−140.8
Na_2O	61.98	Sodium oxide	−99.9
NiO	74.71	Nickel monoxide	−57.3
P_4O_6	219.88	Phosphorus trioxide	−392.0
PbO	223.19	Lead monoxide	−52.3
PbO_2	239.19	Lead dioxide	−66.3
Pt_3O_4	649.27	Platinum oxide	−39.0
PuO_2	274.00	Plutonium dioxide	−252.9
RaO	242.00	Radium oxide	−125.0
RhO	118.91	Rhodium monoxide	92.0
RhO_2	134.91	Rhodium dioxide	44.0
SO_3	80.06	Sulfur trioxide	−108.6
SiO_2	60.09	Silicon dioxide	−217.3
SnO	134.69	Tin monoxide	−68.3
SnO_2	150.69	Tin dioxide	−138.8
SrO	103.62	Strontium monoxide	−141.5
SrO_2	119.62	Strontium peroxide	−151.4
TeO_2	159.60	Tellurium dioxide	−77.1
ThO_2	264.04	Thorium dioxide	−293.2
TiO_2	79.90	Titanium dioxide	−224.6
UO_2	270.03	Uranium dioxide	−259.2
UO_3	286.03	Uranium trioxide	−292.0
VO	66.94	Vanadium monoxide	−103.2
V_2O_5	181.88	Vanadium pentoxide	−370.6
WO_2	215.85	Tungsten dioxide	−140.9
ZnO	81.37	Zinc oxide	−83.2
ZrO_2	123.22	Zirconium dioxide	−263.0

Compiled from References 1, 2, and 3.

Table 14-5. Heat of Formation of Miscellaneous Materials.

Formula	Weight, g/mole	Name	ΔH_f, kcal/mole
CO	28.01	Carbon monoxide, gas	- 26.4
CO_2	44.01	Carbon dioxide, gas	- 94.1
$CaCO_3$	100.09	Calcium carbonate, solid	-289.5
$Ca(OH)_2$	74.10	Calcium hydroxide, solid	-235.6
$Ca(OH)_2$	74.10	Calcium hydroxide, liquid	-239.7
ClO	51.45	Chlorine monoxide, gas	24.3
ClO_2	67.45	Chlorine dioxide, gas	24.5
ClO_3	83.45	Chlorine trioxide, gas	37.0
Cl_2O	86.90	Dichlorine monoxide, gas	19.2
FO	35.00	Fluorine monoxide, gas	- 5.2
F_2O	54.00	Oxygen difluoride, gas	4.3
HCl	36.46	Hydrogen chloride, gas	- 22.1
HCl	36.46	Hydrogen chloride, liquid	- 40.0
H_2O	18.02	Water, liquid	- 68.3
H_2O	18.02	Water, steam	- 57.8
H_2O_2	34.02	Hydrogen peroxide, liquid	- 44.9
H_2O_2	34.02	Hydrogen peroxide, gas	- 32.6
H_2SO_3	82.02	Sulfurous acid, liquid	-146.8
H_2SO_4	98.08	Sulfuric acid, liquid	-193.9
IO	142.90	Iodine monoxide, gas	41.8
MgS	56.37	Magnesium sulfide, solid	- 84.2
$Mg(OH)_2$	58.33	Magnesium hydroxide, solid	-221.9
$MgCO_3$	84.32	Magnesium carbonate, solid	-261.7
$MgSO_4$	120.37	Magnesium sulfate, solid	-304.9
NO	30.01	Nitric oxide, gas	21.6
NO_2	46.01	Nitrogen dioxide, gas	7.9
N_2O	44.02	Nitrous oxide, gas	19.6
Na_2CO_3	105.99	Sodium carbonate, solid	-269.5
Na_2CO_3	105.99	Sodium carbonate, liquid	-275.1
N_2O_5	108.02	Nitrogen pentoxide, gas	2.7
$NaCl$	58.44	Sodium chloride, solid	- 98.2
$NaCl$	58.44	Sodium chloride, aqueous	- 97.3
$NaHCO_3$	84.01	Sodium bicarbonate, solid	-226.0
$NaHCO_3$	84.01	Sodium bicarbonate, liquid	-222.1
NaF	41.99	Sodium flouride, solid	-135.9
NaI	149.89	Sodium iodide, solid	- 69.3
$NaOH$	40.00	Sodium hydroxide, liquid	-112.2
O_3	48.00	Ozone, gas	34.1
SO	48.06	Sulfur monoxide, gas	1.5
SO_2	64.06	Sulfur dioxide, gas	- 70.9
SO_3	80.06	Sulfur trioxide, gas	- 94.6
SbO	137.75	Antimony monoxide, gas	47.7
VO	66.94	Vanadium monoxide, gas	25.0
VO_2	82.94	Vanadium dioxide, gas	- 57.1

Compiled from References 1, 2, and 3.

$$\Delta H_f\ H_2O\ (l) : \quad -68.3 \times 3 = -204.9$$
$$\Delta H_f\ CO_2 : \quad -94.1 \times 6 = -564.6$$
$$\Delta H_f\ C_6H_6 : \quad (-)11.7 \times 1 = -\ 11.7$$

$$\Sigma\ \Delta H_f = -781.2 \text{ kcal/mole}$$

Relating $\Sigma\ \Delta H_f$ to the weight of benzene present:

1/(78.12 gram/mole) • (−781.2) kcal/mole • 1802 = −18020 BTU/lb

Note that in the above examples the heating value is negative, indicative of an exothermic or heat-releasing, reaction. These values, 18020 Btu/lb and 17293 Btu/lb for the high and low heating values of benzene, are consistent with the published values shown in Table 14-1, 18210 Btu/lb and 17480 Btu/lb, respectively.

This method of analysis can be used to determine the loss of heat from ash formation. For instance, if calcium is present as calcium hydroxide (heat of formation, −239.7 kcal/mole) and loses moisture as steam and forms lime, CaO (heat of formation, −151.8 kcal/mole), the endothermic reaction is calculated as follows:

$$Ca(OH)_2 \text{ - - - -> } CaO + H_2O$$

$$\Delta H_f\ H_2O\ (g) : \quad -57.8$$
$$\Delta H_f\ CaO : \quad -151.8$$
$$\Delta H_f\ Ca(OH)_2 : \quad -(-)239.7$$

$$\Sigma\ \Delta H_f = +30.1 \text{ kcal/mole}$$

Relating $\Sigma\ \Delta H_f$ to the weight of calcium hydroxide present:

1/(74.10 gram/mole) • (30.1) kcal/mole • 1802 = 732 BTU/lb

The heat of formation of calcium hydroxide and steam were found in Table 14-5 and for lime in Table 14-4. The resultant "heat of combustion," 730 Btu/lb, is a positive quantity, indicating an endothermic reaction, one in which heat must be added for the reaction to occur. Therefore, when $Ca(OH)_2$ is present in a combustion reaction, 730 Btu must be subtracted from the feed heating value per pound of $Ca(OH)_2$. From their above equations, 56.08 lb of lime (CaO) is equivalent to 74.10 lb of $Ca(OH)_2$. Using this ratio, for each pound of lime present, 732 • 74.10/56.08, or 967 Btu is removed by the lime.

DU LONG'S APPROXIMATION

An empirical method for determining the heating value of coal, the Du Long formula can be used as a rough approximation of the heating value

Table 14-6. Typical Combustion Parameters for Stoichiometric Burning.

Q Btu/lb	C fraction	H_2 fraction	O_2 fraction	O_2 Req'd lb	Air Req'd lb	CO_2 Out lb	N_2 Out lb	Dry Gas lb/10KB	H_2O lb	H_2O lb/10KB
2000	.092	.110	.798	0.32	1.37	0.34	1.05	6.95	0.98	4.90
4000	.159	.118	.723	0.63	2.74	0.58	2.11	6.73	1.05	2.63
6000	.225	.125	.650	0.94	4.05	0.82	3.11	6.55	1.12	1.87
8000	.292	.132	.576	1.25	5.38	1.07	4.13	6.50	1.18	1.48
10000	.359	.140	.501	1.56	6.75	1.31	5.19	6.50	1.25	1.25
12000	.425	.147	.428	1.87	8.06	1.56	6.19	6.46	1.31	1.09
14000	.492	.155	.353	2.18	9.43	1.80	7.25	6.46	1.38	0.99
16000	.558	.162	.280	2.49	10.74	2.04	8.25	6.43	1.45	0.91
18000	.625	.169	.206	2.79	12.07	2.29	9.28	6.43	1.51	0.84
20000	.691	.177	.132	3.11	13.43	2.53	10.32	6.43	1.58	0.79
22000	.758	.184	.058	3.42	14.75	2.77	11.34	6.41	1.64	0.75

Note: KB = thousand Btu

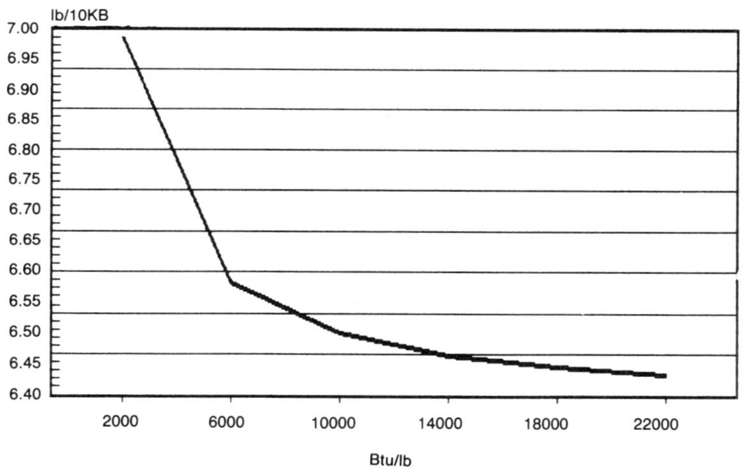

Fig. 14-1. Dry gas generation from stoichiometric burning.

of other carbonaceous materials. This is only an approximation, and it should be used only when other means of calculations do not appear satisfactory:

$$Q = 14544 \times C + 62028 \times (H - 0.125 \times O) + 4050 \times S$$

Q is in Btu/lb and C, H, O, and S are the weight fractions of carbon, hydrogen, oxygen, and sulfur present. The weight fractions should sum to 100 percent unless an inert material, such as nitrogen, is present. With the

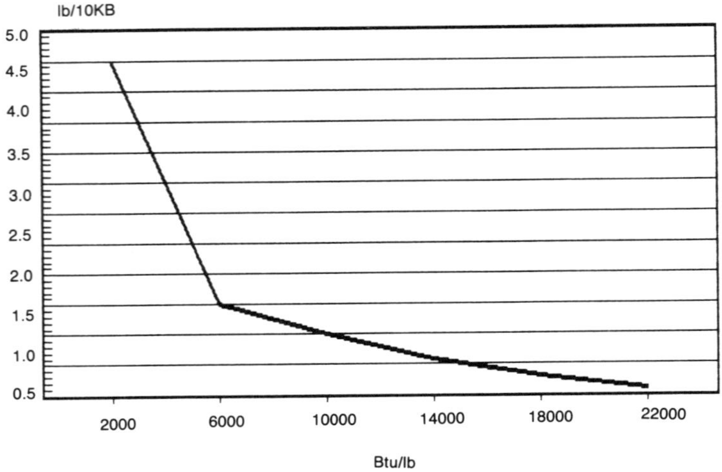

Fig. 14-2. Moisture generation from stoichiometric burning.

presence of nitrogen, for instance, the weight fraction (percent) of C, H_2, O_2, and S will total 100 percent less the fraction of nitrogen present.

Du Long's approximation can be used to estimate waste characteristics when no other data are available. Table 14-6 lists combustion parameters for an organic compound with carbon, hydrogen, and oxygen. When only a heating value is known, this table can be used to estimate the air quantity required for burning, and the products of combustion. Figure 14-1 is a graphic representation of this table for gas generation, and Fig. 14-2 shows moisture of combustion, also taken from this table.

REFERENCES

1. *Thermodynamic Data for Industrial Incinerators*, National Bureau of Standards (Washington, D.C.: Government Printing Office, 1972).

2. D. Stull and H. Prophet, *Thermochemical Tables*, Second Edition (Washington, D.C.: Government Printing Office, June 1971).

3. Perry and Chilton, *Chemical Engineering Handbook*, Fifth Edition (New York: McGraw-Hill).

15

Incinerator Calculations

By its very definition, a waste stream cannot be completely defined. The basic analysis of the stream may be known, but most wastes change in both characteristics and quantity from day to day and from hour to hour. Detailed calculations describing wastes and thermal processes are, at best, approximations. Minor losses in accuracy within assumptions made to expedite calculations are insignificant when compared to the inaccuracies caused by constant changes in the nature of the waste stream.

This chapter addresses specific incinerator design parameters, such as particle properties, thermal destruction criteria, and chamber turbulence. Furnace characteristics relevent to destruction of a hazardous waste constituent are also discussed. Heat-balance techniques and calculations are described in Chapter 16.

DROPLET SIZE

Droplet size is the most important parameter in the evaluation of liquid-waste incineration. Before destruction can occur, the waste must be vaporized. Evaporation rate is primarily a function of droplet size and, to a lesser extent, the material's physical characteristics, pressure, and temperature.

The evaporation rate varies with droplet surface area. Through observation and experimentation, a simplified approximation of evaporation rate has been developed:

(1) $$T = (D^2 - d^2) / (S \times 10^6)$$

T = time required for evaporation, seconds
D = initial mean particle size, microns
d = final particle size, microns
S = evaporation rate constant, mm^2/second

A simplified expression for the evaporation rate constant, S, is given in Equation (2):

(2) $$S = 8 \times K_v \times (T_g - T_e) \times \{1 + 0.276 \times (N_{Re})^{.5} \times (N_{Sc})^{.33}\} / (L \times \epsilon_L)$$

K_v = thermal conductivity of vapor at T_g, kg/(cm – K°)
T_g = gas temperature, K°
T_e = drop surface temperature, K°
N_{Re} = Reynolds number
N_{Sc} = Schmidt number
L = latent heat of evaporation of liquid at T_e, kcal/kg
ϵ_L = liquid density at T_e gm/cm^3

Exact calculation of the evaporation rate constant is extremely difficult. The calculation itself is involved and, more importantly, the parameters upon which this calculation is based are difficult to determine. Both the Reynolds number and the Schmidt number are functions of particle density and viscosity, and these properties change significantly as the droplet undergoes temperature change.

The value of S normally varies from 0.25 to 1.25 mm^2/sec. For water an approximate value for S is 0.33 mm^2/sec.

When $d^2 = 0$ in Equation (1), T becomes the time required for complete evaporation. Table 15-1 lists the time required for evaporation of water as a function of particle diameter D, with d = 0, and this relationship is shown graphically in Fig. 15-1.

PARTICLE SIZE VERSUS DESTRUCTION AND REMOVAL EFFICIENCY

Waste destruction is a function of both evaporation and chemical bond dissolution. Considering only evaporation, however, a relationship can be established between the number of particles present and the ability to meet

Table 15-1. Droplet Evaporation, Water.

Droplet Size Microns	Time Required for Evaporation seconds
< 15	< 0.001
15	.001
25	.002
50	.008
75	.017
100	.030
125	.047
150	.068
200	.121
250	.189
300	.272
350	.371
400	.485
450	.614
500	.758

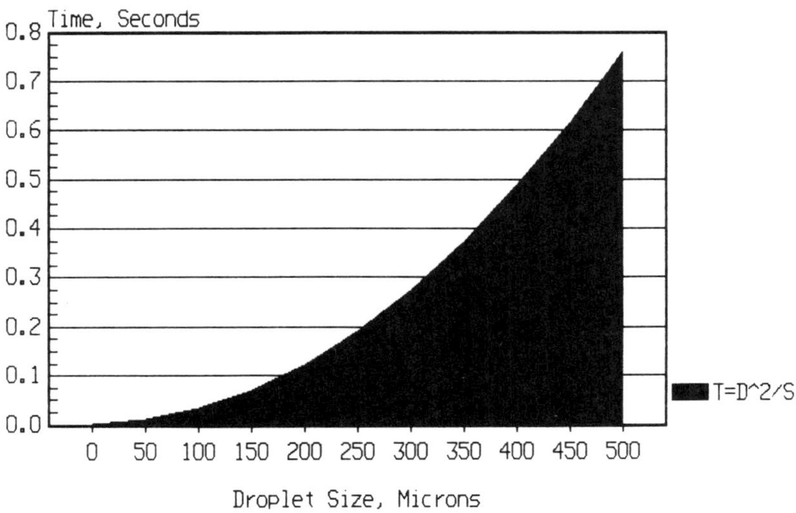

Fig. 15-1. Droplet evaporation rate.

a specific destruction and removal efficiency (DRE). This analysis is as follows:

With N particles of F microns mean particle size, the time for complete evaporation is, from equation (1):

(3) $$T = F^2 / (S \times 10^6)$$

With an additional M particles at a larger size, D micron diameter, what size (d diameter) will these particles be at time T? From equation (1):

(4) $$d = (D^2 - S \times T \times 10^6)^{0.5}$$

Insert T from equation (3) into equation (4):

(5) $$d = (D^2 - F^2)^{0.5}$$

The weight of particles is proportional to particle volume. The volume of particles (spherical) entering this reaction is:

(6) $$V_{in} = (4/3) \times \pi \times (F^3 \times N + D^3 \times M)$$

The volume of particles exiting the reaction is:

(7) $$V_{out} = (4/3) \times \pi \times d^3 \times M$$

Insert into equation (7) the number of particles remaining when the N particles have fully evaporated, from equation (5):

(8) $$V_{out} = (4/3) \times \pi \times (D^2 - F^2)^{1.5} \times M$$

Destruction and removal efficiency (DRE) is defined as follows:

(9) $$DRE = 1 - (W_{out}/W_{in}) = 1 - (V_{out}/V_{in})$$

Combine equations (6) and (8) with equation (9):

(10) $$DRE = 1 - \{(D^2 - F^2)^{1.5} \times M/(F^3 \times N + D^3 \times M)\}$$

Rearranging terms:

(11) $$M/N = (1 - DRE) \times F^3/\{(D^2 - F^2)^{1.5} - (1 - DRE) \times D^3\}$$

The ratio M/N represents the number of larger particles than can be tolerated in waste feed and still allow a specific DRE. For instance, for a DRE of 99.99 percent, with a population of 10000 particles of 100 microns diameter, equation (11) becomes:

(12) $$M = 1{,}000{,}000 / (D^2 - 10{,}000)^{1.5}$$

The values of M (number of larger particles) corresponding to an initial diameter D are listed in Table 15-2 and are graphed in Fig. 15-2 for 99.99 percent DRE.

Table 15-2. Allowable Particles to Maintain 99.99% DRE in a Reaction Containing 10,000 Particles at 100 Microns Equivalent Diameter.

Diameter, Microns	Number of Particles
105	30.5
110	10.4
115	5.5
120	3.4
125	2.4
130	1.7
135	1.3
140	1.1
145	0.9
150	0.7
>150	<0.5

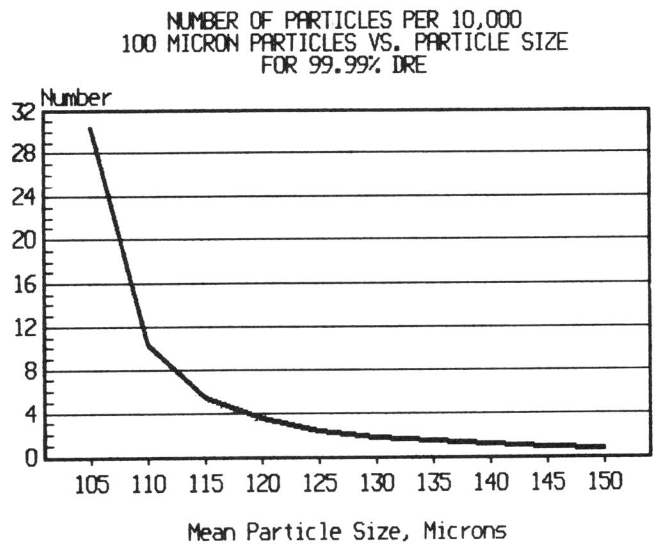

Fig. 15-2. Number of particles versus particle size for 99.99% DRE.

An example of the significance of these values is that if more than 30 particles are present at a diameter in excess of 105 microns, a DRE of 99.99 percent cannot be achieved. The presence of only one particle greater than 150 microns in diameter will prevent reaching a DRE of 99.99 percent, regardless of any other factors of incinerator design or operation.

Evaporative effects are one of the most important considerations in nozzle and atomization design for both the waste stream and the supplemental fuel. Equipment manufacturers normally provide information on droplet-size range or pattern as a function of waste characteristics and atomization type and pressure.

TURBULENCE

One of the most important phenomena contributing to effective combustion is turbulence. Turbulence of an air or gas stream within an incinerator promotes mixing and results in a greater burnout of combustible components.

Turbulence is difficult to quantify. An estimate of the point at which turbulent flow begins (as compared to laminar flow, which precludes mixing) is indicated by the dimensionless Reynolds number (N_{RE}). This change from laminar to turbulent flow is actually not a discrete point or occurrence, but occurs over a range of values of N_{RE}.

N_{RE} is defined as follows:

(13) $$N_{RE} = V \times D / K$$

V = velocity of stream, ft/sec
D = diameter or D_e, equivalent diameter, ft
K = kinematic viscosity, ft²/sec, per Fig. 15-3

For non-circular sections the equivalent diameter is used:

(14) $$D_e = 2 \times a \times b / (a + b)$$

where a and b are the side dimensions, feet.

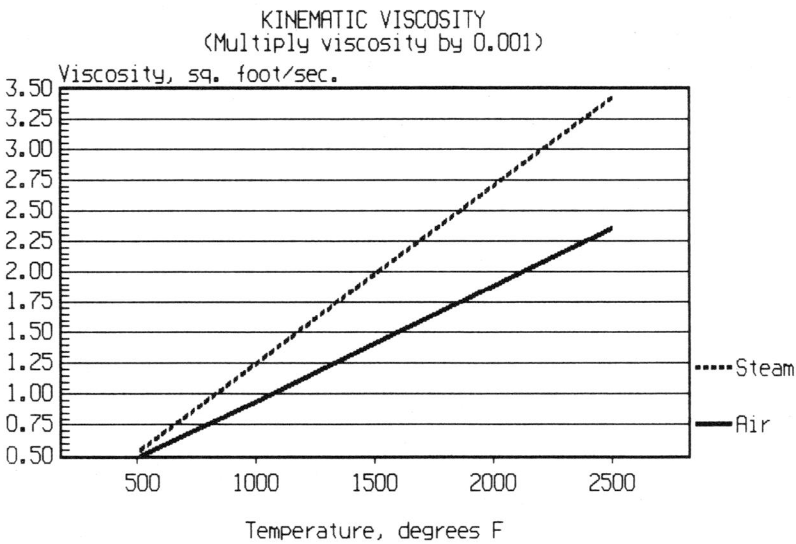

Fig. 15-3. Kinematic viscosity.

The critical N_{RE} (N_{RE} = 2,300) corresponds to that condition where flow passes from the laminar to the turbulent condition. To assure the presence of turbulence, a calculation of N_{RE} should be made. For example, for dry gas flow at 30 ft/sec in a circular flue 6 ft in diameter and at a temperature of 2000° F (corresponding to a kinematic viscosity of 0.0019 ft^2/sec, from Fig. 15-3, assuming dry gas flow equivalent to the flow of dry air), N_{RE} is calculated as follows:

$$(15) \qquad \begin{aligned} N_{RE} &= V \times D/K \\ &= (30\ \text{ft/sec}) \times (2\ \text{ft})/(.0019\ \text{ft}^2/\text{sec}) \\ &= 31{,}579 \end{aligned}$$

This value of N_{RE} is much higher than the critical N_{RE}; therefore, turbulent flow can be assumed.

The use of N_{RE} to indicate turbulence is inexact. Values of N_{RE} significantly above 2,300 can reliably be assumed to indicate the presence of turbulent flow, but when N_{RE} is below 10,000, more detailed analyses must be performed on the stream. These analyses would include a more rigorous analysis of the actual viscosity of the stream and more detailed calculations of the flow parameters, (velocity and equivalent diameter).

ORGANIC DESTRUCTION

The destruction of an organic compound is a function of temperature, residence time at that temperature, and the properties of the compound. The rate of destruction is as follows:

$$(16) \qquad dC/dt = -kC$$

C = concentration at time t (seconds)
k = rate constant

Upon integration of this equation:

$$(17) \qquad t = (1/k) \times \ln (C_o/C)$$

where C_o, destruction, is 0 at t = 0.

Expressing the rate constant, k, in Arrhenius form:

$$(18) \qquad k = V \times \exp \{- E/(R \times T)\}$$

combining equations (17) and (18):

$$(19) \qquad N = 1 - \exp [- V \times t \times \exp\{-E/(R \times T)\}]$$

and solving for incinerator temperature:

(20) $$T = E/\left(R\times \{\ln(t) + \ln[-\ln(1-N)]/V\}\right)$$

where N = destruction efficiency
V = frequency factor (second^{-1})
E = activation energy (cal/g – mole)
R = universal gas constant (1.987 cal/g – mole – ° K)
T = incinerator temperature (° K)

Values for V, the frequency factor, and E, activation energy, are listed in Table 15-3 for selected compounds.

Using equation (20) and the coefficients listed in Table 15-3, destruction was calculated for each of the Table 15-3 compounds. Table 15-4 lists the temperature required for 99.99 percent destruction (not considering droplet evaporation) for maintenance of one-half, one, and two second retention times at the indicated temperature.

As expected, the higher the residence time, the lower the temperature required for destruction is.

BULK-RESIDENCE-TIME CALCULATIONS

A determination of the ability of a furnace to destroy a particular waste requires consideration of the geometry of the furnace and the temperature distribution within the furnace. A plug-flow model can approximate furnace characteristics.

Plug flow assumes that there is no longitudinal mixing of gaseous combustion products. The temperature, composition, and velocity of the gases are uniform at any cross-section of the furnace.

First, a time-temperature relationship must be established for the furnace. In plug flow, without temperature change (isothermal) along the flow path, the mean residence time, t, is calculated as follows:

(21) $$t = (V/Q) \times 60$$

t = mean residence time, seconds
V = furnace volume, cubic feet
Q = volumetric flow rate, scfm

The actual flow rate is proportional to temperature, as follows:

(22) $$Q' = Q \times (T/460)$$

Q′ = actual flow rate, scfm
T = temperature, ° R

combining equations (21) and (22):

(23) $$t = 27{,}600 \times V / (Q \times T)$$

Table 15-3. Destruction Kinetics Parameters.

COMPOUND (Reference)	Frequency Factor (1/second)	Activation Energy (cal/g-mole)
Acetic Anhydride (4)	1.00E+12	34500
Acetonitrile (2)	4.70E+07	40000
Acrylonitrile (1)	2.18E+12	52100
Acrylonitrile (3)	1.30E+06	31000
Aniline (3)	9.30E+15	71000
Azomethane (4)	3.50E+16	52500
Benzene (1)	7.42E+21	95900
Benzene (2)	2.80E+08	38000
Butene (4)	5.00E+12	63000
Carbon Tetrachloride (3)	2.80E+05	26000
Chloroform (2)	2.90E+12	49000
Dichlorobenzene (3)	3.00E+08	39000
Dichloromethane (3)	3.00E+13	64000
Ethane (3)	1.30E+05	24000
Ethyl Chlorocarbonate (4)	9.20E+08	29100
Ethyl Nitrite (4)	1.40E+14	37700
Ethyl Peroxide (4)	5.10E+14	31500
Ethylene Dibutyrate (4)	1.80E+10	33000
Ethylidene dichloride (4)	1.20E+12	49500
Hexachlorobenzene (1)	1.90E+16	72600
Hexachlorobenzene (3)	2.50E+08	41000
Hexachlorobutane (3)	6.30E+12	59000
Hexachloroethane (3)	1.90E+07	29000
Methane (3)	3.50E+09	48000
Methyl Iodide (4)	3.90E+12	43000
Monochlorobenzene (3)	8.00E+04	23000
Nitrobenzene (3)	1.40E+15	64000
Paracetaldehyde (4)	1.30E+15	44200
Pentachlorobiphenyl (1)	1.10E+16	70000
Pentachlorobiphenyl* (1)	7.44E+19*	53600*
Propylene Oxide (4)	1.40E+14	58000
Pyridine (3)	1.10E+05	24000
Tetrachlorobenzene (2)	1.90E+06	30000
Tetrachloroethylene (2)	2.60E+06	33000
Toluene (1)	2.28E+13	56500
Toluene* (1)	2.10E+12*	77500*
Trichlorobenzene (2)	2.20E+08	38000
Trichloroethane (3)	1.90E+08	32000
Vinyl Chloride (1)	3.57E+14	63300

Note:

1. The Frequency Factor (V) is also known as the Arrhenius Factor.
2. Activation Energy (E) is also known as Dissociation Energy.

*Pyrolysis mode of destruction.

Table 15-4. Destruction Kinetics for 99.99% Destruction.

COMPOUND	Temperature of Destruction, °F		
	0.5 Second	1.0 Second	2.0 Second
Acetic Anhydride	805	771	738
Acetonitrile	1997	1887	1786
Acrylonitrile (1)	1392	1343	1296
Acrylonitrile (3)	2056	1909	1778
Aniline	1440	1402	1366
Azomethane	892	866	841
Benzene (1)	1372	1345	1320
Benzene (2)	1622	1538	1461
Butene	1708	1653	1600
Carbon Tetrachloride	1987	1822	1679
Chloroform	1262	1217	1174
Dichlorobenzene	1668	1583	1504
Dichloromethane	1602	1553	1506
Ethane	1994	1816	1662
Ethyl Chlorocarbonate	1028	972	920
Ethyl Nitrite	692	666	641
Ethyl Peroxide	463	442	423
Ethylene Dibutyrate	985	938	894
Ethylidene Dichloride	1341	1293	1246
Hexachlorobenzene (1)	1443	1406	1370
Hexachlorobenzene (3)	1802	1710	1626
Hexachlorobutane	1553	1502	1453
Hexachloroethane	1438	1347	1265
Methane	1822	1742	1667
Methyl Iodide	1034	996	959
Monochlorobenzene	2028	1838	1675
Nitrobenzene	1355	1316	1279
Paracetaldehyde	796	770	744
Pentachlorobiphenyl	1404	1367	1331
Pentachlorobiphenyl*	674	656	638
Propylene Oxide	1312	1272	1233
Pyridine	2041	1856	1697
Tetrachlorobenzene	1895	1761	1642
Tetrachloroethylene	2062	1922	1798
Toluene	1379	1334	1292
Toluene*	2298	2225	2156
Trichlorobenzene	1653	1567	1487
Trichloroethane	1336	1262	1194
Vinyl Chloride	1415	1373	1336

Note:

1. The Frequency Factor (V) is also known as the Arrhenius Factor.
2. Activation Energy (E) is also known as Dissociation Energy.

Plug flow is an ideal flow which is not realized in actual furnaces. There will be some longitudinal mixing, and a portion of the gas may find a faster path than the average gas flow, exiting the furnace in time less than t. This time is defined as the fast-path residence time, τ. The minimum or fast-path residence time has been found to be approximately 0.5 t (Reference 5).

Equation (23) was developed for isothermal flow. With an actual furnace, gas exits at a temperature that is lower than the maximum reaction zone temperature. Furnace operation is not isothermal. A time-temperature relationship must be expanded to describe the temperature profile within the furnace to determine if there is sufficient residence time within the furnace to provide the necessary destruction parameters of a particular hazardous waste compound.

To make this determination the temperature within the furnace must be calculated as a function of the distance from the highest temperature level in the furnace.

The differential mean-residence time, dt, across an element of volume, dV, can be expressed in the differential form of equation (21), as follows:

$$dt = dv/Q' \tag{24}$$

With A being cross-sectional area of the furnace (square feet), which is assumed constant, and dx an element of axial length of the furnace:

$$dV = A \times dx \tag{25}$$

Rewrite equation (23) in the form of equation (26):

$$dt = [(27{,}600 \times A) / (Q \times T)] \times dx \tag{26}$$

An approximate solution of equation (26) can be obtained by assuming a linear gas temperature-axial distance profile between the point of maximum temperature and the furnace exit. If the temperature is T_m at a distance of x_m and the exit temperature is T_e at a distance of x_e, a linear relationship can be expressed as follows for the temperature T at a distance x:

$$(T - T_m)/(x - x_m) = (T_e - T_m)/(x_e - x_m) \tag{27}$$

Rearranging terms:

$$T = \{(T_m - T_e) \times x + T_e \times x_m - T_m \times x_e\}/(x_m - x_e) \tag{28}$$

Substitute this value for T into equation (27) and integrate:

(29) $(t - t_m) = (27{,}600 \times A/Q) \times \ln(T/T_m) \times (x_m - x_e)/(T_m - T_e)$

If the assumption is made that the maximum temperature in the furnace (T_m) occurs close to the furnace entrance, or $x_m << x_e$, at the data point in time which defines $t_m = 0$, and noting that this implies that the product $A \times x_e$ is the furnace volume, V, equation (29) reduces to the following equation:

(30) $t = 27{,}600 \times V \times \ln(T/T_m)/\{Q \times (T_e - T_m)\}$

Equation (30) is a simplified expression for the mean residence time of a gas, t, above a given temperature, T. The mean residence time of the gas stream, t_e, above the furnace exit temperature, T_e, from equation (30), is as follows:

(31) $t_e = 27{,}600 \times V \times \ln(T_e/T_m)/\{Q \times (T_e - T_m)\}$

The fast-path residence time of the gases, τ, is assumed to be ½ the mean residence time. The expression for the fast-path process, from equation (30), is as follows:

(32) $\tau = 13{,}800 \times V \times \ln(T/T_m)/\{Q \times (T_e - T_m)\}$

Equations (30) and (32) provide a relationship between temperature and residence time for a particular furnace, with a geometry represented by Q and V, and temperatures within the furnace and at the furnace exit represented by T_m and T_e. For a particular furnace, the two values of retention time, t and τ, can be plotted as a function of temperature. Figure 15-4 is a plot of temperature vs. residence time for a typical furnace, a chamber with a volume of 5000 ft^3, a gas velocity of 16800 scfm, a maximum temperature of 2700° F and an exit temperature of 1140° F.

One curve is the mean residence time, t, and the other is the fast-path, or minimum residence time, τ. The furnace can be safely said to operate within these two curved lines. If the destruction requirements of a specific waste lie beneath these curves, the waste is a candidate for destruction in this particular furnace. If the waste characteristics are above these curves, then the waste is not a good match for this furnace. If the waste destruction characteristics plot between these two curves, then the waste is likely a candidate for destruction in this furnace, but additional evaluations must be made. Such evaluations include a clear definition of the operating point of the furnace and a more rigorous definition of furnace parameters.

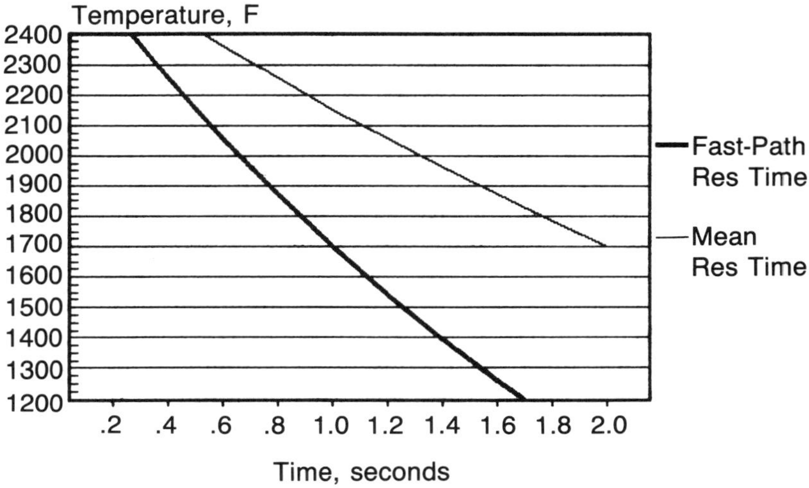

Fig. 15-4. Temperature versus residence time, typical furnace.

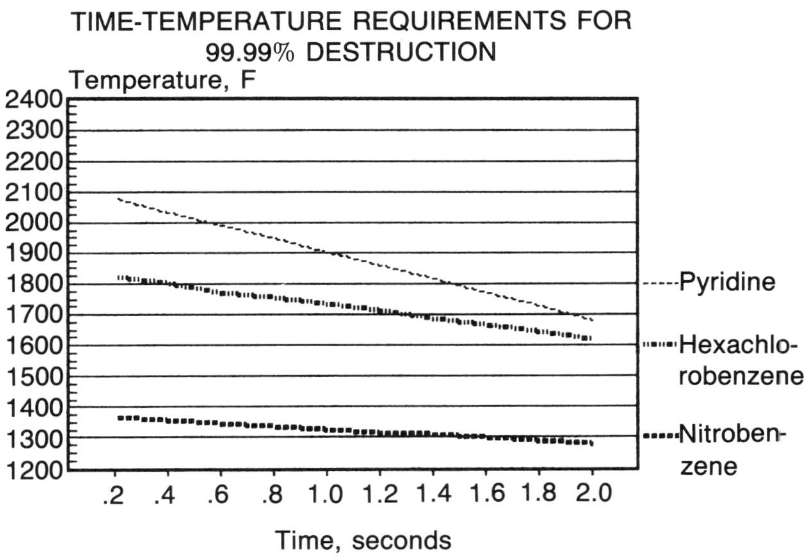

Fig. 15-5. Temperature versus residence time, three hazardous constituents.

Figure 15-5 is a plot of the destruction characteristics of three wastes, using the data from Table 15-4. The first of these wastes is nitrobenzene, which requires relatively low temperatures for destruction. Pyridine requires much higher temperatures for four nines destruction, while hexachlorobenzene lies somewhere between these two others.

Superimposing Fig. 15-4 on Fig. 15-5 yields Fig. 15-6, showing temperature versus residence-time destruction criteria. These three selected

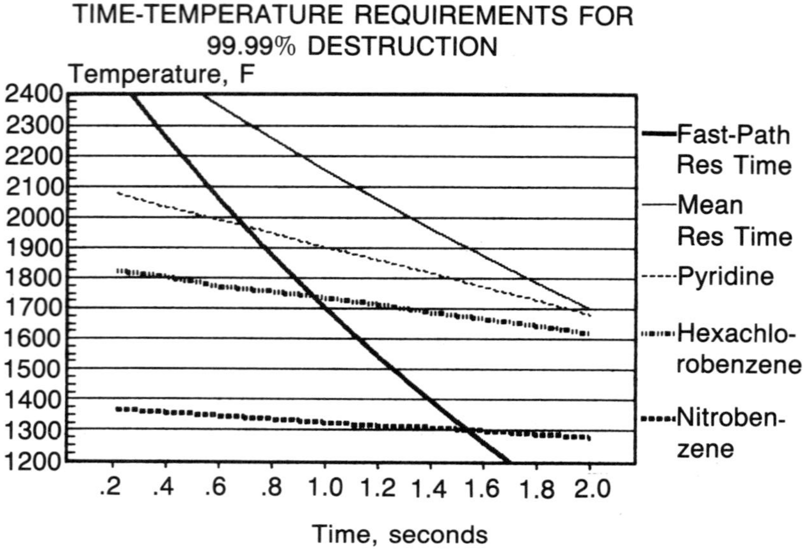

Fig. 15-6. Temperature versus residence time, destruction criteria.

wastes all lie beneath the curve representing mean residence time for this furnace design, which indicates that this furnace may provide the required destruction efficiency. On the other hand, the furnace must be operated above 1980° F to ensure that all of these waste streams will be destroyed (to four nines) even if the fast-path predominates.

These curves can be drawn for any waste stream and for any furnace configuration. In calculating furnace configuration, however, approximations usually must be made on furnace volume and flows because a furnace is rarely of such uniform geometry that these parameters can be readily determined.

REFERENCES

1. D. Duvall, "High Temperature Decomposition of Organic Hazardous Waste," *Proceedings of the Sixth Annual Research Symposium on the Treatment of Hazardous Waste*, EPA-600/9-80-011, United States Environmental Protection Agency, 1980

2. B. Dellinger, "Examination of Fundamental Incinerability Indices for Hazardous Waste Destruction," *Proceedings of the 11th Annual Research Symposium on the Incineration and Treatment of Hazardous Waste*, EPA-600/9-85-028, United States Environmental Protection Agency, September 1985.

3. B. Dellinger, "Determination of the Thermal Decomposition Properties of 20 Selected Hazardous Organic Compounds," *Proceedings of the 10th Annual Research Symposium on the Incineration and Treatment of Hazardous Waste*, EPA-600/9-84-022, United States Environmental Protection Agency, September 1984.

4. R. Perry, *Chemical Engineers' Handbook*, Fifth Edition (New York: McGraw-Hill Book Company).

5. T. D. Ellis, R. B. Dienner, and C. R. Brunner, *Industrial Hazardous Waste Incineration*, AIChE short course manual, 1986.

6. *A Technical Overview of the Concept of Disposing of Hazardous Wastes in an Industrial Boiler*, United States Environmental Protection Agency Industrial Environmental Research Laboratory, Office of Research and Development, 1981.

16

System Calculations

An incinerator system can be analyzed on a gross basis, as can any other self-contained physical or chemical system. In this chapter, mass and heat balance techniques will be developed that will provide an analysis of system inputs and outputs. Additional calculations are included which will aid in the overall development of system parameters.

MASS AND ENERGY BALANCES

The purpose of mass and heat (energy) balance calculations is to determine inputs and outputs of a system. These techniques provide a means of finding, for instance, the air required for a process, the quantity of flue gas generated, the temperature within the system, supplemental fuel requirements, etc. Mass and heat balance calculations are applications of the first law of thermodynamics: in a steady state the inputs to a system are equal to the outputs from that system, i.e., mass and energy are conserved.

Before balance techniques can be defined the concept of *system* must be defined.

SYSTEM BOUNDARY

A system boundary is a definition of the entire system under observation. Of concern is what crosses the system boundary, not what is internal to the system.

As an example, Fig. 16-1 is a diagrammatic representation of a multiple-hearth furnace system. The mass balance is defined as follows: flow into the system is positive and flow leaving the system is negative:

$$M_a + M_f - M_g - M_r = 0$$

Heat flow is also conserved. For a heat of combustion Q_f of the sludge feed and h enthalpy for each designated mass flow, using the convention that heat entering the system boundary is positive and that exiting the boundary is indicated negative:

$$Q_f + M_a h_a + M_f h_f - M_g h_g - M_r h_r = 0$$

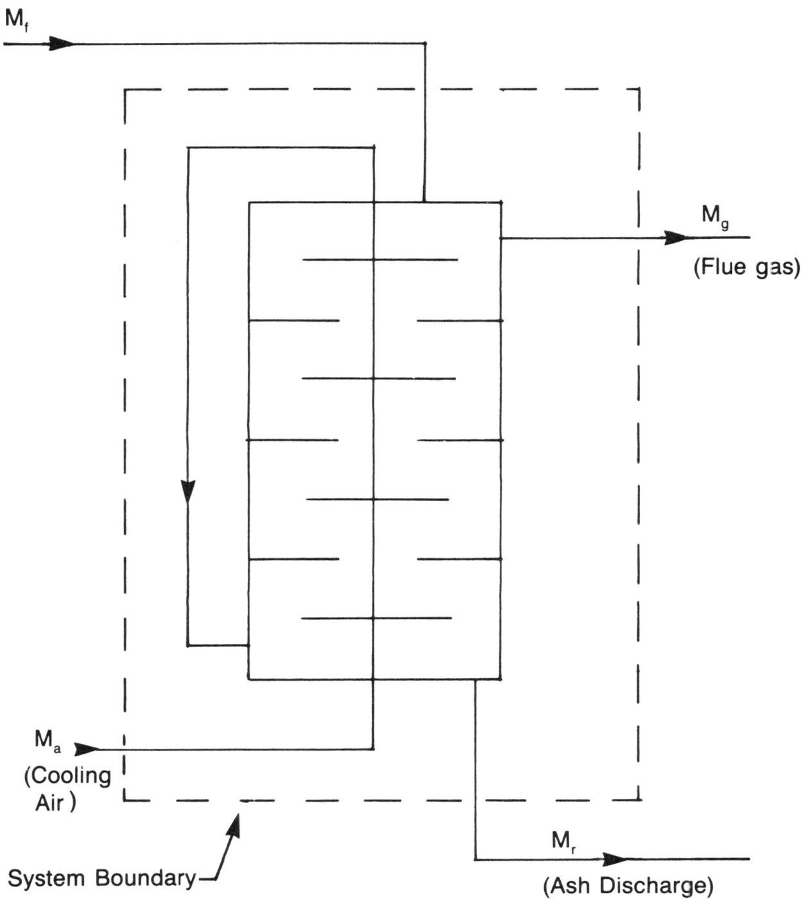

Fig. 16-1. System boundary.

The heat or mass flow exiting the system is not a function of the internal recycle. It is a function of the streams crossing the boundary. The amount of fuel required for this process, for instance, will not depend on the internal temperature of the process, but on the temperature of the flue gas stream exiting the boundary.

UNITS OF ENERGY

The unit of energy commonly used in the United States is the *British thermal unit,* Btu. One Btu is that amount of heat required to raise the temperature of one pound of water one degree Fahrenheit. The *calorie* is an equivalent unit, that amount of heat which will raise the temperature of one gram of water one degree centigrade.

In accordance with the first law of thermodynamics, the energy put into a system must be accounted for within that system. A corollary of that law is the equivalence of heat and work. For instance, a fixed quantity of water stirred by a mixer (mechanical work) will experience an equivalent rise in temperature (heat energy). The constants of conversion of mechanical and heat energy, as well as other thermodynamic conversions, are presented in Table 16-1.

BASIS OF CALCULATIONS

There are a number of assumptions that should be made to expedite the calculation of incinerator system parameters.

- All hydrogen present converts to water vapor, H_2O, unless otherwise noted below.
- All chlorides (or fluorides) convert to hydrogen chloride, HCl (or hydrogen fluoride, HF).
- All carbon converts to carbon dioxide, CO_2.
- All sulfur present converts to sulfur dioxide, SO_2.

Table 16-1. Energy Conversions.

1 Btu	=	778 Foot-Pounds
1 Btu	=	1055 Joules
1 Btu	=	252 calories
1 Btu	=	0.252 Calories (kilocalories)
1 Btu	=	0.0002931 Kilowatt-Hours
1 horsepower-hour	=	2544 Btu
1 kilowatt-hour	=	3412 Btu
1 Calorie	=	3.968 Btu
1 Calorie	=	1 kilocalorie (1000 calories)

- Alkali metals convert to hydroxides: sodium to sodium hydroxide ($2Na + O_2 + H_2$ - -> $2NaOH$) and potassium to potassium hydroxide ($2K + O_2 + H_2$ - -> $2KOH$).
- Non-alkali metals completely oxidize: copper to cupric oxide ($2Cu + O_2$ - -> $2\ CuO$), iron to ferric oxide ($4Fe + 3O_2$ - -> $2Fe_2O_3$).
- All nitrogen from the waste, fuel, or air, will take the form of a diatomic molecule, i.e., nitrogen is present as N_2.

These assumptions are adequate for a heat balance. For instance, SO_2 is present when burning a waste with sulfur at from 15 to 30 times the amount of SO_3 present, and the amount of CO generated is below 150 ppmv, less than 0.3 percent of the CO_2 present.

STANDARDIZED METHODS

A set of calculation sheets have been developed to aid balancing mass and heat. They are organized step-by-step as a mass balance (Table 16-4), heat balance (Table 16-5) and flue gas discharge (Table 16-6).

Table 16-4. Mass Balance.

Step	Description	Unit	Quantity	Derivation
M1	Total flow	lb/hr	5000	known
M2	Moisture	%	55	known
M3		lb/hr	2750	M1-M2/100
M4	Dry feed	lb/hr	2250	M1-M3
M5	Ash	%	30	known
M6		lb/hr	675	M4-M5/100
M7	Volatile	lb/hr	1575	M4-M6
M8		Btu/lb	8000	known
M9		MBH	12.60	M7-M8/1000000
M10	Dry gas (DG)	lb/10KB	6.50	Table 14-6
M11		lb/hr	8190	M9-M10-100
M12	Moisture	lb/10KB	1.48	Table 14-6
M13		lb/hr	1865	M9-M12-100
M14	DG + Moisture	lb/hr	10055	M11 + M13
M15	100% Air	lb/hr	8480	M14-M7
M16	Total Air	Fraction	1.5	Table 2
M17		lb/hr	12720	M15-M16
M18		SCFM	2827	M17/.075-60
M19	Excess Air	lb/hr	4240	M16-M15
M20	Humidity	lb/lb dry air	0.01	Appendix E
M21		lb/hr	127	M17-M21
M22	Total Moisture	lb/hr	4742	M3 + M13 + M21
M23	Total Dry Gas	lb/hr	12430	M11 + M19

Table 16-5. Heat Balance.

Step	Description	Unit	Quantity	Derivation
M6	Ash	lb/hr	675.	M4-M5/100
M9	Volatile	MBH	12.60	M7-M8/1000000
M21	Humidity	lb/hr	127	M17-M21
M22	Total Moisture	lb/hr	4742	M3 + M13 + M21
M23	Total Dry Gas	lb/hr	12430	M11 + M19
H1	Cooling air flow temp.	lb/hr	0	known
H2		°F	–	known
H3		Btu/lb	–	Appendix B
H4		MBH	0	H1-H3
H5	Ash	°F	1000	known
H6		Btu/lb	160	.17-(H5-60)
H7		MBH	0.11	M6-H6/1000000
H8	Radiation	fraction	0.03	Table 3
H9		MBH	0.38	M9-H8
H10	Humidity loss	MBH	−0.12	970-M21/1000000
H11	Total loss	MBH	0.37	H4 + H7 + H9 + (−)H10
H12	Outlet	MBH	12.23	M9-H11
H13		°F	1376	M22, M23, H12
H14	Required Temp.	°F	2000	known
H15	Enthalpy, dry gas	Btu/lb	502.3	Appendix B
H16		MBH	6.24	M23-H15/1000000
H17	Enthalpy, H_2O	Btu/lb	2067.4	Appendix B
H18		MBH	9.80	M22-H17/1000000
H19	Total gas	MBH	16.04	H16 + H18
H20	Deficiency	MBH	3.81	H19-H12
H21	Excess air, FO	%	30	known
H22	Fuel oil (FO)	Btu/gal	54535	Appendix G
H23		gal/hr	69.86	H20-1000000/H22
H24		MBH	9.78	H23-.140
H25	Air req'd, FO	lb/gal	134.41	Appendix G
H26		lb/hr	9390	H23-H25
H27		SCFM	2087	H26/.075-60
H28	Dry gas, FO	lb/gal	133.54	Appendix G
H29		lb/hr	9329	H23-H28
H30	Moisture, FO	lb/gal	8.75	Appendix G
H31		lb/hr	611	H23-H30
H32	Outlet	MBH	22.01	H12 + H24
H33	Outlet dry gas	lb/hr	21759	M23 + H29
H34	Outlet moisture	lb/hr	5353	M22 + H31

It is important to note that these calculations are a first step. They provide good initial estimates of incinerator parameters in a straightforward and deliberate manner. They are estimates in that they are based on consideration of the flue gas stream to consist of just two components:

Table 16-6. Flue Gas Discharge.

Step	Description	Unit	Quantity	Derivation
H32	Outlet	MBH	22.01	H12 + H24
H33	Outlet dry gas	lb/hr	21759	M23 + H29
H34	Outlet moisture	lb/hr	5353	M22 + H31
F1	Saturation	Btu/lb DG	1011.54	H32-1000000/H33
F2		°F	186	Appendix F
F3	Saturation moist.	lb/lb DG	0.8794	Appendix F
F4		lb/hr	19135	H33-F3
F5	H_2O required	lb/hr	13782	F4-H34
F6		gpm	27.56	F5/500
F7	Venturi factor	fraction	8	known
F8	Quench required	lb/hr	110256	F5-F7
F9		gpm	221	F8/500
F10	Sump temperature	°F	155	F2/1.2
F11	Temp. difference	°F	95	F10-60
F12	Discharge temp.	°F	140	known
F13	Discharge	Btu/lb DG	178.82	Appendix F
F14		MBH	3.89	H33-F13/1000000
F15	Cooling load	MBH	18.12	H32-F14
F16	Cooling H_2O	lb/hr	190737	F15-1000000/F11
F17		gpm	381	F16/500
F18	Discharge H_2O	lb/lb DG	0.1530	Appendix F
F19		lb/hr	3329	H33-F18
F20	Drain w/o recirc	lb/hr	303017	H34 + F8 + F16 − F19
F21		gpm	606	F20/500
F22	Drain w/recirc	lb/hr	192761	H34 + F16 − F19
F2		gpm	386	F22/500
F24	Outlet gas flow	CF/lb DG	18.82	Appendix F
F25		CFM	6825	H33-F24/60
F26	Fan pressure	in. WC	40	known
27	Fan correction	fraction	1.11	407/(407 − F26)
F28	Corrected outlet	ACFM	7576	F25-F27

moisture and dry gas. The dry gas component is considered to have the properties of dry air. These are enthalpy and saturation properties.

After an initial set of calculations are completed, based on dry air and moisture, a similar set of calculations can be run based on the actual gas discharge. The off-gas will probably contain moisture, oxygen, nitrogen, carbon dioxide, hydrogen chloride, and sulfur dioxide. By applying the techniques presented here for dry gas and moisture, a more accurate determination of incinerator parameters can be made. Enthalpy lists and equations are included in Appendix C for these five gases.

Further considering the nature of incinerator operation, air in excess of the stoichiometric requirement is required to ensure that there is suffi-

Table 16-2. Excess Air Values.

Incinerator Type	Excess Air Range
Fluid Bed Furnace	40% – 60%
Infra-Red Incinerator	20% – 80%
Liquid Waste Incinerator	10% – 60%
Multiple Hearth Incinerator	75% – 150%
Rotary Kiln	75% – 200%

cient air to combust the waste stream. The amount of excess air required is a function of the incinerator as well as other factors. As indicated in Table 16-2, excess air can range from 10 to 200 percent. The air in excess of that required in the combustion process itself exits the incinerator in the flue gas. The flue gas, therefore, will have a dry gas component with properties approaching that of air, particularly where the amount of excess air introduced into the incinerator is above 50 percent.

These calculations are based on a 60° F datum. All inputs to the system are assumed to be at 60° F. This includes the waste stream, ambient air, and fuel. Establishing a datum eliminates the need to add terms in the calculation for system heat inputs. The tabular values in the appendix are based on this datum, to expedite calculations.

MASS FLOW

The MASS FLOW calculation sheet (Table 16-4) estimates air consumption and products of combustion. To illustrate the use of this sheet, the following waste firing in a liquid waste incinerator is assumed:

- 5000 lb/hr liquid waste feed
- 55 percent moisture
- 30 percent of solids are non-combustible
- 8000 Btu/lb of volatiles

Following the MASS FLOW sheet step-by-step:

M1: Insert total flow, 5000 lb/hr, which is a known quantity from the statement of the example.

M2: Insert moisture percent, 55, which is a known quantity from the statement of the example.

M3: Moisture contained within the waste, 55 percent of the total flow or $(55/100) \times 5000$ lb/hr = 2750 lb/hr.

M4: Dry feed into the furnace is equal to the total waste feed less moisture; 5000 lb/hr − 2750 lb/hr = 2250 lb/hr.

M5: Non-combustible or ash content, 30 percent of dry feed, is assumed in the example.

M6: Ash lb/hr from the waste, (30/100)×2250 lb/hr = 675 lb/hr.

M7: Volatile content. The waste solid feed is considered to have two components, ash and volatiles. The volatiles are, therefore, the difference between the total solids content and the ash content, 2250 lb/hr − 675 lb/hr = 1575 lb/hr volatiles.

M8: Volatile heating value is stated by the problem to be 8000 Btu/lb.

M9: The heat content of the feed stream is expressed in MBH, million Btu per hour. In this case 1575 Btu/lb volatiles × 8000 Btu/lb volatile/1000000 = 12.60 MBH.

M10: Dry gas generated by the burning of the waste volatile content. This generation rate can be determined from the chemical constituents of the waste (1.00 lb of carbon generates 3.66 lb of CO_2 and 8.77 lb of N_2; 1.00 lb of hydrogen generates 8.92 lb of H_2O and 26.08 lb of N_2). When the chemical make-up of the waste is unknown, an estimate of gas generation can be taken from Table 14-6. In this case, Table 14-6 was used. At a volatile heating value of 8000 Btu/lb, 6.50 lb/10KB is generated as off-gas. Note the convention: value per 10KB, or 10000 Btu. This method of presenting these values is an outgrowth of calculations based on coal burning technology conventions, wherein the heating value of coal volatiles is roughly 10000 Btu/lb.

M11: Dry gas generation is the generation rate multiplied by the waste heating content, 6.5 lb dry gas/10KB × 12.60 MBH × 100 = 8190 lb/hr. The factor 100 is a conversion factor (100 × 10KB = MB).

M12: Moisture generation from the combustion of waste volatiles is determined in a manner similar to that discussed in M12. From Table 14-6, at 8000 Btu/lb volatile, 1.48 lb/10KB moisture of combustion is generated.

M13: Moisture generation is equal to the generation rate multiplied by the waste heat content, 1.48 lb moisture/10KB × 12.60 MBH × 100 = 1865 lb/hr.

M14: To determine the amount of air required for this process, the products of combustion of the volatiles must first be totaled, 8190 lb/hr dry gas + 1865 lb/hr moisture = 10055 lb/hr.

M15: Air required for burning the waste stream is equal to the products of combustion less the waste feed rate. In other words, the air quantity plus the volatile quantity is equal to the sum of the products of combustion: dry gas and moisture of combustion. The air required for this reaction is therefore 10055 lb/hr products of combustion − 1575 lb/hr volatiles = 8480 lb/hr air. This is the stoichiometric air requirement for the waste stream.

M16: The total air requirement is equal to the excess air requirement plus the stoichiometric requirement, or 1 + excess air fraction. From the statement of this example, the furnace in question is a liquid-waste incinerator. From Table 16-2, an excess air rate of 50 percent is assumed. This entry, therefore, is 1.5, the total air requirement for burning this waste stream.

M17: The total air flow is the total air requirement multiplied by the stoichiometric air flow, 1.5 × 8480 lb/hr stoichiometric air = 12720 lb/hr total air for waste combustion.

M18: The air requirement for waste combustion is often provided by a single fan. The equivalent volumetric flow at standard conditions is based on a rate per minute, at air density of 0.075 lb/ft^3. 12720 lb/hr air / (60 × 0.075 lb/ft^3) = 2827 scfm, the waste combustion air fan requirement.

M19: The excess air quantity is that air not required for combustion. This air will pass through the process, unreacted. Excess air is equal to total air less the stoichiometric air requirement, 12720 lb/hr total air − 8480 lb/hr stoichiometric air requirement = 4240 lb/hr.

M20: Humidity is the amount of water vapor in the air and is a function of ambient temperature. From the psychrometric chart, Appendix E, a value of humidity of 0.01 lb water vapor per lb of dry air is chosen for this example, based on air at 60° F and 90 percent relative humidity.

M21: The moisture present due to humidity is the humidity multiplied by the dry gas flow, 0.01 lb moisture/lb dry gas (air) × 12720 lb/hr dry gas = 127 lb/hr moisture from humidity.

M22: The total moisture from combustion of the waste stream is equal to the sum of the moisture fraction of the waste entering the process, the moisture of combustion, and the humidity; 2750 lb/hr with the waste + 1865 lb/hr moisture of combustion + 127 lb/hr humidity = 4742 lb/hr total moisture.

M23: The total dry gas is equal to the dry gas generated from combustion of the waste plus the excess air passing through the process; 8190 lb/hr dry gas from combustion + 4240 lb/hr excess air = 12430 lb/hr total dry gas.

Now that the products of combustion of the waste have been determined, a heat balance can be calculated.

HEAT BALANCE

The Heat Balance, Table 16-5, will develop the temperature of the incinerator chamber, establish the need for supplemental fuel, will determine the amount of fuel required and will generate flue gas quantities.

To continue the process developed through the MASS FLOW calculations, note the following system characteristics:

- no cooling air flow
- ash exits at 1000° F from the process
- the required temperature in the incinerator is 2000° F
- fuel oil will be fired with 30 percent excess air

Following the Heat Balance sheet step-by-step:

H1: Insert cooling-air flow. In this example cooling-air flow = 0. Cooling air is that air flow which picks up heat from the process and then crosses the process boundary, i.e., is discharged to the atmosphere.

H2: Insert cooling-air temperature. With no cooling-air flow, a cooling-air temperature is not applicable. This temperature is the temperature of the cooling-air flow as it exits the system boundary, for instance, if the cooling air from the center shaft of a multiple-hearth incinerator were to be discharged to the atmosphere, this is the temperature of that discharge, normally in the range of 250° to 450° F.

H3: Insert the enthalpy of the air exiting the system, from Appendix B.

H4: This is the heat loss represented by the cooling-air discharge, the cooling-air flow multiplied by the enthalpy of the stream.

H5: Ash represents a loss of heat from the system. The temperature of the ash discharge from the incinerator chamber is stated as a given, 1000° F.

H6: Heat content of the ash discharge. Generally, ash contains heat in proportion to its temperature. Soils have a heat capacity of approximately 0.17 Btu/(lb – ° F), and this number is used in this calculation. Methods from Chapter 14 can be used to determine the heat of formation removed by ash from the incineration process when the constituent analysis of the ash is known. In this example the ash heat content is calculated as 0.17 Btu/(lb – ° F) × (1000° F – 60° F datum) = 160 Btu/lb.

H7: The total heat loss from the system due to ash discharge is the ash heat value multiplied by the ash quantity, 160 Btu/lb × 675 lb/hr = 0.11 MBH.

H8: Radiation loss is the loss from a furnace by radiation from its surface. This can be approximated by a value based on the heat released within the chamber, listed in Table 16-3, or it can be calculated based on the external surface area of furnace. In this example, a value of 3 percent (0.03) was selected, based on Table 16-3.

Table 16-3. Radiation Loss Factors.

Heat Input MBH	Expected Loss From Radiation Percent of Heat Input
< 15	3.00
20	2.75
25	2.50
30	2.00
40	1.75
>50	1.50

H9: The heat lost by radiation is equal to the radiation loss multiplied by the heat release, 0.03 × 12.60 MBH = 0.38 MBH. (Note that if this loss is approximated by a surface heat loss factor, this calculation would be based on surface area, not on heat-release rate.)

H10: Humidity enters the process in combustion air. Humidity is a vapor, not a liquid. Its enthalpy is higher than if it entered the process as a liquid because it has already absorbed heat in being raised to the vapor (or gaseous) state. This means that if it were included in calculations as "liquid water", a correction must be made to represent its elevated energy level. In the mass balance, humidity was considered "water" in step M22, when the humidity was added to moisture of combustion (which originates as a liquid at 60° F) and water in the waste, which is also an input to the system as a liquid at 60° F. The correction is the heat of vaporization of water at one atmosphere, 970 Btu/lb. Since this is an energy credit to the system, it is given a negative sign. In this example, the humidity correction is the humidity moisture flow multiplied by the heat of vaporization, 127 lb/hr × (−) 970 Btu/lb / 1000000 = (−) 0.12 MBH.

H11: The temperature of the flue gas is to be determined. The total loss from the system, i.e., the heat generated by the process, but not exiting in the flue gas, must be determined. These losses are the sum of the cooling-air loss, the loss from ash discharge, and radiation loss, less the humidity correction; 0 Btu/hr cooling air + 0.11 Btu/hr ash + 0.38 Btu/hr radiation − 0.12 Btu/hr humidity correction = 0.37 MBH loss from the flue gas.

H12: The incinerator outlet represents the flue gas outlet of the incinerator, the heat input less losses; 12.60 MBH input − 0.37 MBH loss = 12.23 MBH.

H13: The temperature within an incinerator chamber is not uniform. The temperature is highest near a burner, and the pattern of temperature within the chamber will vary as a function of air flow quantity and direction, waste introduction, and chamber geometry. The temperature calculated here is the temperature at the outlet of the incinerator chamber, where the gases within the chamber have had an opportunity to mix. This is not the highest temperature within the process.

The outlet temperature is calculated as follows:

Find the temperature, t, at which the enthalpy of the gas equals the heat available. The heat available, from H12, is 12.23 MBH. The enthalpy of the gas, calculated with the data from Appendix B, is obtained by trial and error.

(a) Temp. ° F	(b) h_{air} Btu/lb	(c) $12430 \times h_{air}$ MBH	(d) h_m Btu/lb	(e) $4742 \times h_m$ MBH	(f) MBH Total
1300	312.3	3.88	1665.1	7.90	11.78
t					12.23
1400	338.9	4.21	1719.8	8.16	12.37

By interpolation,

$$t = 1300 + (1400 - 1300) \times (12.23 - 11.78)/(12.37 - 11.78)$$
$$t = 1300 + 100 \times 0.76$$
$$t = 1376\ ^\circ\ \mathrm{F}$$

with:
(a) assumed temperature
(b) enthalpy of air from Appendix B
(c) heat in air at assumed temperature
(d) enthalpy of moisture from Appendix B
(e) heat in moisture at assumed temperature
(f) total heat in stream, (f) = (c) + (e)

Note that the total heat in the gas stream at the assumed temperature of 1300° F is less than the heat available, 12.23 MBH. Therefore, the second temperature assumed was higher, 1400° F. The heat in the gas stream was higher at 1400° F than what was available. The actual temperature of the stream must be between these two temperatures, i.e., between 1300° F and 1400° F. By interpolation, this temperature was found to be 1376° F.

H14: The required temperature is 2000° F, in the statement of the example. The temperature of the gas stream has been found to be 1376° F, which is lower than the required temperature. The temperature of the gas must be increased, and this means that supplemental fuel is required. If the required temperature would have been found to be less than the actual temperature, there would be at least four courses of action: increase the excess air requirement of the system; increase the water content of the feed; discharge cooling air; or do nothing. It may be that a process requires a minimum temperature and as long as the heat generated within that process generates a temperature higher than this minimum, no action is to be taken. When the waste is able to burn at a desired temperature without the addition of supplemental fuel, the waste is said to burn autogenously.

H15: The enthalpy of the dry gas component of the flue gas at the desired temperature, 2000° F, should be inserted here, from Appendix B, 502.3 Btu/lb.

H16: The heat content of the dry gas component of the flue gas is the product of the dry gas flow multiplied by its enthalpy; 12430 lb/hr dry gas × 502.3 Btu/lb/1000000 = 6.24 MBH.

H17: The enthalpy of the moisture component of the flue gas at the desired temperature, 2000° F, should be inserted here, from Appendix B, 2067.4 Btu/lb.

H18: The heat content of the moisture component of the flue gas is the product of the moisture in the gas stream and its enthalpy; 4742 lb/hr moisture × 2067.4 Btu/lb/1000000 = 9.80 MBH.

H19: The enthalpy of the gas stream at 2000° F is the sum of the dry gas and moisture component enthalpies; 6.24 MBH dry gas + 9.80 MBH moisture = 16.04 MBH.

H20: As discussed above, in step H14, supplemental fuel is required to bring the temperature of the gas stream to the required level. The heat deficiency that must be satisfied is the difference in heat in the gas as generated compared to the required heat level, 16.04 MBH required − 12.23 MBH existing = 3.81 MBH deficiency.

H21: Fuel oil requires a source of air to burn as a supplemental fuel. The amount of excess air to be used to burn fuel oil is to be inserted here, 30 percent from the example statement.

H22: Fuel oil will release its own products of combustion when it burns. Heat is required to raise the temperature of these products of combustion (moisture and dry gas) to the desired process temperature. The tables in Appendix G list the net heat available from #2 fuel oil as a function of process temperature and excess air. Note that as the temperature of the process rises, less heat is available from the fuel. For this example, at 30 percent excess air and 2000° F, 54535 Btu is available from a gallon of #2 fuel oil.

H23: The fuel oil quantity required to provide the deficiency need to bring the process to the desired temperature is the heat deficiency divided by the net fuel heating value; 3.81 MBH × 1000000/54535 Btu/gal = 69.86 gal/hr #2 fuel oil.

H24: The gross heat input from one gallon of fuel oil s 140000 Btu. The gross heat input from the required fuel quantity is, therefore, 69.86 gal/hr #2 fuel oil × 140000 Btu/gal #2 fuel oil/1000000 =9.78 MBH.

H25: The air required for combustion of the fuel oil is, from Appendix G, 134.41 lb/gal.

H26: Air flow for combustion of the fuel oil is the specific air requirement multiplied by the fuel quantity; 134.41 lb/gal fuel oil × 69.86 gal/hr #2 fuel oil = 9390 lb/hr air.

H27: The blower providing air to the supplemental fuel system is rated in volumetric flow, 9390 lb/hr / (0.075 lb/ft^3 × 60 min/hr) = 2087 scfm, supplemental fuel combustion air blower.

H28: The dry gas generated from the fuel oil is, from Appendix G, 133.54 lb/gal.

H29: The dry gas flow from fuel combustion is the fuel-oil quantity multiplied by the gas flow rate, 69.86 gal/hr fuel oil × 133.54 lb/gal = 9329 lb/hr dry gas from fuel-oil combustion.

H30: The moisture generated from the combustion of fuel is 8.75 lb/gal, from Appendix G.

H31: The moisture flow from fuel combustion is the product of the fuel oil rate and the moisture rate, 69.86 gal/hr fuel oil × 8.75 lb/gal = 611 lb/hr moisture of combustion.

H32: The heat within the flue gas at the outlet of the incinerator, including the heat generated from supplemental fuel, is the sum of the net heat from combustion of the wastes plus the gross heat from combustion of the supplemental fuel; 12.23 MBH from the waste + 9.78 MBH from the combustion of the fuel = 22.01 MBH exiting the furnace.

H33: The total dry gas flow exiting the incinerator is the sum of the dry gas flow generated from combustion of the waste and the dry gas flow from fuel combustion; 12430 lb/hr dry gas from waste + 9329 lb/hr dry gas from fuel = 21759 lb/hr total dry gas flow at outlet.

H34: The total moisture flow exiting the incinerator is the sum of the moisture flow generated from combustion of the waste plus the moisture flow from fuel combustion, 4742 lb/hr moisture from waste + 661 lb/hr moisture from fuel = 5353 lb/hr moisture flow at the incinerator outlet.

The parameters of the incineration process have now been determined. Conditions and requirements of the scrubber system can now be established.

FLUE GAS DISCHARGE

The Flue Gas Discharge, Table 16-6, develops characteristics and parameters of a flue-gas stream passing through a wet scrubber on its way to discharge to the atmosphere. Scrubber water requirements and induced draft fan requirements are determined.

Parameters associated with the scrubbing system are as follows:

- venturi factor of eight
- 140° F exhaust gas discharge temperature
- 40 inches WC ID fan pressure

To continue the process developed through the Mass Flow and Heat Balance calculations, note the following system characteristics:

F1: The hot gas stream will initially be quenched. Water will be added to the gas flow in sufficient quantity to saturate the gas stream i.e., the saturation moisture is the greatest amount of water that a gas (air) can hold. Saturation is a function of the temperature of the gas stream. The saturation temperature corresponds to the enthalpy of the gas, which is calculated by dividing the heat content of the gas stream by the dry gas flow; 22.01 MBH × 1000000/21759 lb/hr dry gas = 1011.54 Btu/lb dry gas.

F2: Knowing the enthalpy of the gas stream, the temperature corresponding to this value is taken from Appendix F, which is 186° F. This is the adiabatic saturation temperature of the gas stream. It defines a quench process. Just enough water is added to the gas stream to saturate it; there is no transfer of heat. The lowering of the 2000° F temperature of the incoming gas stream to 186° F results from a release of latent heat to the water, not from heat transfer across the system boundary.

F3: To determine the amount of water required to provide saturation, insert the saturation humidity from Appendix F, 0.8794 lb moisture/lb dry gas.

F4: The saturation moisture is the dry gas flow multiplied by the saturation humidity, 21759 lb/hr dry gas × 0.8794 lb moisture/lb dry gas = 19135 lb/hr saturation moisture.

F5: The moisture required to quench, or saturate the gas stream, is equal to the saturation moisture less the moisture initially carried along in the gas stream, 19135 lb saturation moisture/hr − 5353 lb inlet moisture/hr = 13782 lb/hr water required.

F6: One gallon per minute is equal to 500 lb/hr; 13782 lb/hr / 500 lb/(hr − gpm) = 27.56 gpm, the water required to quench the hot gas flow.

F7: Going through the scrubber, the water injected into the gas must quench the gas stream and provide sufficient turbulence to remove particulate matter from the gas stream. The amount of water required for this process is a function of the quench-water quantity. A rule-of-thumb is to inject eight times the quench water requirement to provide the gas cleaning required. There is relatively poor heat transfer in a venturi scrubber and this additional water flow will not result in significant cooling below the adiabatic quench temperature.

F8: The quench water supply necessary for quenching and particulate removal is the venturi factor multiplied by the ideal quench flow, 8 × 13782 lb quench water/hr = 110256 lb/hr quench water requirement.

F9: One gallon per minute equals 500 lb/hr: 110256 lb/hr/500 lb/(hr – gpm) = 221 gpm quench water.

F10: The gas stream is generally cooled below 150° F before discharge to the atmosphere. This lower temperature reduces the volume of the stream, decreasing the power requirement of the induced-draft (ID) fan. This reduced temperature also decreases the moisture in the flue gas, which reduces the occurrence of a visible plume at the stack discharge. The calculation of gas cooling requires a determination of the temperature of the water exiting the gas cooler, which is often a tray tower as shown on the right side of Fig. 12-11. The temperature of this discharge is the adiabatic temperature divided by the empirical factor 1.2, 186° F / 1.2 = 155° F, the cooling tower water discharge temperature.

F11: The temperature difference is the difference in cooling water temperature between the cooling tower discharge and the inlet of the cooling tower. The temperature of the water entering the cooling tower (or tray tower) is the datum, 60° F. The temperature difference is 155° F sump temperature – 60° F datum = 95° F.

F12: The discharge temperature is the temperature of the gas stream exiting the cooling tower, 140° F by the statement of this example.

F13: The enthalpy of the gas stream exiting the scrubbing system is the enthalpy of the saturated gas stream at the exit temperature, 140° F. From Appendix F, the discharge enthalpy is 178.82 Btu/lb dry gas.

F14: The heat content of the gas stream exiting the scrubbing system is the dry gas flow multiplied by the enthalpy of the gas stream; 21759 lb/hr dry gas × 178.82 Btu/lb dry gas / 1000000 = 3.89 MBH.

F15: The cooling load is the amount of heat that must be removed from the gas stream to reduce its temperature to the gas discharge temperature, 140° F. This load is the heat entering the scrubber less the heat exiting the scrubber in the gas stream; 22.01 MBH scrubber inlet – 3.89 MBH system outlet = 18.12 MBH cooling load.

F16: Cooling water requirement is the amount of water required to satisfy the cooling load. This calculation is based on the following equation:

$$Q = m \times c_p \times \Delta t$$

Q is the heat transfer; m, the water flow; c_p, the specific heat of water (1 Btu/lb – ° F), and Δt, the rise in temperature of the cooling water. In this illustration, the water quantity is as follows: 18.12 MBH × 1000000 / (1 Btu/lb – ° F × 95° F) = 190737 lb/hr cooling water.

F17: The cooling water flow can be expressed as 190737 lb/hr / 500 lb/hr – gpm) = 381 gpm.

F18: The amount of water exiting the scrubbing system in the exhaust gas is a function of the exhaust gas temperature, 140° F. From Appendix F, the discharge moisture of saturation is 0.1530 lb moisture/lb dry gas.

F19: Moisture flow exiting the scrubbing system in the exhaust is the dry gas flow multiplied by the discharge moisture of saturation; 21759 lb/hr dry gas flow × 0.1530 lb moisture/lb dry gas = 3329 lb/hr discharge moisture.

F20: The amount of water exiting the system in the scrubber drain is the sum of the water entering the scrubber in the incinerator gas stream, the water injected into the quench/venturi system, and the cooling water pumped to the tray tower; minus the water exiting the system in the exhaust gas stream. In this example, 5353 lb/hr incinerator exit + 110256 lb/hr quench + 190737 lb/hr cooling – 3329 lb/hr exhaust = 303017 lb/hr to drain (without quench water recirculation).

F21: The water flow can be expressed as 303017 lb/hr / 500 lb/hr – gpm = 606 gpm.

F22: The quench-water requirement is practically independent of water temperature, unlike the requirement for cooling water, which is a direct function of the cooling-water temperature. Quenching is a function of the latent heat of the water, not its temperature. A substantial savings in water usage can be realized by recirculation of water from the tray tower (spent cooling water). The savings in cooling water is the quench-water flow, 221 gpm in this example. Likewise, the drain water quantity is substantially reduced, to the sum of water flow at the incinerator exit plus the cooling water flow, less the moisture in the exhaust discharge. For this illustration, which includes recirculation, 5353 lb/hr incinerator outlet + 190737 lb/hr cooling − 3329 lb/hr exhaust = 192761 lb/hr to drain.

F23: The water flow can be expressed as 192761 lb/hr / 500 lb/(hr − gpm) = 386 gpm.

F24: The volumetric gas flow must be determined to size the induced-draft (ID) fan. From Appendix F, the volume of gas at the outlet temperature of 140° F is 18.82 cubic feet per lb of dry gas.

F25: The flow exiting the scrubber is the specific flow rate multiplied by the dry gas flow, 18.82 ft^3/lb dry gas × 21759 lb/hr dry gas / 60 min/hr = 6825 cfm outlet gas flow.

F26: The ID fan inlet flow is a function of the pressure across the fan as well as the upstream flow rate. The fan pressure, from the example statement, is 40 inches water column.

F27: The ID fan inlet is at a pressure lower than atmospheric pressure. This decreased pressure introduces an expansion of the volumetric flow entering the fan, equivalent to an increased volumetric flow rate. Atmospheric pressure (14.7 psi) is equivalent to a water column of 407 inches. The correction factor to be applied to the flow at the fan inlet is 407 inches / (407 inches − fan pressure), in this case 407/(407-40) = 1.11 fan correction factor.

F28: The actual volumetric flow rate into the ID fan is the flow exiting the tray tower (cooling tower) multiplied by the fan correction factor, 6825 cfm × 1.11 = 7576 acfm corrected outlet flow.

SUMMARY

The calculations presented in this chapter provide gross parameters of operation. With additional information from Chapter 15 on calculating system parameters, and Appendix D, a listing of specific volume of gases, other parameters of incinerator operation and design, such as residence time, destruction requirements, turbulence, etc., can be calculated. Calculations of heat recovery from incineration processes are presented in Chapter 17.

17

Energy Recovery

The purpose of a hazardous-waste incinerator is to safely dispose of hazardous waste. If there is a market for recovered heat, or if the heat can be effectively used to reduce incinerator fuel consumption, then energy recovery should be considered. An incinerator with energy recovery, however, is first of all an incinerator, which happens to generate energy. It should not be looked upon as an energy generator (such as a boiler) that happens to burn hazardous waste.

In this chapter techniques are developed for evaluating the potential heat recovery from an incineration process.

ENERGY-RECOVERY RATIONALE

Provision of energy-recovery equipment in an incineration system, such as the boiler shown in Fig. 17-1, is costly, representing a relatively high capital investment and increased operating costs. The inclusion of heat recovery in an incineration system must be evaluated to justify its cost. There are many reasons to consider energy recovery. Some of them are:

- Economics: to generate revenue or to save energy costs;
- Conservation: to reduce the quantity of supplemental fuel required, decreasing dependence on external supplies;

Fig. 17-1. Waste-heat boiler installation. Copyright © C. R. Brunner, 1984.

- Gas cooling: to reduce the temperature of the flue gas exiting the incinerator prior to entering the air emissions control system;
- Community relations: to obtain public acceptance of an incinerator project. The public may be more favorably disposed towards a system that recovers or saves energy than one that, in its perception, only incinerates hazardous wastes.

While the first reason may appear to be the most logical and important, often, it is the last consideration that controls the decision to install heat recovery systems and equipment.

Table 17-1 lists advantages and disadvantages of heat recovery in an incineration system.

EXTERNAL USE OF RECOVERED ENERGY

External use of energy is the use of recovered heat outside the incinerator or incinerator process. This is normally steam, hot water, or electric power generated from steam.

Figure 17-2 shows a steam turbine installed at an incinerator facility. The turbine generates electricity for in-plant use and the steam exiting the turbine (it is a non-condensing unit) is market outside the plant.

External heat recovery is only possible where there is a use for the recovered energy. The generation of steam, for instance, is not viable if there is no use for the steam in the plant or at a nearby location. Steam may be required only during winter months, but not on a yearly basis, and this limited, seasonal market may not be sufficient to justify the cost of energy-recovery equipment.

Table 17-1. Heat Recovery versus Wasting of Heat.

WITH HEAT RECOVERY	WITHOUT HEAT RECOVERY
Reduced gas temperatures and volumes due to absorption of heat by heat recovery system.	Hotter gas temperatures.
Moderate excess air.	High excess air required to control furnace temperatures.
Moderate size combustion chamber.	Large refractory-lined combustion chamber to handle high gas flow.
Smaller air and induced draft fans required for smaller gas volume.	Higher gas volumes due to higher temperatures, requiring larger air and gas flow equipment.
Steam facilities including boiler drums and boiler auxiliary equipment required.	No steam equipment required.
Operations involve boiler system monitoring, adjustment for steam demand, etc.	Relatively simple operating procedures.
Steam tube corrosion is possible as well as corrosion within exhaust gas train.	Corrosion possible in the exhaust gas train.
Licensed boiler operators may be required to operate the incinerator.	Conventional operators are satisfactory.
Considerable steam credits are possible, including in-plant energy savings in addition to salvage.	Only credits are possible salvage of the equipment after its useful life.

The generation of electricity is a less efficient mode of energy recovery than steam or hot water generation because the generation of electricity requires that steam be produced. Electricity is generated from this steam flow. On the other hand, electricity is the most marketable energy form. Electric power can be used within a plant, or it can be sold to the local power company. The sale of electric power to a power company is subject to a complex rate structure and set of regulations. Generally, power can be sold to a utility by an incineration facility at 50 to 80 percent of the cost of that power if purchased from that utility, based on PURPA statutes.

Fig. 17-2. Steam turbine at incinerator facility. Copyright © C. R. Brunner, 1979.

PUBLIC UTILITIES POLICY ACT

The Public Utilities Policy Act of 1978 (PURPA), includes regulations relevant to hazardous-waste incineration-facility energy recovery through electric power. A goal of PURPA is to encourage cogeneration and small-scale power production. *Cogeneration* refers to equipment that generates both electricity and useful thermal energy (i.e., heat or steam). In an effort to implement this goal the Federal Energy Regulatory Commission (FERC) has adopted regulations that provide the following:

- Utilities must purchase electricity from qualifying cogeneration and small power-production facilities at an appropriate rate (to be determined by state regulatory agencies), incorporating cost elements of capital offset and fuel replacement.
- Qualifying facilities that produce and sell electric energy are exempt from certain federal and state regulations pertaining to electric utilities.
- Utilities must provide qualifying facilities with electric energy and other types of services, which may be requested by a qualifying facility to supplement or back-up that facility's own generation.
- Implementation of the PURPA regulations is the responsibility of state regulatory authorities and non-regulated electric utilities.

Qualifying facilities fall into two categories: small power-production facilities and cogeneration facilities. To qualify as a small power-production facility a generator must meet the following criteria:

- Power production capacity must be less than 80 megawatts. (Note, however, that to qualify for an exemption from the Federal Power Act, a small production facility's capacity must be less than 30 megawatts).
- Biomass, waste, renewable resources, or any combination thereof must account for greater than 50 percent of the facility's total energy source. Oil, natural gas, and coal, in aggregate, may not exceed 25 percent of the total energy input of the facility during any one year.
- Not more than 50 percent of equity interest in the facility may be owned by an electric utility, utilities, or subsidiary of an electric utility.

To qualify as a cogeneration facility a generator must meet the following criteria:

- The same ownership criteria as small power-production facilities.
- It may not be a new (installed after March 13, 1980) diesel cogeneration facility.
- If it is a topping-cycle facility (first producing electric power, then thermal energy), it must meet the following operating and efficiency standards:
    - ☐ The useful thermal energy (heat or steam) must equal 5 percent or more of the total energy output.
    - ☐ If natural gas or oil are not used for energy input and installation began prior to March 13, 1980, there is no efficiency standard.
    - ☐ If natural gas or oil are used for energy input and installation began on or after March 13, 1980, then the following efficiency standards apply: If thermal-energy output is less than 15 percent of the total energy output, the useful-power output plus one half of the thermal output must be no less than 45 percent of the total energy input of natural gas and oil to the facility. If thermal-energy output is equal to or greater than 15 percent of the total energy output then the useful-power output plus one half the thermal output must be no less than 42.5 percent of the total energy input of natural gas and oil to the facility.
- If it is a bottoming-cycle facility (first producing thermal energy, then electric power), it must meet the following efficiency standards:
    - ☐ If natural gas or oil are not used for any of the energy input as supplementary firing, and installation began prior to March 13, 1980, then there is no efficiency standard.

- ☐ If natural gas or oil are used for any of the energy input as supplementary firing, and installation began on or after March 13, 1980, then the following efficiency standard applies: The power output of the facility during any calendar year must be no less than 45 percent of the energy input of natural gas and oil for supplementary firing.

An additional consideration, subject to Section 292.205(c) of the PURPA regulations, is that a cogeneration facility using natural gas may be exempted from incremental pricing.

Once it has been determined that a facility qualifies, it is eligible for the benefits available to it through PURPA. The benefits include the following obligations of the electric utility to the generator:

- The purchase of any energy and capacity made available by a qualifying facility.
- The sale of any energy and capacity requested by a qualifying facility to that facility.
- Interconnection with a qualifying facility when necessary to accomplish purchases or sales. This obligation is waived if, solely by reason of the purchase or sales over the interconnect, the electric utility would become subject to regulations as a public utility under Part II of the Federal Power Act.

Interconnection costs are to be paid by the qualifying facility and are to be set at a level comparable to the interconnection costs of other customers with similar load characteristics. Rates for purchases and sales are determined by a complex set of rules which require that a utility pay for energy from qualifying facilities based on "avoided costs." These are the costs that the utility saves by not having to generate the energy provided by the qualifying facility. The avoided costs of a facility are usually its highest costs since the cost of the last generating equipment to be brought on line is the avoided cost, and this is the utility's most expensive equipment.

These rules also cover the utility paying for "avoided capacity cost." This relates to the determination of how much of the qualifying facility's power capability is considered reliable enough to displace the need for new power plants. In Florida, the Public Service Commission adopted rules deeming a qualifying facility as eligible for a capacity credit if it provides energy to a utility at 70 percent equivalent capability.

In all cases wherein utilities are selling power to a qualifying facility, rates for sales are to be comparable to those charged other customers served by the utility.

INCINERATOR PROCESS ENERGY RECOVERY

There are a number of potential uses of recovered energy within an incineration process, including:

Direct Air Preheat

The most common instances of the use of direct air heating is for preheat of combustion air. Fluid-bed incinerators are often provided with an

Fig. 17-3. Air preheater. Copyright © C. R. Brunner, 1983.

air pre-heater or recuperator. This equipment, normally an air-to-gas heat exchanger, as shown in Fig. 17-3, draws hot gases from the incinerator through its tubes. Air is directed through the shell of the heat exchanger, and absorbs heat from the hot incinerator exhaust gases passing through the heat exchanger tubes.

Indirect Preheat

These systems, frequently used in Europe, but relatively uncommon in the United States, utilize a heat-transfer fluid. An insertion-type heat exchanger placed in the incinerator exhaust-gas stream absorbs heat in a circulating high-pressure fluid. The fluid is pressurized to maintain it as a liquid throughout the heat-transfer cycle. This fluid usually passes through one or two heat exchangers, where it pre-heats combustion air, re-heats the stack discharge, etc.

Steam-Driven Fans and Pumps

Steam turbines can drive large rotating equipment, often at 50 horsepower or more. In utilizing steam-driven equipment, consideration must be given to plant start-up. If steam is not available at a facility until the incinerator is operating and up to temperature, an electric drive must be provided for start-up. The cost of a dual drive and associated switching equipment (a clutch, at minimum) can be relatively high.

FIRESIDE CORROSION

Energy recovery is limited by the physical properties of materials within the recovery system. The critical item of a heat recovery system is the boiler, which is immersed in the path of the hot, erosive and corrosive exhaust gas stream.

There are two modes of corrosion that affect boiler tubes. Low-temperature, or *dewpoint*, corrosion is metal wastage caused by sulfuric or hydrochloric acid condensation. Chlorides and sulfides in the waste partially convert to free chlorine, hydrogen chloride, sulfur dioxide, and sulfur trioxide in the gas stream. Hydrogen chloride and sulfuric acid (sulfur trioxide readily combines with moisture in the gas stream to form sulfuric acid) condense at temperatures below 300° F, and their condensate attacks steel.

Boiler tubes are, by ASME code requirements, constructed of steel, and it is important to keep their temperature above 300° F. This temperature is a function of the temperature of the hot water or steam generated.

The boiler tubes will be at a temperature close to that of the circulating fluid, and this 300° F limit is also the minimum temperature of the steam or hot water generated. For saturated steam, the pressure corresponding to a temperature of 300° F is approximately 50 psig.

High-temperature corrosion is a more complex phenomenon. At temperatures exceeding 700° F, a complex reaction takes place between the

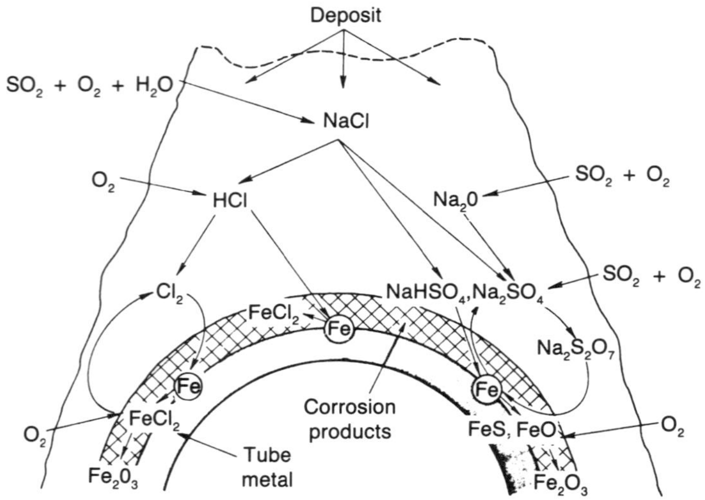

Fig. 17-4. Sequence of chemical reactions explaining corrosion on incinerator boiler tube. Source: Reference 1.

sulfide- and chlorine/chloride-bearing flue gas and the steel boiler-water tubes, as illustrated in Fig. 17-4. Chlorine reacts with the iron in the tube wall to produce ferrous chloride, which, upon contact with oxygen in the flue gas, converts to iron oxide. The iron oxide (rust) is friable and leaves the surface of the steel, causing wastage of the boiler-tube surface. Other components of the waste which become airborne, such as alkali salts, promote this corrosion.

The mechanism of high-temperature corrosion is illustrated in Fig. 17-5. Generally, chlorine is present in incinerator off-gas at a much lower concentration than HCl. HCl is the more significant component, and its corrosive effect increases dramatically as the gas temperature rises above 700° F.

BOILER DESIGN PARAMETERS

In accordance with consideration of heat recovery as ancillary to the basic function of an incinerator, the boiler should be designed to provide maximum reliability, not maximum heat recovery. Careful attention must be focused on the tube spacing and boiler tube design.

In conventional water-tube boilers, the tube spacing is such that the gas velocity across the face of the tubes is in the range of 100 to 120 feet per second. This is the velocity across the first set of tubes in the boiler, where the gas temperature is greatest. Downstream of this point, the gas loses heat to the tubes, which results in a lower temperature and volumetric gas-flow rate, which corresponds to a lower gas-stream velocity across the tube face.

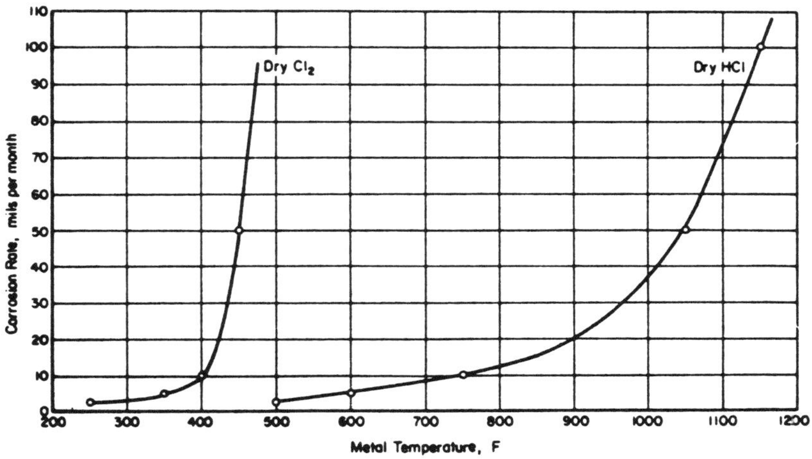

Fig. 17-5. Corrosion of carbon steel in chloride and hydrogen chloride. Source: Reference 1.

Incinerator flue gas, particularly that which results from the burning of solids or sludges, contains particulate matter which tends to erode boiler tubes. To minimize this effect, the velocity of the gas stream in contact with the boiler tubes should be below 40 feet per second.

The second parameter concerns the surface of the boiler tubes. Tube surfaces should be smooth. The provision of fins, studs, or other protuberances on the tube surface increases the effective tube surface and heat transfer. However, these designs encourage particulate matter to collect on these extended surfaces, and to scale and clog the tubes. This action can cause local hot spots which will burn out the boiler tube, and such build-ups decrease heat transfer through the tubes, generating less steam.

These two features, decreased gas velocity and smoother tube surface, illustrate the concept that an incinerator is primarily a waste disposer, not a steam generator. Although higher gas velocities and extended tube surfaces may appear to increase the efficiency of the system in the generation of steam (or hot water), they decrease the reliability of the incinerator. System reliability should have a higher priority than steam generation efficiency.

STEAM VERSUS HOT WATER

Steam is used as a heat recovery medium in incinerator applications far more frequently than hot water. Steam is more versatile in its application, and one pound of steam contains significantly more energy than one pound of water. Whereas hot water is normally only useful for heating buildings during winter months, or for feedwater heating of larger steam systems, steam can be used for process requirements and for equipment loads, which are often year-round requirements. Steam can be converted to hot water when the need arises. Within limitations, steam can also be utilized to generate electric power, which is universally marketable.

The advantage of hot water is its relative simplicity. Steam systems require a constant supply of make-up water, but hot water systems utilize closed-loops and do not require either continuous de-mineralized water supplies, nor do they generate significant wastewater.

STEAM GENERATION

Figure 17-6 is a simplified flow diagram showing the generation and supply of steam from a waste heat boiler. Make-up water is raised to feedwater temperature by a steam flow side-stream from the main flow loop. Heat contained in the condensate is also used to heat feedwater.

Besides raising the feedwater temperature prior to injection into the boiler, the de-aerator acts to help release dissolved oxygen from feedwater. Additional feedwater treatment is usually employed to reduce or prevent scaling and corrosion of exposed boiler surfaces in contact with the feedwater and steam. Water softeners are used to remove most of the calcium and magnesium hardness from the raw water. Chemicals are also added, typically:

Sodium Sulfite

This is an oxygen-scavenging chemical that chemically removes the dissolved oxygen residual not removed in the de-aerator. Hydrazine is another oxygen scavenger that is used in high pressure (over 1200 psia) boiler applications.

Amines

There are a number of amines used for feedwater treatment. They are used for boiler pH or alkalinity control. Excess alkalinity (pH greater

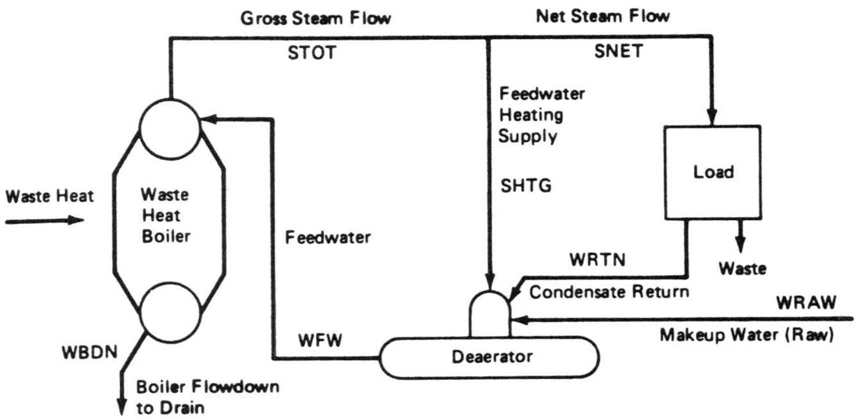

Fig. 17-6. Waste-heat boiler, steam flow.

than 11) permits accelerated scale build-up; low pH (below 6) can cause excessive boiler tube corrosion. Normally, boiler water pH is maintained in the range of 8.0 to 9.5 (slightly alkaline).

Phosphates

This treatment is used to precipitate residual calcium and magnesium hardness remaining in the feedwater after softening. Certain phosphates act as dispersants, preventing adhesion of the precipitate to tube walls.

These chemicals form sludge which accumulates in the lower drum of a boiler. The boiler water must have a blowdown on a regular basis to prevent build-up of sludge in the "mud drum." This blowdown normally represents from 2 to 5 percent of boiler steam generation.

CALCULATING AVAILABLE HEAT

Available heat is that amount of heat that is available to heat the energy medium: to convert water to either steam or hot water, or to heat an air stream. Calculating this number requires a detailed knowledge of the boiler parameters, such as number, size, and spacing of boiler tubes; location of these tubes in the gas stream (waterwall or immersion type tube system) and the nature of the gas stream (constituent gases, gas temperatures, and other physical properties). All of these parameters are rarely, if ever, known for a waste stream. In many cases, the purpose of the calculations for energy recovery is to establish boiler size

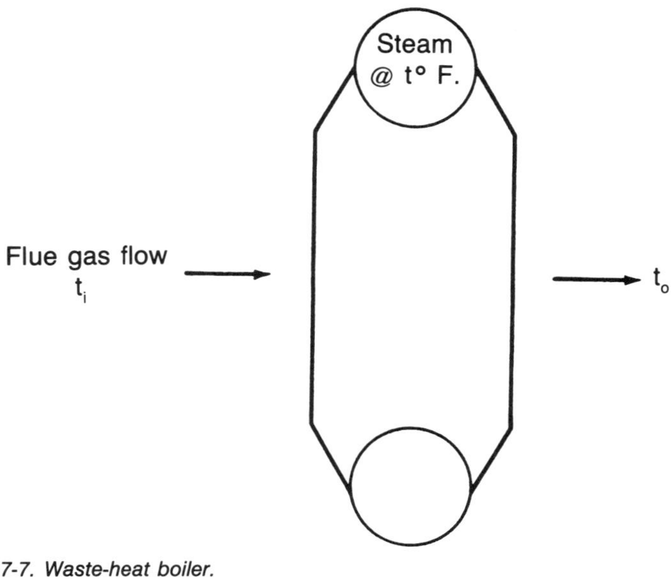

Fig. 17-7. Waste-heat boiler.

parameters. Assumptions have to be made in order to proceed with this calculation.

A more accurate method of determining available heat is to examine current boiler designs. In all boilers, the temperature of the flue gas exiting the boiler tube section is higher than the temperature of the generated steam. In a perfect system, the temperature of the steam will be equal to the temperature of the gas stream exiting the boiler. The approach temperature, t_x is defined as the difference between the steam temperature and the temperature of the flue gases exiting the boiler. For the ideal system, $t_x = 0$. An infinitely large boiler would be required to achieve this efficiency. Real systems are finite, and their approach temperature normally ranges from 30° F to 150° F. By selecting an approach temperature, the available heat is readily calculated, as follows:

$$Q = W \times h@t_1 - h@t_2$$

Referring to Fig. 17-7,

$$t_2 = t_x + t$$

t_x	=	approach temperature, ° F
t_1	=	temperature of hot flue gas, ° F, usually known
t_2	=	temperature of exiting flue gas, ° F, unknown
t	=	steam temperature, ° F, usually known
$h(t)$	=	enthalpy of gas stream, Btu/lb
W	=	gas flow rate, lb/hr

Using the example of Chapter 16, Table 16-5, for incinerator operating parameters:

W_{dg}	=	21759	lb/hr	(dry gas flow)
W_{H_2O}	=	5353	lb/hr	(moisture flow)
Q_1	=	22.01	MBH	(exit heating value)
t_1	=	2000	° F	(temperature entering boiler)

The available heat will be calculated for a number of steam pressure values, using an approach temperature of 75° F.

In Table 17-2, for the generation of steam at 100 psia D&S (dry and saturated), from Appendix H, the steam temperature is 328° F. With an approach temperature of 75° F, the temperature of the gas stream exiting the boiler is 328° F + 75° F = 403° F. At this temperature, the heat in the exhaust-gas stream leaving the boiler is calculated by interpolating enthalpy from Appendix B. At 100 psia the heat in the gas stream exiting the boiler is 8.28 MBH. This represents a 13.73 MBH transfer of energy to the boiler. (The stream entering the boiler has a heat content of 22.01

Table 17-2. Available Heat.

Steam		Gas Exit	Enthalpy		Gas Stream MBH			Available Heat
Pres.	Temp.	Temp.	DG	H_2O	$21759h_{dg}$	$5353h_m$	Total	22.01-Total
100	328	403	82.1	1213.2	1.79	6.49	8.28	13.73
200	383	458	95.8	1261.1	2.08	6.75	8.83	13.18
500	467	542	116.7	1279.3	2.54	6.85	9.39	12.62

MBH. The heat available is 22.01 MBH minus 8.28 MBH exiting the boiler.) The available heat is also calculated for 200 and 500 psia. These values are 13.18 MBH and 12.62 MBH respectively.

The available heat is greatest at the lowest steam pressure. The lower the steam pressure, the lower its temperature, and the lower the temperature of the gas stream exiting the boiler. This allows a greater heat transfer to the boiler.

REFERENCES

1. P. Miller, *Corrosion in Municipal Incinerators*, United States Environmental Protection Agency SW-72-3-3, 1972.

2. C. R. Brunner, *Incineration Systems: Selection & Design* (Reston, VA: Incinerator Consultants Incorporated, 1988.)

APPENDIX A
Table of the Elements

Element	Symbol	Atomic No.	Atomic Weight
Actinium	Ac	89	227.
Aluminum	Al	13	26.98
Americium	Am	95	243.
Antimony	Sb	51	121.75
Argon	Ar	18	39.95
Arsenic	As	33	74.92
Astatine	At	85	210.
Barium	Ba	56	137.34
Berkelium	Bk	97	247.
Beryllium	Be	4	9.01
Bismuth	Bi	83	208.98
Boron	B	5	10.81
Bromine	Br	35	79.90
Cadmium	Cd	48	112.40
Calcium	Ca	20	40.08
Californium	Cf	98	251.
Carbon	C	6	12.01
Cerium	Ce	58	140.12
Cesium	Cs	55	132.91
Chlorine	Cl	17	35.45
Chromium	Cr	24	52.00
Cobalt	Co	27	58.93
Columbium (*see* Niobium)			

APPENDIX A—Continued

Element	Symbol	Atomic No.	Atomic Weight
Copper	Cu	29	63.55
Curium	Cm	96	247.
Dysprosium	Dy	66	162.50
Einsteinium	Es	99	254.
Erbium	Er	68	167.26
Europium	Eu	63	151.96
Fermium	Fm	100	253.
Fluorine	F	9	19.00
Francium	Fr	87	223.
Gadolinium	Gd	64	157.25
Gallium	Ga	31	69.72
Germanium	Ge	32	72.59
Gold	Au	79	196.97
Hafnium	Hf	72	178.49
Helium	He	2	4.00
Holmium	Ho	67	164.93
Hydrogen	H	1	1.01
Indium	In	49	114.82
Iodine	I	53	126.90
Iridium	Ir	77	192.2
Iron	Fe	26	55.85
Krypton	Kr	36	83.80
Lanthanum	La	57	138.91
Lead	Pb	82	207.19
Lithium	Li	3	6.94
Lutetium	Lu	71	174.97
Magnesium	Mg	12	24.31
Manganese	Mn	25	54.94
Mendelevium	Md	101	256.
Mercury	Hg	80	200.59
Molybdenum	Mo	42	95.94
Neodymium	Nd	60	144.24
Neon	Ne	10	20.18
Neptunium	Np	93	237.
Nickel	Ni	28	58.71
Niobium	Nb	41	92.91
Nitrogen	N	7	14.01
Nobelium	No	102	254.
Osmium	Os	76	190.2
Oxygen	O	8	16.00
Palladium	Pd	46	106.4
Phosphorus	P	15	30.97
Platinum	Pt	78	195.09
Plutonium	Pu	94	242.
Polonium	Po	84	210.
Potassium	K	19	39.10

APPENDIX A—Continued

Element	Symbol	Atomic No.	Atomic Weight
Praseodymium	Pr	59	140.91
Promethium	Pm	61	147.
Protactinium	Pa	91	231.
Radium	Ra	88	226.
Radon	Rn	86	222.
Rhenium	Re	75	186.2
Rhodium	Rh	45	102.91
Rubidium	Rb	37	85.47
Ruthenium	Ru	44	101.07
Samarium	Sm	62	150.53
Scandium	Sc	21	44.96
Selenium	Se	34	78.96
Silicon	Si	14	28.09
Silver	Ag	47	107.87
Sodium	Na	11	22.99
Strontium	Sr	38	87.62
Sulphur	S	16	32.06
Tantalum	Ta	73	180.95
Technetium	Tc	43	99.
Tellurium	Te	52	127.60
Terbium	Tb	65	158.92
Thallium	Tl	81	204.37
Thorium	Th	90	232.04
Thulium	Tm	69	168.93
Tin	Sn	50	118.69
Titanium	Ti	22	47.90
Tungsten	W	74	183.85
Uranium	U	92	238.03
Vanadium	V	23	50.94
Xenon	Xe	54	131.30
Ytterbium	Yb	70	173.04
Yttrium	Y	39	88.91
Zinc	Zn	30	65.37
Zirconium	Zr	40	91.22

APPENDIX B
Enthalpy, Moisture and Dry Gas

	Btu/lb (Relative to 60° Datum)	
Temperature, °F	**Air Enthalpy**	**Moisture Enthalpy**
60	0.0	0.0
150	21.0	1091.9
200	32.9	1116.6
250	44.9	1140.7
300	57.0	1164.5
350	69.2	1188.2
400	81.4	1211.8
450	93.8	1235.5
500	106.2	1259.2
550	118.7	1283.1
600	131.2	1307.1
650	143.8	1331.3
700	156.5	1355.7
750	169.2	1380.3
800	181.9	1405.0
850	194.7	1430.0
900	207.6	1455.3
950	220.5	1480.7
1000	233.5	1506.4

APPENDIX B—Continued

Temperature, °F	Btu/lb (Relative to 60° Datum) Air Enthalpy	Moisture Enthalpy
1050	246.5	1532.4
1100	259.6	1558.3
1150	272.7	1584.8
1200	285.8	1611.2
1250	299.0	1638.3
1300	312.3	1665.1
1350	325.6	1692.2
1400	338.9	1719.8
1450	352.3	1747.7
1500	365.7	1775.5
1600	392.7	1832.1
1700	419.8	1890.1
1800	447.1	1948.0
1900	474.6	2007.2
2000	502.3	2067.4
2100	530.1	2128.7
2200	558.1	2189.9
2300	586.2	2252.6
2400	614.6	2315.3
2500	643.1	2377.8
2600	671.7	2443.3
2700	700.5	2511.9

APPENDIX C
Enthalpy, Gases

Reference: Chemical Engineers' Handbook, Perry & Chilton Fifth Edition, 1973, Page 3-119

This table is based on the following relationships:

Carbon dioxide (44.01 lb/mole):
$$H = 10.34 \times T + 0.00137 \times T^2 + 195500/T$$

Hydrogen chloride (36.46 lb/mole):
$$H = 6.7 \times T + 0.00042 \times T^2$$

Nitrogen (28.02 lb/mole):
$$H = 6.5 \times T + 0.0005 \times T^2$$

Oxygen (32.00 lb/mole):
$$H = 8.27 \times T + 0.000129 \times T^2 + 187700/T$$

Sulfur dioxide (64.06 lb/mole):
$$H = 7.7 \times T + 0.00265 \times T^2 + 0.00000083/T$$

Air: $H = 0.2315 \times H_{oxygen} + 0.7685 \times H_{nitrogen}$

where: T = temperature, °K
H = enthalpy, kcal/kg-mole

APPENDIX C—Continued

To convert to Btu/lb (h):
h = H / (0.252 kcal/Btu × 2.2046 lb/kg × molecular weight)

Temp., ° F	CO_2	HC1	N_2	O_2	SO_2	Air
150	18.8	17.8	21.9	18.1	13.2	21.0
200	29.9	27.8	34.1	28.8	20.6	32.9
250	41.3	37.9	46.4	39.8	28.2	44.9
300	52.9	48.0	58.7	51.1	35.9	57.0
350	64.9	58.2	71.1	62.7	43.7	69.2
400	77.0	68.5	83.5	74.4	51.6	81.4
450	89.4	78.8	96.0	86.3	59.7	93.8
500	101.9	89.2	108.5	98.4	67.8	106.2
550	114.6	99.6	121.1	110.5	76.1	118.7
600	127.5	110.1	133.7	122.8	84.5	131.2
650	140.5	120.7	146.4	135.2	93.0	143.8
700	153.6	131.4	159.1	147.6	101.6	156.5
750	166.9	142.1	171.9	160.1	110.4	169.2
800	180.3	152.8	184.7	172.7	119.2	181.9
850	193.9	163.7	197.6	185.3	128.2	194.7
900	207.5	174.6	210.5	198.0	137.2	207.6
950	221.3	185.5	223.5	210.7	146.4	220.5
1000	235.2	196.5	236.5	223.5	155.7	233.5
1050	249.2	207.6	249.6	236.4	165.2	246.5
1100	263.3	218.8	262.7	249.2	174.7	259.6
1150	277.5	230.0	275.9	262.1	184.3	272.7
1200	291.8	241.3	289.1	275.0	194.1	285.8
1250	306.2	252.6	302.3	288.0	204.0	299.0
1300	320.7	264.0	315.7	301.0	214.0	312.3
1350	335.3	275.5	329.0	314.0	224.1	325.6
1400	350.0	287.0	342.4	327.1	234.3	338.9
1450	364.8	298.6	355.9	340.2	244.7	352.3
1500	379.7	310.3	369.4	353.3	255.1	365.7
1600	409.8	333.8	396.6	379.5	276.4	392.7
1700	440.2	357.6	424.0	405.9	298.1	419.8
1800	471.1	381.6	451.6	432.3	320.3	447.1
1900	502.3	405.9	479.3	458.9	342.9	474.6
2000	533.9	430.5	507.3	485.4	366.0	502.3
2100	565.8	455.3	535.5	512.1	389.5	530.1

APPENDIX C—Continued

Temp., ° F	CO_2	HC1	N_2	O_2	SO_2	Air
2200	598.1	480.3	563.9	538.8	413.6	558.1
2300	630.8	505.6	592.4	565.6	438.0	586.2
2400	663.8	531.2	621.2	592.5	463.0	614.6
2500	697.2	557.0	650.2	619.4	488.4	643.1
2600	731.0	583.1	679.3	646.4	514.2	671.7
2700	765.1	609.4	708.7	673.4	540.5	700.5

APPENDIX D
Specific Volume

Temp., °F	Air, ft^3/lb	H_2O, ft^3/lb
70	13.0	21.1
100	14.1	22.7
200	16.6	26.8
300	19.1	30.8
400	21.7	34.9
500	24.2	38.9
600	26.7	43.0
700	29.2	47.0
800	31.7	51.1
900	34.2	55.1
1000	36.8	59.2
1100	39.3	63.3
1200	41.8	67.3
1300	44.3	71.4
1400	46.8	75.4
1500	49.4	79.5
1600	51.9	83.5
1700	54.4	87.6
1800	56.9	91.6
1900	59.4	95.7

APPENDIX D—Continued

Temp., °F	Air, ft^3/lb	H_2O, ft^3/lb
2000	61.9	99.7
2100	64.5	103.8
2200	67.0	107.9
2300	69.5	111.9
2400	72.0	116.0
2500	74.5	120.0
2600	77.0	124.1
2700	79.6	128.1
2800	82.1	132.2
2900	84.6	136.2
3000	87.1	140.3
3100	89.6	144.3
3200	92.2	148.2
3300	94.7	152.5
3400	97.2	156.5
3500	99.7	160.6
3600	102.2	164.6

APPENDIX E
Psychrometric Chart

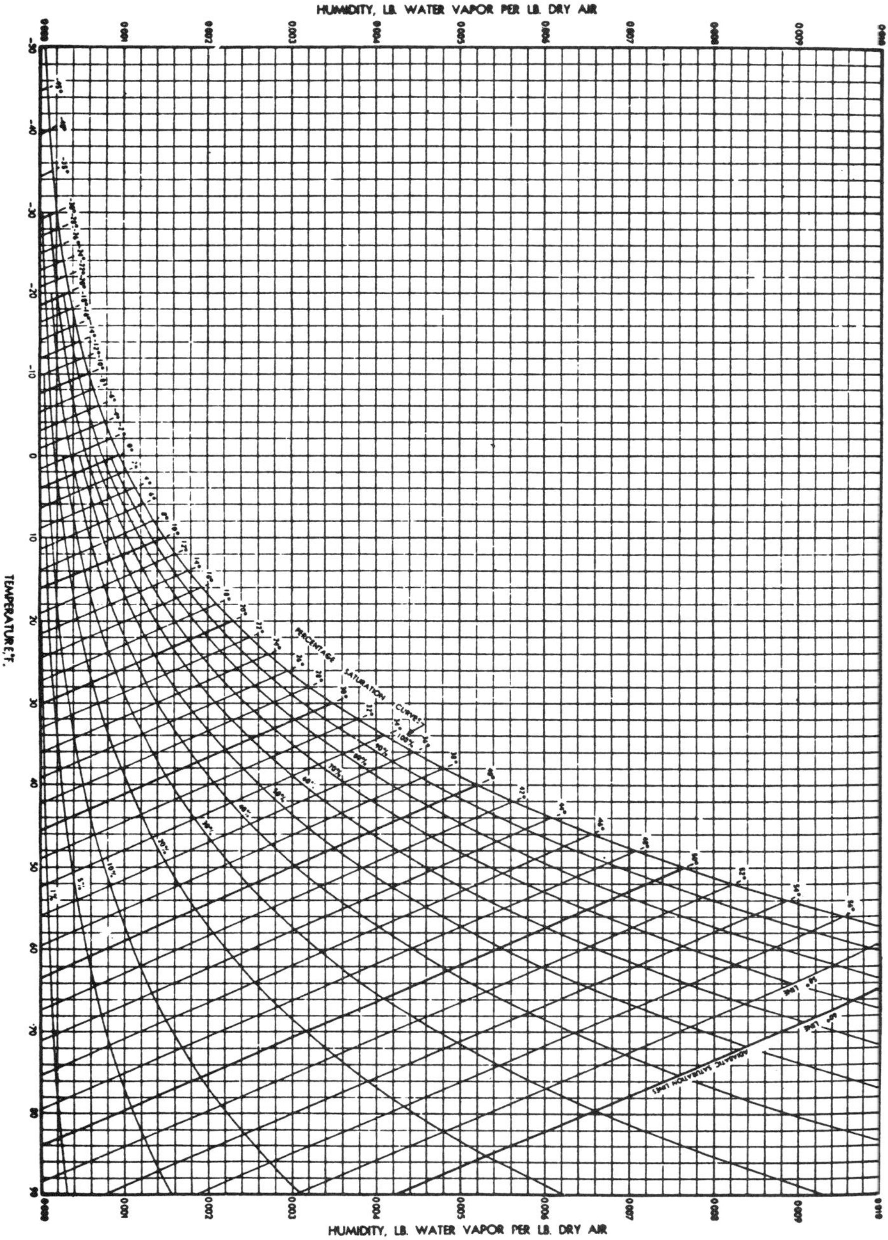

APPENDIX F

Saturation of Dry Air with Moisture

Temp. °F	Humidity lb H_2O/lb DA	Enthalpy Btu mixture/lb DA	Volume ft^3 mixture/lb DA
60	0.01108	0.000	13.329
61	0.01149	0.690	13.363
62	0.01191	1.393	13.398
63	0.01234	2.114	13.433
64	0.01279	2.849	13.468
65	0.01326	3.602	13.504
66	0.01374	4.369	13.540
67	0.01423	5.155	13.680
68	0.01474	5.958	13.613
69	0.01527	6.781	13.650
70	0.01581	7.621	13.688
71	0.01638	8.484	13.762
72	0.01700	9.366	13.764
73	0.01755	10.266	13.803
74	0.01817	11.191	13.842
75	0.01881	12.136	13.882
76	0.01946	13.094	13.922
77	0.02014	14.094	13.963

APPENDIX F—Continued

Temp. °F	Humidity lb H_2O/lb DA	Enthalpy Btu mixture/lb DA	Volume ft^3 mixture/lb DA
78	0.02084	15.109	14.004
79	0.02156	16.149	14.046
80	0.02231	17.214	14.088
81	0.02308	18.306	14.131
82	0.02387	19.426	14.175
83	0.02468	20.571	14.219
84	0.02552	21.744	14.264
85	0.02639	22.947	14.309
86	0.02728	24.181	14.355
87	0.02821	25.445	14.402
88	0.02916	26.744	14.449
89	0.03014	28.074	14.497
90	0.03115	29.441	14.547
91	0.03219	30.841	14.597
92	0.03326	32.279	14.647
93	0.03437	33.751	14.699
94	0.03551	35.266	14.751
95	0.03668	36.815	14.804
96	0.03789	38.408	14.854
97	0.03914	40.039	14.913
98	0.04043	41.715	14.968
99	0.04175	43.438	15.025
100	0.04312	45.209	15.083
101	0.04453	47.023	15.142
102	0.04498	48.886	15.202
103	0.04748	50.806	15.263
104	0.04902	52.770	15.325
105	0.05061	54.792	15.389
106	0.05225	56.868	15.453
107	0.05394	59.000	15.519
108	0.05568	61.190	15.587
109	0.05747	63.444	15.655
110	0.05932	65.764	15.725
111	0.06123	68.144	15.796
112	0.06319	70.589	15.869
113	0.06522	73.102	15.944
114	0.06731	75.690	16.020
115	0.06946	78.357	16.098
116	0.07168	81.095	16.178
117	0.07397	83.197	16.259
118	0.07633	86.815	16.343

APPENDIX F—Continued

Temp. °F	Humidity lb H_2O/lb DA	Enthalpy Btu mixture/lb DA	Volume ft^3 mixture/lb DA
119	0.07877	89.800	16.428
120	0.08128	92.880	16.515
121	0.08388	96.840	16.603
122	0.08655	99.300	16.695
123	0.08931	102.65	16.789
124	0.09216	106.11	16.885
125	0.09511	109.67	16.983
126	0.09815	113.34	17.084
127	0.10129	117.14	17.187
128	0.10453	121.04	17.293
129	0.10788	125.06	17.402
130	0.11130	129.22	17.514
131	0.11490	133.50	17.628
132	0.11860	137.87	17.746
133	0.12240	142.46	17.867
134	0.12640	147.18	17.991
135	0.13040	152.02	18.119
136	0.13460	157.04	18.251
137	0.13900	162.24	18.386
138	0.14350	167.58	18.525
139	0.14820	173.11	18.669
140	0.15300	178.82	18.816
141	0.15800	184.73	18.969
142	0.16320	190.85	19.126
143	0.16850	197.17	19.288
144	0.17410	203.72	19.454
145	0.17980	210.49	19.626
146	0.18580	217.52	19.804
147	0.19200	224.82	19.987
148	0.19840	232.35	20.176
149	0.20510	240.17	20.374
150	0.22120	248.29	20.576
151	0.21920	256.71	20.786
152	0.22670	265.45	21.004
153	0.23440	274.53	21.229
154	0.24250	283.96	21.462
155	0.25090	293.78	21.704
156	0.25960	303.98	21.955
157	0.26880	314.64	22.216
158	0.27820	325.69	22.487
159	0.28810	337.23	22.769

APPENDIX F—Continued

Temp. °F	Humidity lb H_2O/lb DA	Enthalpy Btu mixture/lb DA	Volume ft^3 mixture/lb DA
160	0.29850	349.36	23.063
161	0.30920	361.79	23.368
162	0.32050	374.90	23.685
163	0.33230	388.58	24.017
164	0.34460	402.89	24.365
165	0.35750	417.84	24.725
166	0.37100	433.53	25.102
167	0.38510	449.95	25.492
168	0.40000	467.18	25.914
169	0.41560	485.28	26.347
170	0.43200	504.29	26.804
171	0.44930	524.31	27.272
172	0.46750	545.38	27.787
173	0.48670	567.61	28.315
174	0.50700	591.07	28.876
175	0.52840	615.86	29.465
176	0.55110	642.11	30.089
177	0.57520	669.92	30.749
178	0.60080	699.48	31.449
179	0.62790	730.86	32.193
180	0.65600	764.31	32.984
181	0.68780	800.01	33.829
182	0.72090	838.20	34.731
183	0.75630	879.02	35.694
184	0.79430	922.91	36.728
185	0.83520	970.11	37.839
186	0.87940	1021.0	39.037
187	0.92710	1076.1	40.332
188	0.97900	1135.9	41.737
189	1.03550	1201.0	43.265
190	1.09700	1272.2	44.935
191	1.16500	1350.7	46.764
192	1.24000	1436.3	48.780
193	1.32200	1531.6	51.011
194	1.41400	1637.4	53.488
195	1.51700	1756.2	56.265
196	1.63300	1889.5	59.381
197	1.76500	2040.8	62.918
198	1.91500	2214.0	66.963
199	2.08900	2413.9	71.630
200	2.29200	2648.4	77.102

APPENDIX F—Continued

Temp. °F	Humidity lb H_2O/lb DA	Enthalpy Btu mixture/lb DA	Volume ft^3 mixture/lb DA
201	2.53200	2924.2	83.543
202	2.82000	3255.4	91.270
203	3.17300	3660.8	100.750
204	3.61400	4169.8	112.590
205	4.18100	4821.7	127.800
206	4.93900	5694.0	147.600
207	6.00000	6913.0	176.560
208	7.59400	8748.0	219.300
209	10.248	11802	290.440
210	15.54	17887	432.250
211	31.49	36241	859.820

Note: DA = Dry Air

Source: Derived from O. Zimmerman, I. Lavine, Psychrometric Tables and Charts, Industrial Research Service, Dover, NH, First Edition.

APPENDIX G

Net Fuel Heat Value Parameters

This appendix is based on the following fuel characteristics:

- Number 2 fuel oil
- 19430 Btu/lb
- 7.21 lb/gallon
- 140000 Btu/gal
- 87.46 percent carbon
- 12.54 percent hydrogen

Combustion characteristics were calculated with the following parameters, with X = excess air fraction:

- Humidity = 0.005 lb H_2O/lb dry air
- Carbon dioxide out = 3.20 lb/lb fuel
- H_2O from fuel = 1.12 lb/lb fuel
- Air required = $4.3197 \times 3.32 \times (1 + X)$ lb/lb fuel
- Nitrogen residual = $3.3197 \times 3.32 \times X$ lb/lb fuel
- Oxygen residual = $3.32 \times X$ lb/lb fuel
- Dry gas out = nitrogen + oxygen residual + CO_2
- H_2O out = H_2O from fuel + humidity

APPENDIX G—Continued

0 percent Excess Air		10 percent Excess Air	
Air/gal = 144.75 lb		Air/gal = 155.09 lb	
Dry Gas/gal = 143.88 lb		Dry Gas/gal = 154.22 lb	
H_2O/gal = 8.80 lb		H_2O/gal = 8.85 lb	
T, °F	**Net Btu/gallon**	**T, °F**	**Net Btu/gallon**
200	127542	200	127186
300	124654	300	124046
400	121748	400	120885
500	118818	500	117699
600	115854	600	114474
700	112855	700	111212
800	109303	800	107343
900	106735	900	104557
1000	103615	1000	101164
1100	100457	1100	97730
1200	97261	1200	94255
1300	94029	1300	90740
1400	90761	1400	87187
1500	87461	1500	83600
1600	84124	1600	79973
1700	80765	1700	76322
1800	77380	1800	72643
1900	73967	1900	68934
2000	70527	2000	65197
2100	67064	2100	61433
2200	63585	2200	57654
2300	60080	2300	53847
2400	56561	2400	50023
2500	53032	2500	46190
2600	49464	2600	42316
2700	45860	2700	38404

APPENDIX G—Continued

20 percent Excess Air		30 percent Excess Air	
Air/gal	= 124.07 lb	Air/gal	= 134.41 lb
Dry Gas/gal	= 123.20 lb	Dry Gas/gal	= 133.54 lb
H_2O/gal	= 8.70 lb	H_2O/gal	= 8.75 lb

T, °F	Net Btu/gallon	T, °F	Net Btu/gallon
200	126830	200	126474
300	123437	300	122829
400	120022	400	119159
500	116579	500	115460
600	113095	600	111715
700	109570	700	107928
800	105382	800	103422
900	102379	900	100200
1000	98712	1000	96261
1100	95002	1100	92275
1200	91248	1200	88241
1300	87452	1300	84163
1400	83614	1400	80040
1500	79739	1500	75878
1600	75821	1600	71670
1700	71879	1700	67435
1800	67906	1800	63168
1900	63901	1900	58868
2000	59866	2000	54535
2100	55803	2100	50173
2200	51722	2200	45791
2300	47613	2300	41379
2400	43486	2400	36949
2500	39348	2500	32505
2600	35167	2600	28018
2700	30947	2700	23491

APPENDIX G—Continued

40 percent Excess Air		50 percent Excess Air	
Air/gal	= 144.75 lb	Air/gal	= 155.09 lb
Dry gas/gal	= 143.88 lb	Dry gas/gal	= 154.22 lb
H_2O/gal	= 8.80 lb	H_2O/gal	= 8.85 lb

T, °F	Net Btu/gallon	T, °F	Net Btu/gallon
200	126118	200	125762
300	122221	300	121612
400	118297	400	117434
500	114340	500	113220
600	110336	600	108956
700	106285	700	104643
800	101462	800	99501
900	98022	900	95844
1000	93810	1000	91358
1100	89547	1100	86820
1200	85235	1200	82228
1300	80874	1300	77585
1400	76466	1400	72892
1500	72016	1500	68155
1600	67518	1600	63367
1700	62992	1700	58549
1800	58431	1800	53694
1900	53835	1900	48801
2000	49204	2000	43873
2100	44542	2100	38912
2200	39860	2200	33929
2300	35146	2300	28912
2400	30411	2400	23874
2500	25663	2500	18821
2600	20869	2600	13721
2700	16034	2700	8578

APPENDIX G—Continued

60 percent Excess Air

Air/gal = 165.43 lb
Dry gas/gal = 164.56 lb
H_2O/gal = 8.90 lb

T, °F	Net Btu/gallon
200	125406
300	121004
400	116571
500	112101
600	107577
700	103001
800	97541
900	93665
1000	88907
1100	84093
1200	79221
1300	74297
1400	69319
1500	64294
1600	59215
1700	54106
1800	48957
1900	43768
2000	38542
2100	33282
2200	27997
2300	22679
2400	17336
2500	11978
2600	6572
2700	1122

70 percent Excess Air

Air/gal = 175.77 lb
Dry gas/gal = 174.90 lb
H_2O/gal = 8.95 lb

T, °F	Net Btu/gallon
200	125050
300	120396
400	115708
500	110981
600	106197
700	101358
800	95581
900	91487
1000	86456
1100	81365
1200	76215
1300	71008
1400	65745
1500	60433
1600	55064
1700	49663
1800	44220
1900	38735
2000	33211
2100	27651
2200	22066
2300	16445
2400	10799
2500	5136
2600	——
2700	——

APPENDIX G—Continued

80 percent Excess Air				90 percent Excess Air			
Air/gal	=	186.10 lb		Air/gal	=	196.44 lb	
Dry gas/gal	=	185.24 lb		Dry gas/gal	=	195.58 lb	
H_2O/gal	=	9.01 lb		H_2O/gal	=	9.06 lb	

T, °F	Net Btu/gallon	T, °F	Net Btu/gallon
200	124694	200	124338
300	119788	300	119179
400	114845	400	113983
500	109862	500	108742
600	104818	600	103438
700	99716	700	98074
800	93620	800	91660
900	89309	900	87130
1000	84005	1000	81553
1100	78638	1100	75910
1200	73208	1200	70201
1300	67719	1300	64430
1400	62171	1400	58598
1500	56572	1500	52711
1600	50913	1600	46761
1700	45220	1700	40777
1800	39482	1800	34745
1900	33702	1900	28669
2000	27880	2000	22549
2100	22021	2100	16391
2200	16135	2200	10204
2300	10211	2300	3978
2400	4262	2400	——
2500	——	2500	——
2600	——	2600	——
2700	——	2700	——

APPENDIX G—Continued

100 percent Excess Air		**100 percent Excess Air**	
Air/gal	= 206.78 lb	Air/gal	= 217.12 lb
Dry gas/gal	= 205.92 lb	Dry gas/gal	= 216.26 lb
H_2O/gal	= 9.11 lb	H_2O/gal	= 9.16 lb

T, °F	**Net Btu/gallon**	**T, °F**	**Net Btu/gallon**
200	123982	200	123626
300	118571	300	117963
400	113120	400	112257
500	107623	500	106503
600	102059	600	100679
700	96431	700	94789
800	89700	800	87739
900	84952	900	82774
1000	79102	1000	76651
1100	73183	1100	70455
1200	67194	1200	64188
1300	61142	1300	57853
1400	55024	1400	51450
1500	48850	1500	44989
1600	42610	1600	38458
1700	36334	1700	31891
1800	30008	1800	25271
1900	23636	1900	18602
2000	17218	2000	11887
2100	10760	2100	5130
2200	4272	2200	——-
2300	——	2300	——
2400	——	2400	——
2500	——	2500	——
2600	——	2600	——
2700	——	2700	——

APPENDIX G—Continued

120 percent Excess Air		130 percent Excess Air	
Air/gal = 227.46 lb		Air/gal = 237.80 lb	
Dry gas/gal = 226.60 lb		Dry gas/gal = 236.94 lb	
H_2O/gal = 9.21 lb		H_2O/gal = 9.26 lb	
T, °F	**Net Btu/gallon**	**T, °F**	**Net Btu/gallon**
200	123270	200	122914
300	117354	300	116746
400	111394	400	110532
500	105383	500	104264
600	99300	600	97920
700	93146	700	91504
800	85779	800	83818
900	80595	900	78417
1000	74199	1000	71748
1100	67728	1100	65000
1200	61181	1200	58174
1300	54564	1300	51275
1400	47876	1400	44303
1500	41127	1500	37266
1600	34307	1600	30155
1700	27448	1700	23005
1800	20534	1800	15796
1900	13569	1900	8536
2000	6556	2000	1225
2100	——	2100	——
2200	——	2200	——
2300	——	2300	——
2400	——	2400	——
2500	——	2500	——
2600	——	2600	——
2700	——	2700	——

APPENDIX G—Continued

140 percent Excess Air		150 percent Excess Air	
Air/gal = 248.14 lb		Air/gal = 258.48 lb	
Dry gas/gal = 247.27 lb		Dry gas/gal = 257.61 lb	
H_2O/gal = 9.32 lb		H_2O/gal = 9.37 lb	
T, °F	**Net Btu/gallon**	**T, °F**	**Net Btu/gallon**
200	122558	200	122202
300	116138	300	115530
400	109669	400	108806
500	103144	500	102025
600	96541	600	95161
700	89862	700	88219
800	81858	800	79898
900	76239	900	74060
1000	69297	1000	66845
1100	62273	1100	59545
1200	55168	1200	52161
1300	47987	1300	44698
1400	40729	1400	37155
1500	33405	1500	29544
1600	26004	1600	21852
1700	18562	1700	14119
1800	11059	1800	6322
1900	3503	1900	——
2000	——	2000	——
2100	——	2100	——
2200	——	2200	——
2300	——	2300	——
2400	——	2400	——
2500	——	2500	——
2600	——	2600	——
2700	——	2700	——

APPENDIX H
Steam Enthalpy

Source: J. Keenan, F. Keyes, Thermodynamic Properties of Steam
John Wiley & Sons, New York, NY, 1957

Pressure	Temp.	Saturated	Steam Temperature, Degrees Farenheit						
psia	°F	Vapor	400	500	600	700	800	1000	1200
14.7	212	1150.4	1239.9	1287.1	1334.8	1383.2	1432.3	1533.1	1637.5
20	228	1156.3	1239.2	1286.6	1334.4	1382.9	1432.1	1533.0	1637.4
30	250	1164.1	1237.9	1285.7	1333.8	1382.4	1431.7	1532.7	1637.2
40	267	1169.7	1236.5	1284.8	1333.1	1381.9	1431.3	1532.4	1637.0
50	281	1174.1	1235.1	1283.9	1332.5	1381.4	1430.9	1532.1	1636.8
60	293	1177.6	1233.6	1283.0	1331.8	1380.9	1430.5	1531.9	1636.6
70	303	1180.6	1232.1	1282.0	1331.1	1380.4	1430.2	1531.6	1636.3
80	312	1183.1	1230.7	1281.1	1330.5	1379.9	1429.7	1531.3	1636.2
90	320	1185.3	1229.1	1280.1	1329.8	1379.4	1429.3	1531.0	1635.9
100	328	1187.2	1227.6	1279.1	1329.1	1378.9	1428.9	1530.8	1635.7
120	341	1190.4	1224.4	1277.2	1327.7	1377.8	1428.1	1530.2	1635.3

APPENDIX H—Continued

Pressure	Temp.	Saturated	Steam Temperature, Degrees Farenheit						
psia	°F	Vapor	400	500	600	700	800	1000	1200
140	353	1193.0	1221.1	1275.2	1326.4	1376.8	1427.3	1529.7	1634.9
160	364	1195.1	1217.6	1273.1	1325.0	1375.7	1426.4	1529.1	1634.5
180	373	1196.9	1214.0	1271.0	1323.5	1374.7	1425.6	1528.6	1634.1
200	382	1198.4	1210.3	1268.9	1322.1	1373.6	1424.8	1528.0	1633.7
220	390	1199.6	1206.5	1266.7	1320.7	1372.6	1424.0	1527.5	1633.3
240	397	1200.6	1202.5	1264.5	1319.2	1371.5	1423.2	1526.9	1632.9
260	404	1201.5	-	1262.3	1317.7	1370.4	1422.3	1526.3	1632.5
280	411	1202.3	-	1260.0	1316.2	1369.4	1421.5	1525.8	1632.1
300	417	1202.8	-	1257.6	1314.7	1368.3	1420.6	1525.2	1631.7
325	424	1203.5	-	1254.6	1312.8	1366.9	1419.6	1524.5	1631.2
350	432	1203.9	-	1251.5	1310.9	1365.5	1418.5	1523.8	1630.7
375	438	1204.3	-	1248.4	1308.9	1364.1	1417.5	1523.1	1630.2
400	445	1204.5	-	1245.1	1306.9	1362.7	1416.4	1522.4	1629.6
450	456	1204.6	-	1238.4	1302.8	1359.9	1414.3	1521.0	1628.6
500	467	1204.4	-	1231.3	1298.6	1357.0	1412.1	1519.6	1627.6
550	477	1203.8	-	1223.7	1294.3	1354.0	1409.9	1518.2	1626.6
600	486	1203.2	-	1215.7	1289.9	1351.1	1407.7	1516.7	1625.5
650	495	1202.3	-	1207.2	1285.3	1348.0	1405.4	1515.3	1624.5
700	503	1201.2	-	-	1280.6	1345.0	1403.2	1513.9	1623.5
750	511	1200.0	-	-	1275.7	1341.8	1400.9	1512.4	1622.4
800	518	1198.6	-	-	1270.7	1338.6	1398.6	1511.0	1621.4
900	532	1195.4	-	-	1260.1	1332.1	1393.9	1508.1	1619.3
1000	545	1191.8	-	-	1248.8	1325.3	1389.2	1505.1	1617.3
1100	556	1187.8	-	-	1236.7	1318.3	1384.3	1502.2	1615.2
1200	567	1183.4	-	-	1223.5	1311.0	1379.3	1499.2	1613.1

Note: Enthalpy in Btu/lb.

Glossary

abrasion—The wearing away of surface material by the scouring action of moving solids, liquids, or gas.

absorption—The penetration of one substance into or through another.

activated carbon—A highly adsorbent form of carbon used to remove odors and toxic substances from gaseous emissions or to remove dissolved organic material from wastewater.

acute LC(50)—A concentration of a substance, expressed in parts weight per million parts of medium, that is lethal to 50 percent of the test population of animals under specific test conditions.

acute toxicity—Any poisonous effect produced within a short period of time, usually 24 to 96 hours, resulting in severe biological harm and often death.

adhesion—Molecular attraction which holds the surfaces of two substances together.

adsorption—The attachment of the molecules of liquid or gaseous substances to the surface of a solid.

aerosol—A particle of liquid or solid matter that can remain suspended in the air because of its small size.

afterburner—A device that includes an auxiliary fuel burner and combustion chamber to incinerate combustible gas contaminants.

air emissions—For stationary sources, the release or discharge of a pollutant by an owner or operator into the ambient air, either by means of a stack or as fugitive dust, mist, or vapor, as a result inherent to the process.

air heater—A heat exchanger through which air passes and is heated by a medium of a higher temperature, such as hot combustion gases in metal tubes.

air pollutant—Dust, fumes, smoke and other particulate matter, vapor, gas, odorous substances, or any combination thereof. Also, any air pollution agent or combination of such agents including any physical, biological, or radioactive substance which is emitted into or otherwise enters the ambient air.

air pollution—The presence of any air pollutant in sufficient quantities and of sufficient characteristics and duration as to be, or likely to be, injurious to the health or welfare of animal or plant life, or as to interfere with the enjoyment of life or property.

air preheater—An air heater which raises the temperature of an air stream entering a process.

ambient air—That portion of the atmosphere external to buildings to which the general public has access.

AQCR—Air quality control region.

ash—Inorganic residue remaining after the burning of combustible substances.

autogenous combustion—The burning of organic material in which the gross heat of combustion is at such a level that it permits burning without the addition of supplementary fuel.

autothermic combustion—See "autogenous combustion."

BACT—Best available control technology.

baffles—Deflector vanes, guides, grids, grating, or similar devices constructed or placed in air or gas flow, flowing water, or slurry flow, to effect a more uniform distribution of velocities; absorb energy; divert, guide or agitate fluids; and check eddies.

baghouse—An air emissions control device used to trap particulate by filtering gas streams through large fabric bags, usually made of cloth or glass fibers.

barometric damper—A hinged or pivoted plate that automatically regulates the amount of air entering a duct or flue to maintain a constant draft within an incinerator.

batch-fed incinerator—An incinerator that is periodically charged with waste, and in which one charge is allowed to burn out before another is added.

biomedical waste—Waste derived from hospital, laboratory, and health-care facility operation.

blowdown—The minimum discharge of recirculating water for the purpose of discharging materials contained in the process, the further buildup of which would cause concentrations or amounts exceeding limits established by best engineering practice.

blower—A fan used to force air or gas under pressure.

bottom ash—The solid material that remains on a hearth or falls off the grate of an incinerator after burning is complete.

breeching—A passage that conducts the products of combustion to a stack or chimney.

bridge wall—A partition between chambers over which the products of combustion pass.

burning area—The horizontal projection of a grate or hearth.

burning hearth—A solid surface to support the solid fuel or solid waste in a furnace, upon which materials are placed for combustion.

burning rate—The rate at which solid waste is incinerated or heat is released during incineration. The burning rate is usually expressed in pounds of waste per square foot of burning area per hour or in Btu per cubic foot of furnace volume per hour.
butterfly damper—A plate or blade installed in a duct, flue, breeching, or stack that rotates on an axis to regulate the flow of gases.

CAA—Clean Air Act.
cake—The solids discharged from a dewatering apparatus.
calcination—The process of heating a waste material to a high temperature without fusing in order to produce a useful change.
carbonaceous matter—Pure carbon or carbon compounds present in the waste or residue of a combustion process.
carcinogenic—Capable of causing the cells of an organism to react in such a way as to cause cancer.
catalytic combustion system—A process in which a substance is introduced into an exhaust gas stream to encourage the burning or oxidation of vaporized hydrocarbons; the substance itself remains intact.
caustic soda—Sodium hydroxide, a strong alkaline substance used to scrub acid from flue gases.
centrifugal collector—A mechanical system using centrifugal force to remove aerosols from a gas stream or to dewater sludge.
CERCLA—Comprehensive Environmental Response, Compensation, and Liability Act.
chamber—An enclosed space inside an incinerator.
charging chute—An overhead passage from which waste materials drop into an incinerator.
checker work—A pattern of multiple openings through which the products of combustion pass to accelerate the turbulent mixing of gases.
classification—The separation and re-arrangement of waste materials according to composition (e.g., organic or inorganic), size, weight, color, shape, etc., using specialized equipment.
clinkers—Hard sintered or fused pieces of residue formed in a furnace by the agglomeration of ash, metals, glass, and ceramics from the waste stream.
combustibles—Materials that can be ignited at a specific temperature in the presence of air to release heat.
combustion—The production of heat and light through a chemical process, usually oxidation.
combustion air—The air used for burning a fuel (or waste).
combustion gases—The mixture of gases and vapors produced by burning.
conduction—The transfer of heat by physical contact between substances.
continuous-feed incinerator—An incinerator into which waste is charged continuously to maintain a steady rate of burning.
controlled-air incinerator—An incinerator with two or more combustion chambers in which the amounts and distribution of air are controlled. Combustion takes place in the first zone and gases are burned in a subsequent zone or zones.
corrosion—The gradual wearing-away of a substance by chemical action.

curtain wall—A refractory construction within a furnace that deflects combustion gases downward.

cyclone separator—A separator that uses a swirling air flow to sort mixed materials according to the size, weight, and density of the pieces.

decontamination/detoxification—Processes which convert pesticides into nontoxic compounds, or the selective removal of radioactive material from a surface or from within another material.

deep-well injection—Disposal of raw or treated hazardous wastes by pumping them into deep wells for filtration through porous or permeable subsurface rock and containment within surrounded layers of impermeable rock or clay.

dewatering—A physical process which removes sufficient water from sludge so that its physical form is changed from a fluid to a damp solid.

dispersion technique—The use of dilution to attain ambient air quality levels, including any intermittent or supplemental control of air pollutants varying with atmospheric conditions.

draft—The difference between the pressure within an incinerator and that in the atmosphere.

drag conveyor—A conveyor that uses vertical steel plates fastened between two continuous chains to drag material across a smooth surface.

DRE—Destruction and removal efficiency.

drying hearth—A solid surface in an incinerator upon which wet waste materials, liquids, or waste matter that may liquify before burning are placed to dry, or to burn with the help of hot combustion gases.

duct—A conduit, usually metal or fiberglass, round or rectangular in cross section, used for conveyance of air.

dust—Fine-grained particles, light enough to be suspended in air.

electrostatic precipitator—An air emissions control device that imparts an electric charge to particles in a gas stream, causing them to collect on an electrode.

emission rate—The amount of pollutant emitted per unit of time.

ESP—Electrostatic precipitator.

exhaust system—The system comprising a combination of components which provide for enclosed flow of exhaust gas from the furnace exhaust port to the atmosphere.

extraction test procedure—A series of laboratory operations and analyses designed to determine whether, under severe conditions, a solid waste, stabilized waste, or landfill material can yield a hazardous leachate.

FERC—Federal Energy Regulatory Commission.

filter—A porous device through which a gas or liquid is passed to remove suspended particles or dust.

firebrick—Refractory brick made from fireclay.

fireclay—A sedimentary clay containing only small amounts of fluxing impurities, high in hydrous aluminum silicate, and capable of withstanding high temperatures.

fire point—The lowest temperature at which an oil vaporizes readily enough to burn at least five seconds after ignition.

fixed carbon—The ash-free carbonaceous material that remains after volatile matter is driven off a dry waste sample.

flammable waste—A waste capable of igniting easily and burning rapidly.

flash drying—The process of drying a wet organic material by passing it through a high-temperature zone at such a rate that the water is rapidly evaporated, but the organic material, protected by the boiling point of water, is not overheated.

flash point—The minimum temperature at which a liquid or solid gives off sufficient vapor to form an ignitable vapor-air mixture near the surface of the liquid or solid.

flue—Any passage designed to carry combustion gases and entrained particulate.

flue gas—The products of combustion, including pollutants, emitted to the air after a production process or combustion takes place.

fluidized-bed combustion—Oxidation of combustible material within a bed of solid, inert (non-combustible) particles which, due to the action of vertical hot air flow, act as a fluid.

fly ash—The solid airborne combustion residue from burning fuel (waste).

forced draft—The positive pressure created by the action of a fan or blower which supplies the primary or secondary combustion air in an incinerator.

fugitive emissions—Emissions other than those from stacks or vents.

fume—Solid particles under 1 μ in diameter, formed as vapors condense or as chemical reactions take place.

furnace—A combustion chamber; an enclosed structure in which heat is produced.

fusion point—The temperature at which a particular complex mixture of minerals can flow under its own weight.

grain loading—The rate at which solids are emitted from a pollution source, in grains per cubic foot of gas emitted, where 7000 grains weigh one pound.

grate—A piece of furnace equipment used to support solid waste or solid fuel during drying, igniting, or burning.

grease—A group of substances including fats, waxes, free fatty acids, calcium and magnesium soaps, mineral oils, and certain other non-fatty materials.

guillotine damper—An adjustable plate used to regulate the flow of gases, installed vertically in a breeching.

hardness—A characteristic of water, imparted by salts of calcium, magnesium, and iron, that causes curdling of soap, deposition of scale in boilers and damage in some industrial processes.

hazardous waste—A waste, or combination of wastes, that may cause or significantly contribute to an increase in mortality or an increase in serious irreversible, or incapacitating reversible, illness, or that pose a substantial present or potential hazard to human health or the environment when improperly treated, stored, transported, disposed of, or otherwise managed. A legal definition of hazardous waste is included in the provisions of RCRA.

hearth—The bottom of a furnace, upon which waste materials are exposed to the flame.

heat balance—An accounting of the distribution of the heat input and output of an incinerator or a boiler, usually on a time basis.

heavy metals—Metallic elements, such as mercury, lead, and cadmium, with high atomic weights, which tend to accumulate in the food chain.
HEPA—High efficiency particulate air filter.
hydrocarbon—Any of a vast family of compounds containing carbon and hydrogen in various combinations, found particularly in fossil fuels and industrial waste streams.

ignitability—The quality of having a flash point of less than 140° F.
incineration—An engineered process using controlled flame combustion to thermally degrade waste materials.
incinerator stoker—A mechanically operated moving grate for supporting, burning, or transporting solid waste in a furnace and discharging the residue.
induced draft—The negative pressure created by the action of a fan, blower, or other gas-moving device located between an incinerator and a stack.
infiltration air—Air that leaks into the chambers, ducts, or flues of an incinerator.
inorganic matter—Chemical substances of mineral origin, not containing carbon-to-carbon bonding.
isokinetic sampling—Sampling in which the linear velocity of the gas entering the sampling nozzle is equal to that of the undisturbed gas stream at the sample point.

KBH—Thousand Btu per hour.
kraft paper—A comparatively coarse paper noted for its strength and used primarily as a wrapper or as packaging material.

LAER—Lowest achievable emission rate.
landfill—A land disposal site employing an engineered method of disposing of wastes on land that minimizes environmental hazards by spreading wastes in thin layers, compacting the wastes to the smallest practical volume, and applying cover materials at the end of each operating day.
LEL—Lower explosive limit.
lime—Any of a family of chemicals consisting essentially of calcium oxide or hydroxide made from limestone (calcite).
LUST—Leaking underground storage tank.

masking—Blocking out one sight, sound, or smell with another.
MBH—Million Btu per hour.
MCU—Modular combustion unit.
mist—Liquid particles, measuring 40 μ to 500 μ, in diameter that are formed by condensation of vapor.
mixing chamber—A chamber usually placed between the primary and secondary combustion chambers and in which the products of combustion are thoroughly mixed by turbulence that is created by increased velocities of gases, checker work, or turns in the direction of gas flow.
multiple-chamber incinerator—An incinerator that consists of two or more chambers, arranged as in-line or retort types, interconnected by gas-passage parts or flues.

mutagenic—The property of a substance or mixture of substances which, when it interacts with a living organism, causes the genetic characteristics of the organism to change and its offspring to have a decreased life expectancy.

NAAQMD—National ambient air quality standards.
natural draft—The negative pressure created by the height of a stack or chimney and the difference in temperature between flue gases and the atmosphere.
NCP—National Contingency Plan.
NEHAPS—National Emission Standard for Hazardous Air Pollutants.
NIOSH—National Institute for Occupational Safety and Health.
NPDES—National Pollutant Discharge Elimination System.
NPL—National priorities list.
NSR—New source review.

odor threshold—The lowest concentration of an airborne odor that a human being can detect.
opacity—Degree of obscuration of light; for example, a window has zero opacity; a wall has 100 percent opacity.
open-pit incinerator—A burning apparatus that has an open top and a system of closely spaced nozzles that place a stream of high-velocity air over the burning zone.
organic matter—Chemical substances comprised mainly of covalently bonded carbon, which may have their origin in animal or plant life, coal, petroleum, or laboratory synthesis.
organic nitrogen—Nitrogen combined in organic molecules, such as protein, amines, and amino acids.
overfire air—Air under control as to quantity and direction, introduced above and beyond a fuel (waste) bed by induced or forced draft.
oxidant—A substance that reacts chemically to increase the valence of another substance.

packed tower—A pollution-control device that forces dirty gases through a tower packed with crushed rock, wood chips, or other packing while liquid is sprayed over the packing material. Pollutants in the gas stream either dissolve in or chemically react with the liquid.
particulate—Fine liquid or solid particles, such as dust, smoke, mist, fumes, or smog, found in the air or in emissions.
particulate matter—Any material, except water in uncombined form, that is or has been airborne and exists as a liquid or a solid at standard conditions.
pathogenic waste—Discarded materials that contain organisms capable of causing disease.
PICs—Products of incomplete combustion.
plume—Visible emissions from a flue or chimney.
POHC—Principal organic hazardous constituent.
pollution—The presence of matter or energy whose nature, location, or quantity produces undesirable environmental effects. Also, the man-made or man-induced alteration of the chemical, physical, biological, and radiological integrity of the water, land, or air.

pour point—The lowest temperature at which an oil will flow or can be poured under specified conditions of test.

precipitators—Air emissions control devices that collect particles from air emissions by electrical or mechanical means.

precursor—A pollutant that takes part in a chemical reaction resulting in the formation of one or more new pollutants.

primary combustion air—The air admitted to a combustion system when the fuel is first oxidized.

primary pollutant—A pollutant emitted directly from a polluting stack.

primary standard—A national air emissions standard intended to establish a level of air quality that, with an adequate margin of error, will protect public health.

process waste—Any designated toxic pollutant which is inherent to or the unavoidable result of any manufacturing process, including that which comes into direct contact with or results from the production or use of any raw material, intermediate product, finished product, by-product, or waste product.

proximate analysis—The analysis of a fuel (or waste) to determine, on a percentage basis, how much moisture, volatile matter, and ash the sample contains, in addition to its heating value.

PRP—Potentially responsible party.

PSD—Prevention of significant deterioration.

PURPA—Public Utilities Regulatory Policies Act.

pyrolysis—The chemical decomposition of organic matter through the application of heat in an oxygen-deficient atmosphere.

RACT—Reasonable available control technology.

RCRA—Resource Conservation and Recovery Act.

reactivity—The tendency to create vigorous reactions with air or water; tendency to explode; to exhibit thermal instability with regard to shock; or ready reaction to generate toxic gases.

refractory material—Nonmetallic substances used to line furnaces because they can endure high temperatures and resist abrasion, spalling, and slagging.

residue—Solid or semi-solid materials such as, but not limited to, ash, ceramics, glass, metal, and organic substances remaining after incineration or processing.

RI/FS—Remedial investigation/feasibility study.

ROD—Record of decision.

SAC—Starved air combustion.

SARA—Superfund Amendments and Re-authorization Act.

scrubbing—The removal of impurities from a gas stream by the spraying of a fluid.

secondary burner—A burner installed in the secondary combustion chamber of an incinerator to maintain a minimum temperature and to complete the combustion of incompletely burned gas.

secondary combustion air—The air introduced above or below the fuel (waste) bed by a natural, induced, or forced draft.

secondary pollutant—A pollutant formed in the atmosphere by chemical changes taking place between primary pollutants and other substances present in the air.

secondary standard—A national air quality standard that establishes that ambient concentration of a pollutant that, with an adequate margin of safety, will protect the public welfare (all parts of the environment except human health) from adverse impacts.

settling chamber—Any chamber designed to reduce the velocity of the products of combustion and thus to promote the settling of fly ash from the gas stream before it is discharged to the next process or to the environment.

slag—The fused and vitrified matter separated during the reduction of a metal from its ore, or formed in a furnace, usually as a result of excessive operating temperatures.

slagging—Destructive chemical action that forms slag on refractory materials subjected to high temperatures; also a molten or viscous coating produced on refractory materials by ash particles.

sludge—Any solid, semisolid, or liquid waste generated from a municipal, commercial, or industrial wastewater treatment plant, water supply treatment plant, or air emissions control facility, or any other such waste having similar characteristics and effects.

smoke—Particles suspended in air after incomplete combustion of materials containing carbon.

soot—Carbon dust formed by incomplete combustion.

spray chamber—A chamber equipped with water sprays that cool and clean the combustion products passing through it.

stack—Any chimney, flue, vent, roof monitor, or conduit arranged to discharge emissions to the atmosphere.

stationary source—Any building, structure, facility, or installation which emits or may emit air pollutant.

stoichiometric combustion—Combustion with the theoretical air quantity.

stoker—A mechanical device to feed solid fuel or solid waste to a furnace.

teratogenic—Affecting the genetic characteristics of an organism so as to cause its offspring to be misshapen or malformed.

theoretical air—The quantity of air, calculated from the chemical composition of a waste, that is required to burn the waste completely.

thermal efficiency—The ratio of heat used to total heat generated.

toxic substance—A chemical or mixture that may present an unreasonable risk of injury to health or to the environment.

TSCA—Toxic Substances Control Act.

tuyeres—Openings or ports in a grate through which air can be directed to improve combustion.

UEL—Upper explosive limit.

ultimate analysis—The chemical analysis of a solid, liquid, or gaseous fuel.

underfire air—Forced or induced combustion air (quantity and direction are controlled) introduced under a grate to promote burning within a fuel bed.

UST—Underground storage tank.

vapor—The gaseous phase of substances that are liquid or solid at atmospheric temperature and pressure.

vapor plume—The stack discharge consisting of flue gases made visible by condensed water droplets or mist.

volatile—Any substance that evaporates at low temperature.
volatility—The property of a substance or substances to convert to vapor or gas without chemical change.

waste—Unwanted materials left over from manufacturing processes, or refuse from places of human or natural habitation.
wastewater—Water carrying dissolved or suspended solids from homes, farms, businesses, institutions, and industries.
waterwall incinerator—An incinerator whose furnace walls consist of vertically arranged metal tubes through which water passes and absorbs the radiant energy from burning solid waste.
windbox—A chamber below a furnace grate or surrounding a burner, through which air is supplied under pressure to burn the fuel.

Index

COMPUTER PROGRAM

The mass flow, heat balance and flue gas discharge analyses presented in the System Calculations chapter of this handbook have been developed into a computer program written in the basic language and menu driven. The program accepts up to four different waste streams as feed concurrently. It considers individual gas components, and has primary chamber, afterburner, waste heat boiler and scrubber systems options. The program displays and prints out more comprehensive information than is immediately available from the analysis sheets in this chapter. Mail the label below to Incinerator Consultants Incorporated for additional information.

ICI INCINERATOR CONSULTANTS INCORPORATED

INCINERATOR CONSULTANTS INCORPORATED
11204 Longwood Grove Drive
Reston, Virginia 22094
703/437-1790

NAME :__________________________

ORGANIZATION :__________________________

STREET :__________________________

CITY, STATE, ZIP :__________________________

Help Us Help You

So that we can better provide you with the practical information you need, please take a moment to complete and return this card.

1. I am interested in books on the following subjects:

- ☐ architecture & design
- ☐ automotive
- ☐ aviation
- ☐ business & finance
- ☐ computer, mini & mainframe
- ☐ computer, micros
- ☐ electronics
- ☐ engineering
- ☐ hobbies & crafts
- ☐ how-to, do-it-yourself
- ☐ military history
- ☐ nautical
- ☐ other ______

2. I own/use a computer:

- ☐ Apple/Macintosh ______
- ☐ Commodore ______
- ☐ IBM ______
- ☐ Other ______

3. This card came from TAB book (no. or title):

______

4. I purchase books from/by:

- ☐ general bookstores
- ☐ technical bookstores
- ☐ college bookstores
- ☐ mail
- ☐ telephone
- ☐ electronic mail
- ☐ hobby stores
- ☐ art materials stores

Comments ______

______

______

Name ______

Address ______

City ______

State/Zip ______

TAB BOOKS Inc.

NO POSTAGE
NECESSARY
IF MAILED
IN THE
UNITED STATES

BUSINESS REPLY MAIL

FIRST CLASS PERMIT NO. 9 BLUE RIDGE SUMMIT, PA 17214

POSTAGE WILL BE PAID BY ADDRESSEE

TAB BOOKS Inc.
Blue Ridge Summit, PA 17214-9988